D0908845

Emergency Incident
Management Systems

Emergency Incident Management Systems

Fundamentals and Applications

LOUIS N. MOLINO, Sr.

WILEY-INTERSCIENCE

A JOHN WILEY & SONS, INC., PUBLICATION

Library of Congress Cataloging-in-Publication Data:

Molino, Louis N.
   Emergency incident management systems : fundamentals and applications / Louis N. Molino.
     p. cm.
   Includes index.
   ISBN-13: 978-0-471-45564-6 (cloth)
   ISBN-10: 0-471-45564-4 (cloth)
   1. Emergency management—United States.   2. Crisis management—United States.   3. Disaster
   relief—United States.   4. Risk management—United States.   I. Title.
   HV551.3.M65 2006
   363.34'80973—dc22   2006000883

10 9 8 7 6 5 4

The Authors and Contributors would like to dedicate this book
to the members of all emergency services disciplines
that have died or been injured in the line of duty in service
to the citizens of the United States of America.

ACKNOWLEDGMENTS

In a project of the magnitude such as this book there is no "one man band" at work. More often than not it is a culmination of the work of many people. Some do small things like read one chapter; others are there for all of the steps that it takes to get a book to the presses today. I'd like to thank all of them here and individually but that would double the size of the book overall so I'll try and thank those that played major roles in this project:

To my co-author Paul Hannemann who wrote much of several of the chapters of this book, thanks for sharing your vast knowledge and experiences from the wildland and Incident Management Team worlds.

To David and Lynn White, thanks for the introduction to Paul but, moreover, thanks for the everyday encouragement and all the help you've been since I met you here in Texas. You truly define "Texan" to this "Damn Yankee"

To Jenny Holderby who was my sounding board for the whole of the project (starting with the proposal process), who is my first choice for Technical Editor, cheerleader, and is likely the best research librarian on the Internet. This book would not be possible without your input.

To my subject matter experts, Kim Jones, RN, Brandon Graham, and Ed Smith that wrote the chapters dealing with Hospital Emergency Incident Command Systems, Law enforcement-based IMS and Corporate Incident management and Business Continuity Planning (respectively), your expertise in your areas showed from draft one and you worked with me to get this done. For that, I owe you one.

To Dr's. Bryan Bledsoe, D.O. and Chuck Stewart, D.O., thanks for helping me break into the "writing business".

To the folks that put up with me at Lone Wolf Enterprises, Ltd. and Media Professionals International, Inc. what can I say but thanks for making this book a reality.

To all of the people that have taught me and that I have taught over the years that I have been in emergency services and to those that mentored me and were mentored by me, even those that I am at times seemingly at odds with. Thanks for helping me learn the art of persuasion by professional argument and the art of compromise by disagreement with professionalism.

And lastly to my kids, just thanks for being my kids.

ABOUT THE AUTHOR

Louis N. Molino, Sr. is a twenty-four year veteran emergency services provider with both fire and EMS experience in rural, suburban, and urban environments. He has been published in a number of national fire and EMS magazines and has also been a chapter contributor for a number of major fire and EMS texts. He currently serves as a member of the Board of Directors of the National Fire Academy Alumni Association for Region VI and as a Member of the Board of Directors of the Emergency Medical Services Association of Texas. In addition he is a member of several other professional associations at all levels. He is currently a fire and EMS consultant and emergency services instructor based in College Station, Texas where he resides with his four children. He can be reached at LNMolino@aol.com and enjoys lively discussions on any topic related to the emergency services world.

CONTENTS

PREFACE

Since its beginning after September 11, 2001 this book project has been a series of roller coaster rides. Sometimes the project was going up hill and sometimes it was going down hill (seemingly very fast), but this was a result of the fact that on that date the emergency services world changed forever almost in a blink of an eye. The world of emergency services Incident Management Systems got very complicated on that day and it has changed in many ways. But, when examined closely, it has remained more like it was before that date than it has changed and that in itself is a credit to the men and women that developed the systems in place for such incident management before that day. The Emergency Incident Management System of today and those that develop tomorrow are based on those earlier systems and all of the sound footwork that was done long before that second day that will live in infamy. This book is by no means the "be all end all" text on the topic as there could never be such a text due to the dynamics of the topic. This book is hopefully one more brick in a solid wall that builds an even more capable Emergency Incident Management System of the future.

Chapter 1

INTRODUCTION AND HISTORY OF INCIDENT MANAGEMENT SYSTEMS

It has been said, "Necessity is the mother of all invention". As with all great inventions such as the telephone or the automobile, the needs of select groups of people have led to the invention of what we now call "Incident management Systems". Unlike any physical invention, the Incident Management System, or rather its precursor the Incident Command System, is not a physical thing that one can pick up. It is rather a concept or more correctly stated, a set of ideas, policies, procedures and or ways of "doing things" that will, when employed properly, bring control to chaotic emergencies of all types.

Many people, including this author, have attempted to date with a degree of specificity the development of the Incident Command System. This task has proven nearly impossible as many different concepts have evolved over the past forty years, not in a linear format but rather more concurrently, leading emergency services to this point in the developmental of emergency Incident Management Systems. The unthinkable occurred on September 11, 2001, changing the world, as we knew it forever.

This set of concepts, known collectively as Incident Management has become even more necessary due to events of September 11, 2001 both in the United States and throughout the world. The concepts are still, and dare it be said, will forever be developing in fact due to those events and are in some ways developing even more rapidly than before because of it.

THE MILITARY CONNECTION

Since the dawn of man, men have had disagreements. Those disagreements have grown into fights. The fights have become wars. The desire of making war has led man on a seemingly endless quest to better his ability to engage in and win such battles. In that quest, the science of war has evolved. In some circles, it has become known as the "art of war".

1

Figure 1.1 World Trade Center collapse

Much of the "art of war" revolves around four elements: command, control, coordination and communications, collectively known as "C4". It should be noted that the military often uses the acronym "C4I". The "I" represents the element of Intelligence. Many would argue that in the post 9/11 era this element is also an emergency services concern and that will be discussed later in this work. These same elements are needed to tackle and conquer nearly any form of emergency response situations.

The United States military has evolved from men with muskets and swords on horseback to what is very likely by far the most precise, sophisticated military force on the planet. Emergency services in the United States have also evolved, from the days of a single constable and a bucket brigade in a large city into modern-day police forces and the American Fire Service, who with other contemporary delivery systems, provide emergency response services of all forms to all citizens. Each utilizes a great deal of technological marvels in their work, not unlike that of our military forces.

The technological revolution has begun to spiral at a seemingly amazing rate in today's digital age both in the military and in the emergency services communities. The fact remains that technology by itself does not win war nor does it respond to every day emergencies or once in a millennium events like September 11, 2001. It is not technology but rather men and women who use that technology who perform the tasks needing to be done be it in war, or at an emergency response incident. In order for those men and women to do their jobs safely, efficiently, and often correctly, they need tools. Not all tools are forged out of steel. Some are concepts and operations systems that will allow them to bring control to the chaos whether they respond in war or a disaster. Their primary tool is not a weapon nor is it a fire engine but rather concepts that have become known as an Incident Management System.

We have proven repeatedly that our military might is likely the best on the planet. We have both the tactics and the technology that leave other nations envious of our power, yet the true reason for the United States' military success is not technology but rather a mission-oriented, goal-driven mind-set. This mind-set is also reflected in personnel of the emergency services community in the United States and abroad.

While there is no doubt that a good deal of the reason that the United States military enjoys such success in the modern world is due to the massive technology at our disposal. Remember, technology is only a tool in a collective toolbox. Much of that same technology is used on a daily basis in emergency response activities. The military of the 21st Century has had to deal with technological jumps that are unprecedented in history. These jumps or leaps have undoubtedly caused many headaches and unseen problems that to past military leaders were unknown. These same technological advances have been the cause of similar problems in respect to emergency response. Modern warfare requires technology to become bigger and better, faster and stronger. Yet, the warrior on the ground remains the most important part of war. Technology will allow him to win war with greater speed and ease. Still, in the end, it is the soldier on the ground fighting and often dying in war. The same is true in respect to emergency responders. It is not the soldier but the firefighter, the police

Figure 1.2 Command and control

Figure 1.3 Modern day military warfare control room

officers and the paramedic that serve their country on the home front. Technology will not and cannot do their jobs for them, yet it allows them to do their jobs better. The men and women of emergency services make up the front line of Homeland Defense.

Many of the "fathers of incident command/management" (use of the plural intentional) had military backgrounds. They could see the obvious and often very unobvious parallels of the needs of their emergency response agencies. They began to mold and modify military command and control structures learned serving their country into systems that would allow their respective agencies to better respond to those needing assistance in times of crisis.

These men first adopted then adapted military command philosophy for use in their day-to-day response activities. This adaptation process was not seamless nor was it an overnight success. To this day it still remains a never-ending process as the modern emergency services community looks to the military for guidance in ever-increasing demands for response to incidents which in the past have not been an issue for their systems. The concepts of both emergency services incident management theory and in military sciences are evolving. Often those evolutions are on similar tracks.

Thus the beginnings of the modern day Incident Management Systems were born almost simultaneously in vastly different areas, for a wide and somewhat diverse set of reasons, but for one common goal, to better serve the needs of the community and to do the job of saving lives and protecting property in better ways.

THE BIRTH OF IMS: FIRECSOPE

When a student of emergency services begins to study emergency management theory, they will quickly see that the real birth of the concept we know today as IMS was also found in the late 1960's in California after devastating wildfires ravaged much of the southern part of that state. The fire service in California, along with other state and local governmental agencies, knew that they had to find ways to overcome a series of repeated events and shortcomings that occurred during these large-scale statewide emergency operations.

The agencies quickly began to identify common failures in their own response to these events specifically when a multitude of agencies responded to the same incident. These shortcomings were seen time and time again in operations. The main ones were:

- There was no answer to the question of "Who is in charge?"
- There was no clearly identifiable incident leader or commander, nor formal protocol or legal statute clarifying the responsibility of such a position at these incidents. Many laws could be cited but none positively had the legal impact necessary to clarify a single person or agency in charge and further it was often subject to varying interpretation based on the perspective of the agency(s) or individual(s) involved.
- Many conflicts arose among fire chiefs, police chiefs and other "in charge" official types at various levels of government in the affected jurisdictions.

Figure 1.4 Large scale wildfire aerial shot showing magnitude

- Further complications arose in the multitude of agencies from the federal, state, and a myriad of local and elected officials who seemed to want to have a piece of the "command pie", often without the responsibility.
- No attempt at any formal or informal basis was made to form a type of collaborative organizational structure for the purposes at hand.
- The established chain of command structures in each agency were rigidly enforced each by their own respective agencies, for a variety of reasons including traditions and turf war issues. There was no way to establish any type of leadership or command ladder using anyone from the various agencies represented regardless of any qualifications for the leadership role.
- The use of appropriate span of control was not considered.
- Many operations were undertaken with only one leader or supervisor and, due to the inherent complexity or more often than not the shear danger of the operations being undertaken, required much supervision. The military's use of the platoon system was never considered. On many incidents, companies would freelance. Since there was no one keeping track of who was doing what or where, a catastrophic incident could occur on the response scene since there was no accountability for anyone operating in any capacity.
- No interagency operations were made. If such operations did occur between two or more agencies, it was often done on the fly with no real planning or inter-agency integration before undertaking inherently dangerous and often risky operations.
- There was no common terminology. Each response group had its own self-evolved professional vocabulary or lingo, including acronyms and the use of "non-words". This not only lead to confusion but in some cases one agency's use of a term, acronym, or other vernacular would be in direct contrast with another agency which could often lead to misunderstandings. Operational level people were put at risk due to these miscommunications or misunderstandings.
- There was no joint communications system and no inter-operability was provided. In many cases, communication system incompatibilities were evident. Even where sophisticated, modern communications systems were in place for each agency or each jurisdiction, both areas suffered incompatibilities and agency-or-jurisdiction-specific problems occurred in terms of inter-operability and inter-agency communications. Often the existing systems were technologically incompatible and the use of uncommon terminology was more of an issue jurisdictionally or agency-based. On some incidents, response forces were able to physically "see" one another from a distance, yet had no interagency capability to communicate with them.
- There was no formal logistics control in place. A large-scale or complex incident demanded more resources than often was available at a given point in time. There was no way to systematically and appropriately ration resources or equipment any jurisdiction or agency might have at its disposal. Often in post-incident critiques it would be learned that one area of the incident needed a particular resource thought to be unavailable, yet that resource or equipment was in fact

available for use unbeknownst to the agency needing it at the time and therefore operating without said resources needlessly.

California's solution to the question of how to solve these and other problems was a project called "Firefighting Resources of California Organized for Potential Emergencies", which became better known as FIRECSOPE. FIRECSOPE then developed the Incident Command System (ICS) by way of an interagency task force of local, state, and federal agencies.

In the history of FIRECSOPE, it is stated that the task of "Designing a standardized emergency management system to remedy the problems listed above took several years and extensive field testing." This very accurate statement could be followed up with one regarding the ongoing nature of the process and still another regarding the system never being totally refined to any given end point due to the ever changing nature of the environment that emergency responders operate in, not unlike their military counterparts.

The same document further states that early in the Incident Command System development process, four essential requirements became clear:

- The system must be organizationally flexible to meet the needs of incidents of any kind and size.
- Agencies must be able to use the system on a day-to-day basis for routine situations as well as for major emergencies.
- The system must be sufficiently standard to allow personnel from a variety of agencies and diverse geographic locations to rapidly meld into a common management structure.
- The system must be cost-effective.

Once the framework of the system then known as Incident Command System or just Incident Command was laid, the system began to evolve and spread much like the wildfires it was designed to combat in the first place.

Many larger agencies and eventually states themselves began to adopt, refine, and adapt the work done in California for use in their own agencies.

Evolution of "Big Three" IMS Systems

As was stated earlier, the birth of the Incident Command System in the early part of the 1970s was a result of a series of fires that taxed the California response community to the point of nearly breaking. Like the invention of a more physical object, the very nature of history tends to focus on the products rather then the true evolutionary development process needed for all such endeavors.

History tells us that the light bulb was "invented" by Thomas Edison in 1879, yet a simple Internet search would show that many prior inventors had a hand in the development process to the end point of a commercially deployable incandescent light bulb.

The modern day IMS that the emergency services community uses was a functional outgrowth or spin-off of the original Incident Command System, developed for the specific needs of large-scale wildland fires. It became evident to users of that early Incident Command System that the system itself was well suited for use in other emergencies including natural and technological disasters.

Undoubtedly those who assisted in the early expansion of the Incident Command System to include non-wildfire situations were also Californians, as the state seems to have a propensity for all forms of disasters which occur regularly on a large scale if not seemingly daily basis.

The reason for this outgrowth was simple. The emergency management community quickly realized that the underlying problems that led the wild land community to develop the original Incident Command System were also common to other incidents both in terms of complexity and scope.

They were and still are:

- Large spans of control causing too many people to be reporting to one supervisor
- Multiple types of organizational structures among response agencies at a variety of levels of government
- No formal way to reliably share incident information
- Incompatible and inadequate communication systems and procedures
- No formal method of coordinated planning among agencies
- Severe misunderstandings of lines of authority
- Major terminology differences between agencies and response disciplines
- Lack of formal or in some cases unspecified incident objectives
- Lack of any incident action planning or backup plans

Common factors in all disaster or emergency response situations are that they may occur with no advance warning or notice. They may then develop rapidly from an incident as small as a grass fire that can evolve into a major wildland conflagration, or one that may occur almost instantaneously on a wide-scale covering many jurisdictions at once such as an earthquake or tornado.

If left to their own devices unchecked, they may grow in size, proportion, or complexity increasing personal risk for response personnel and civilians who may be affected by the event. The risk of life and property loss can be extremely high.

A great example of this was the wildland fires in Yellowstone National Park. In a single week in the summer of 1988 fires within the park alone encompassed more than nearly 99,000 acres, and by the end of the month, dry fuels and high winds combined to make the large fires nearly uncontrollable. On the worst single day, August 20, 1988, tremendous winds pushed fire across more than 150,000 acres requiring a massive national level response.

Figure 1.5 Yellowstone fires 1988

There are often several agencies that will respond to these events with some on-scene responsibility. These events can quickly become multi-jurisdictional in nature, and often have high public and media visibility.

Lastly, the cost of response and eventual mitigation is always a major consideration which may be complicated, due in part to the vast array of funding available at a multitude of governmental and private sector levels for response and or restoration of the community.

These factors led to the development of three "manifestations" of the Incident Command System. Each was by nature of its visibility and funding, therefore, the "Big Three" of Incident Command Systems in terms of the evolutionary process. All other later Incident Command System were outgrowths of theses three systems. They are:

- The Wildfire Incident Command System (FIRECSOPE). The system born from FIRECSOPE which then evolved into the system most widely used by the wild-land firefighting system in the United States.
- The National Fire Academy (NFA) Incident Command System. In 1980 the Federal government recognized the Incident Command System as the model for emergency management of incident scenes but with a bit of an "east coast" flavor to its manifestation which further expanded to include more palatable non-wild land fire applications which are the bulk of emergency services incidents faced by urban responders.
- The Fire Ground Commander System (Phoenix Fire Department). Developed by the Phoenix Fire Department and made famous by their Chief Alan Brunicini. This system used the same basic or underlying principles as the FIRECSOPE model but was made to be more practical for structural fire-based incidents.

Later in time, all these systems would meld in many ways and some would argue that their differences are only semantic in nature. This will be discussed later in this work.

THE MELDING OF THE IMS CONCEPTS OF TODAY

Once the Incident Command System was widely used and accepted throughout the United States by fire agencies, it very rapidly became used more and more by law enforcement, hospitals, and other public safety applications, and for emergency and event management.

In 1980, the original FIRECSOPE system and other existing IMS systems transitioned into a national program called the National Interagency Incident Management System (NIIMS). NIIMS became the backbone of a wider-based system for all federal agencies having a role in wildland fire management. This was initially and is to this day is administered by the United States Department of the Interior through its National Wildfire Coordinating Group (NWCG) based in Boise, Idaho.

The following agencies and entities, among others, have at some point endorsed the use of the concepts of the Incident Command System and are participants in the NIIMS process:

- The Federal Emergency Management Agency's (FEMA) , National Curriculum Advisory Committee on Incident Command Systems/Emergency Operations Management Systems recommends adoption of the Incident Command System as a multi-hazard all-agency system.
- FEMA's National Fire Academy (NFA) has adopted the Incident Command System as a model system for fire services. NFA serves as a focal point for federally-based incident command and management training though a variety of models of training delivery both on its Emmitsburg, Maryland campus and through its off campus training delivery system.
- FEMA's Urban Search and Rescue (USAR) Response System, a component of the Federal Response Plan, uses ICS as the basis of its onsite management structure and to allow for seamless interface with the local Incident Management System.
- The National Fire Protection Association (NFPA). Many NFPA Standards reference or in some cases directly call for the development and use of an incident management system.

THE UNITED STATES COAST GUARD (USCG)

The USCG incorporated basic incident command and management structures and principles into the National Response System (NRS) used for oil and hazardous material pollution response. These are documented in the National Response Plan currently under revision along with other federal planning documents related to emergency management. These revisions are being undertaken under the auspices of the Secretary of the Department of Homeland Security (DHS).

THE OCCUPATIONAL SAFETY AND HEALTH ADMINISTRATION (OSHA)

OSHA requires that all governmental and private organizations that in any way respond to or otherwise handle incidents that may involve any form of hazardous materials use ICS. This is contained in 29 CFR 1910.120(q).

THE ENVIRONMENTAL PROJECTION AGENCY (EPA)

EPA requires that any employer in a non-OSHA state also comply with the provisions of 29 CFR 1910.120 (q) and that they make use of ICS at any type hazardous materials incidents. This document is in Section 311 of the EPA's regulations.

Figure 1.6 United States Coast Guard Logo

Figure 1.7 Occupational Safety and Health Administration Logo

Other Agencies

Some states now require the use of an emergency management system based on ICS. Many businesses are implementing Comprehensive Emergency Management plans (CEMP) in the post 9/11 era. Most are adopting many of the principles of ICS, as they are universal in nature and meld well with local CEMP's programs already put in place by the response community.

In March of 2003, President George W. Bush by executive order required that all federal agencies complete reviews and that revisions be made to existing emergency response plans at all levels of the federal government. This effort was to be coordinated by the Secretary of the Department of Homeland Defense.

Figure 1.8 Environmental Protection Agency Logo

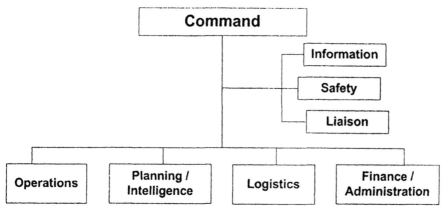

Figure 1.9 Top ICS Command Chart (CFLOP)

CONCLUSION

It has been said recently by those with an 'inside track" to developments at the highest levels of government in reference to the Incident Management System that the systems we "once knew" would "be changed forever in short order".

The drive for these changes, of course, are the hijacking events of September, 2001 and later in that same year the Anthrax attacks in the Northeast as well as other credible real-world threats to the long term national security of the United States.

The events referenced proved that we can no longer work in vacuums in the public safety community. Agencies that were once thought to never have the need to work together found that they must indeed work with each other. Responders from every corner of the United States and some from around the world found themselves working side-by-side, hand-in-hand at incidents seemingly beyond even Hollywood's imagination.

If we are to expect to work together, we must create a unified command system for managing people, resources and equipment that will be necessary for the successful outcome of any incident, however large or small it may be.

Use of Incident Management Systems allows that to happen with both speed and efficacy but only if it is implemented from the "top down" and the "bottom up" nearly simultaneously and only if all of the "players" in the "Public Safety Game" are willing to make it work both for the "big ones" as well as on the every day "mundane" emergency responses. In essence, we need to bring the chaos under control.

Chapter 2

THE FIVE "Cs" OF COMMAND

INTRODUCTION

If one were to conduct an extensive study of the literature written on the control and management of emergency incidents, one would find a common pattern of five concepts repeated throughout: command, control, communications, coordination, and cooperation. We shall call these concepts the Five "C's."

While I have not read every piece of material written to date concerning this topic, I have, in researching this book and as a diligent student of the subject in general, read many of the major works available on the subject. I have also read and written many articles on the various subtopics related to the complex subject of emergency services, particularly from the management standpoint, and specifically concerning command and control.

In studying this material, it is evident, since the Five C's recur repeatedly, that they bear a high level of importance in respect to the overall topic itself. It is therefore a safe conclusion that they are the basic concepts of the system that we have come to call the Incident Management System (IMS) or Incident Command System (ICS).

The dictionary definitions are as follows:

- Command

  a: to direct authoritatively: <to order>

  b: to exercise a dominating influence over: <to have command of>

  c: to have at one's immediate disposal: <he has the money>

  d: to demand or receive as one's due: <she commands a high fee>

  e: to overlook or dominate from or as if from a strategic position: <to have a broad view>

  f: to have military command of as senior officer: <he is in charge>

- Control
  - a: to exercise restraining or directing influence over : <to regulate>
  - b: to have power over : <to rule>
  - c: to reduce the incidence or severity of especially to innocuous levels <control a disease>
- Communications
  - a: an act or instance of transmitting: <as in talking>
  - b: information communicated by a verbal or written message: <as in a letter>
  - c: a process by which information is exchanged between individuals through a common system of symbols, signs, or behavior <as in language>
  - d: a technique for expressing ideas effectively: <as in speech>
- Coordination
  - a: the act or action of coordinating
  - b: the harmonious functioning of parts for effective results
- Cooperation
  - a: the action of cooperating : <common effort>
  - b: association of persons for common benefit

Definitions are general in nature. They cannot fully describe the implied contexts of specific uses of words either independently or, perhaps more important, interdependently with one another. To fully understand these words in terms that are meaningful and relative to the subject at hand, one must consider their meanings and more often than not their multiple meanings as they apply to the topics and subtopics of emergency incident management.

Command

While no one of the Five "Cs" has greater importance than another in terms of incident-management, Command is often seen as the "key word" or linchpin. Without effective command you cannot gain control of the incident scene. This has been proven time and time again in the history of emergency response. Effective command requires an authoritative presence, one person who is in charge or takes charge, and situational awareness of the incident scene and its requirements, as well as any area or larger concerns related to response (such as the events of September 11, 2001). Using this situational awareness, the commanding officer establishes strategic goals and operational objectives, gives orders and/or instructions, and assigns tasks related to the goals and objectives of the emergency-response operation.

Control

To gain control of a chaotic situation such as an emergency response, the first step is to eliminate undesirable factors such as fires, leaks, and hazards that can cause injury to

Figure 2.1 Chief Officer at Incident Command Post

civilians and responders alike. In order for response forces to accomplish this goal, they must set incident priorities. For example, if a fire on the third floor of an 80-story high-rise building can be controlled in five minutes, there will be no need to evacuate the residents of the upper floors, as the fire will never present a hazard to them. Therefore, fire suppression becomes the top-priority operation. However, should the fire begin to spread large volumes of smoke to the upper floors of the building, presenting an immediate danger to those residents, rescue and or evacuation will become a higher priority. If the fire were caused by a major explosion on a middle floor, generating large volumes of toxic smoke, fire suppression will be an even lower priority for the first-arriving emergency response forces. The priorities for emergency-response personnel are referred to as "LIP," an acronym for:

- Life safety
- Incident stabilization
- Property conservation

These are the three primary considerations for any emergency-response situation in generic terms, regardless of the genesis of the emergency or the background of a responder. They make good common sense: save lives, control the incident whatever its type, and only then worry about property conservation. By controlling individual elements of the situation, response forces will have the ability to regulate the factors that determine how an incident plays out.

Communications

Clear and concise communications at an emergency incident may well make the difference between life and death! This applies to both responders and civilians. In order to facilitate effective communications, a variety of methods are generally applied. Such methods include but are by no means limited to:

- Face-to-face discussions
- Radio
- Telephone (land line or cellular)
- Written messages including paper and e-mail
- Instant-messaging systems
- High-tech communications systems such as satellite and other space-based systems

Figure 2.2 Face-to face communications at an Incident Command Post

Figure 2.3 Portable radio attached to the belt

Figure 2.4 Incident dispatchers talking on the phone

Figure 2.5 Personal computer in use at an Incident Command Post

Figure 2.6 Satellite communications

Communication is by definition a two-way process: there must be both a sender and a receiver of the information being relayed. This is required regardless of the method used to facilitate the communication. Communication has three parts: sending, receiving, and validation or confirmation. The sender transmits the information to the receiver, who processes the information and should repeat it back to the sender to allow the latter to validate or confirm if the former understands the sender's message.

This validation process is even more important when any type of technology (radio, computer, etc.) is used. In the absence of face- to-face discussion, where body language and facial expressions can be used by the sender to validate the receiver's understanding, it is important to make sure that the receiver has understood the meaning and context of the message, because there is no visual affirmation

An example of the hazards of lack of validation is a 9-1-1 call where a sender says that he needs the police but then hangs up. The receiver has no idea why the caller needs the police. Even if the dispatcher can locate the call, he or she has no idea of the nature or severity of the incident. It could be a barking-dog complaint or an armed robbery in progress. The response to these two examples would be markedly different. The lack of validation of the exact nature of the communication could put lives at risk.

Many of the communications that occur during a significant emergency incident are potentially critical in nature. This fact, when coupled with the sheer volume of information sent and received, makes it imperative that all communications be clear and concise to assure that the response forces and command personnel are able to better categorize and then prioritize this information.

It must also be noted that in real-world incidents, as well as in exercises that are designed to test response capability, poor responses are most often related to failures in communications.

Coordination

Coordination is a cornerstone in emergency-management theory. In order for an effective Incident Management System to function, coordination must take place on several levels simultaneously. Coordination ties directly to communication both horizontally and vertically within the chain of command and is often dependent on interagency cooperation to be successful; again this is a frequent cause of failure in both exercises and real-world emergency events.

One level of coordination is at the incident scene itself. Say that a mutual aid resource delivers equipment or personnel to an incident; those resources must be coordinated with the response efforts underway at that time. Coordination is essential regardless of whether the response involves a single agency response or several agencies. Coordination of resources controls confusion, prevents freelancing, and strengthens the overall response.

Another level of coordination is at the regional level. If a major incident occurs that requires a more robust response, some resources could be limited in availability to the Incident Commander. A higher authority such as an Emergency Operations Center (EOC) will need to be contacted to request additional resources. The EOC will then

Figure 2.7 Emergency Operations Center activated

determine how to provide those resources. Regional, state, federal, and/or national level EOC's may need to become involved.

Cooperation

The final element is cooperation. Cooperation involves a working relationship between the parties that are likely to join together in an emergency situation to obtain a positive outcome for all those involved. The time to begin this cooperative effort is not at the time of the emergency. As one wise emergency manager once said, "If the fire chief is handing you his business card for the first time in the middle of the disaster, you are so far behind you'll never catch up."

Cooperation is the same as teamwork. All members of a sports team play together as one unit in order to win the game. Emergency services agencies are in theory no different, with each discipline and sub-discipline bringing its unique skills to the response effort. Without cooperative teamwork the whole effort will fail. In any emergency response, where more than one discipline responds (the vast majority of responses), all disciplines must learn how to work with the others. In the midst of a heated battle such as a firefight or a shooting event, every response agency will do whatever it takes to get the job done. Small incidents are a day-to-day reality in a lot of places in the United States, yet when a major catastrophe occurs we often see communications, command, and control breakdowns. If the groundwork of cooperation has been laid in advance by all agencies involved in a response area, be it a city, a county, a state, a

Figure 2.8 Multi-agency response

region, or an even larger area, the field troops on the ground will have a easier time dealing with the tactical realities of the operation, no matter what the root cause of the incident.

CONCLUSION

The Five "Cs" are Command, Control, Communications, Coordination, and Cooperation. When they are accomplished together, they enable chaos to be converted to control in as effective a manner as possible given any emergency scenario. They are repeated throughout the literature, as they truly form the basic foundation of all emergency-incident-management systems that have proven to be effective over time.

Chapter 3

THE EVOLUTION OF INCIDENT MANAGEMENT SYSTEMS

The emergency services system that is in place today in the United States has become larger, faster, and, perhaps at times, better over the past several decades.

In the years after World War II this country saw a tremendous growth spurt that has become known as the "Baby Boom". This Baby Boom generation along with the subsequent generations since have seen technological advances in all fields in the past 60 years. Advances of all types now take a few years whereas they literally took centuries prior to this point in history. The world of emergency services is no different in many ways and, yet in other ways, it remains the same. Some universal truths like fire can only burn up, over, then down unless it is "helped" and that if left untreated "all bleeding will stop, eventually" remain as true today as they were one hundred or one thousand years ago.

In the United States we often credit Benjamin Franklin as having been the first American Fire Chief. What would Chief Franklin say if he was to set foot in the Philadelphia Fire Department's Alarm Room, or the quarters of "Squirt 8" in his beloved "Olde City" section of the city today? If Chief Franklin was riding with Battalion 4 on a Box Alarm to Independence Hall he'd still see fire burn up, over, and down given a natural event. Since Chief Franklin we've seen massive change in emergency services.

The 1970s saw the advent of Emergency Medical Services (EMS). EMS systems are based on the tactics employed by field medics on the battlefield. EMS systems cut out precious minutes of the so-called "Golden Hour," which was once thought to give severe trauma patients their best chance at long-term recovery. While some modern studies suggest that the theory was flawed, countless lives have been saved by EMS systems modeled after the ones developed here in the United States. Pre-hospital care that was nonexistent fifty years ago can save lives today. Not all that long ago the local mortician may have been the one to get the call to transport those too ill to ride in a car to the hospital using the same hearse that would eventually carry them to their final resting place. There was room for a cot in the back allowing the patient to lie down. In rural areas, the hospital may have been several miles away and transport in a car could have been more traumatizing than the illness or incident.

Today transport to a specialized facility hundreds of miles away is but a few minutes by helicopter or fixed-wing aircraft designed to facilitate such transfers. Simple bandages and tourniquets have been replaced by advanced life-support ambulances staffed by Emergency Medical Technicians, many of whom trained to the paramedic level and are comparable and, in some ways, more advanced than nurses and nearly as skilled as physicians. The advanced equipment onboard these "Mobile Intensive-Care Units" can keep a patient alive in many worst-case scenarios by giving them a "jump start" on the diagnosing and treatment process of their injury or illness. Victims of all types of accidents and illnesses, who in the past would have a slim survival rate, often have a complete recovery due to immediate care that pre-hospital EMS provides. Modern medicine as a whole relies on a system of step-by-step interventions.

In some ways, medicine uses a form of IMS as well. Many times a doctor will be forced to do something based on the priorities of the patient's condition. This is generally pretty obvious. If they are not breathing you need to fix that before you worry about a broken leg. At times the Incident Management System helps managers to determine what the critical needs of the incident are and then allows them to address those critical issues in a logical fashion and in a timely manner.

Emergency services management has learned to work smarter, protect staff better, and refine procedures to maximize effectiveness. As a community and a profession, members place high value on education, training, and real-world expertise. The training of responders on the street, as well as in the executive suite, is literally as important as life and death in many instances.

As a professional community, emergency services management has come to realize that organizing personnel and equipment on an incident begins almost immediately to create control among the chaos.

An Incident Management System is of great value during large-scale incidents as well as everyday responses, as it provides an organizing mechanism. This is due to the federal government mandate that any incident involving a hazardous material apply the basic principles and concepts of the Incident Command System. This mandate came in 1986 with the adoption of 29 CFR 1910.120 by the Department of Labor's Occupational Safety and Health Administration (OSHA).

TAKING CONTROL

Using an Incident Command or Management System starting with the beginning of the response, the initial size-up of the situation, allows the Incident Commander an opportunity to begin to control the emergency situation immediately. With this simple communication the officer will effectively inform dispatch that the apparatus has arrived, that he/she is taking control of the incident and, at the same time, announces to all other emergency responders that he is the Incident Commander and is in charge of this incident at this time.

Usually, and often as a matter of protocol, the officer in charge of the incident is the officer arriving on the first response vehicle. A well-trained officer will have begun preparation enroute to the incident, gathering as much information as possible about the situation from dispatchers and making a mental picture of the area around the incident.

During the initial size-up, the officer will decide if more assistance will be necessary, depending on manpower and equipment needs as he perceives them at that time. This decision should also be based on several factors such as availability of water, air supply, rehab requirements, or other incident needs. In extreme weather conditions more manpower may be needed in order to rotate crews and form specialty teams. The decision to request mutual aid may need to be made immediately. The officer has the option to call more resources of any type if the need arises, again based on his perceptions of the incident and its needs.

As more personnel arrive on the scene, they will ask for direction from the officer in charge (now known as the Incident Commander), who will be making a number of decisions in a short time. At this point, the officer must decide whether he has the training necessary for the situation or whether he needs to call in an officer of higher rank and transfer command to that officer. In most systems when this transfer of command occurs, the first-in officer will become the Operations Section Chief and begin directing the tactical operations as needed, with the higher-ranking or more qualified officer assuming the role of Incident Commander.

Once this occurs, the Incident Commander begins to fill the other General Staff positions as needed by the incident. Once these positions are filled and other critical objectives are met, the planning process begins. Many emergency responders call these meetings huddles, as this is where the strategies for the long haul are developed on both a strategic and a tactical level.

With a structure fire, for example, one firefighter will be assigned to choose a crew and attack the fire (operations) and another will be assigned to equipment (logistics). It is very unlikely that other sections such as planning or finance will need to be filled in such a simple scenario, but ultimately these duties will be assumed by the Incident Commander.

One of the very early actions taken by an Incident Commander is to appoint a Safety Officer or personally assume those duties himself. Again, like all the positions in IMS, it can also be filled later as more personnel arrive. The Safety Officer is not part of General Staff but works directly for the Incident Commander as a part of the Command Staff. At this point the Incident Action Plan (IAP) is purely oral in nature in that it is verbalized to the staff as they arrive and are assigned to the IMS positions that they will be assuming. Many incidents grow rapidly to the point where they become long-duration (generally any event greater then twelve hours is considered long-duration event). At this point the IC will begin to break the day into operational periods, generally twelve hours in length. (This duration can be changed at the discretion of the Incident Commander based on the needs of the incident.)

ADDING TO THE CHAOS

Many of the procedures stated above are second nature to members of the emergency services community, but in extreme situations, such as terrorist activities, volunteers and other people such as public health responders may become involved who are not fully trained in the procedures and concepts of the Incident Command System.

With the threat of terrorism and the consequences of bioterrorism in particular and all that is associated with that monster of modern life many of these "new" responders are being brought into the emergency services "fold" as it were.

This is particularly true in the case of smaller, rural agencies. In these instances the value of mutual aid becomes evident. One department may need to rely on its nearest neighbor to supplement manpower or equipment. Responders as a whole are a proud group of people, but they tend to know when they need help and are always ready to help and lend a hand when called on.

On significantly larger incidents requiring massive mutual aid, responders often create more chaos due to the sheer number of personnel converging on the scene with all types of equipment, each department bringing its own chain of command. This is a dangerous situation and must be rectified. If all agencies are using the same Incident Management System, management becomes a lot easier.

On an everyday "normal emergency response" such as a structure fire, depending on the situation, more than one agency may be required to fulfill the needs of the incident. There is a need for law enforcement to handle the traffic and the crowds; EMS personnel should be on the scene, possibly, along with the power company, the gas company, and the water company. The response to even day-to-day incidents can be made more effective with proper use and implementation of an effective Incident Management System.

During the initial period of amassing the resources needed for the response, communications are often confusing due the high volume of radio traffic within each department and between departments. The matter of communicating with other agencies such as law enforcement, utilities, and medical personnel creates more challenges and more chaos. Interoperability among agencies has come a long way since September 11, yet there is still much to be done.

Perhaps the biggest reason that an Incident Management System needs to be in place at even the smallest of day-to-day emergency events is responder safety. The Incident Commander has little chance of maintaining an effective "span of control" unless she delegates authority and responsibility for the troops that are tactically operational on the incident.

In our simple structure fire example the incident commander needs to make several decisions by this time, including the initial attack, search and rescue, where to place equipment, how many firefighters to send in, and anticipating what additional resources might be needed before the fire is out and all fighters and equipment are back in service. The fire may have escalated; there may be danger of collapse or exposure involvement. At this moment it is not appropriate to try to design a system for the task or come up with a means of organizing this situation. There are many tasks to accomplish before this incident comes to a close, many of which may need to happen quickly, often simultaneously. Having a good organizational method or system can make the difference in how the incident plays out. The Incident Management System is such a system for both the day-to-day events that we respond to on a daily basis or "The Big One."

CONCLUSION

In short, the use of an effective incident management system in the response to day-to-day emergencies is essentially a dress rehearsal for such an application on a major event. It is also a way to bring control to the chaos of emergencies.

Chapter 4

COMMON COMPONENTS OF EMERGENCY INCIDENT MANAGEMENT SYSTEMS

An effective Incident Management System (IMS) provides for a solid management structure and creates a logical system for conducting on-site operations in a manner that is both effective and safe for all people involved. To bring calm to the chaos of crisis is the overall objective. When properly used, the ideal IMS will be applicable to even small-scale daily operational activities in all areas of public safety regardless of which service (e.g., police, fire, EMS, etc.) is using the system. The nature of the specific service or activity will have no particular bearing on how the system is employed at the base level. An Incident Management System seeks to manage, organize, categorize, and simplify any activity at the strategic rather than the tactical level. In some cases the Incident Management System has applicability even in areas not directly involved in the public safety arena such as daily operations of public works entities and the like or other tasks where several people or agencies will need to work in unison yet separately to achieve a specific stated goal.

An effective Incident Management System will bring order to any type of complex situation. This is even truer of chaotic events, which public safety agencies are accustomed to dealing with on a daily basis. This is accomplished because every Incident Management System, no matter what it is called or who designed it, will bring with it a standardized operational structure and common terminology. These common elements provide a useful yet flexible management tool, thereby allowing the public safety agency to concentrate on both strategic and tactical objectives and operations to provide for the highest level of safety and accountability.

The true value of an effective IMS used on a day-to-day basis for small-scale or common incidents is that it can rapidly be expanded and adapted to any situation regardless of its scope or size. The most popular Incident Management Systems used in the United

States over the past thirty-five years are particularly adaptable to incidents involving multi-jurisdictional (cities, states, or even national scope) or multi-disciplinary responses (police, fire, EMS, public works, and private industry). These systems, when properly employed in all the sectors that come together in a community-wide response effort, have been shown through experience to be extremely effective in controlling the short- and long-term consequences of any number of types of disasters, both naturally occurring and man-made.

An effective Incident Management System provides the framework that will allow for a flexible yet rapidly deployable and expandable (as well as collapsible) organizational format to support the tactical functions that need to be performed with any type of incident. The basic concepts of any Incident Management System will not change regardless of the number of agencies responding to the incident, nor will the specific needs of the incident impact greatly on the base concepts of the system.

In an effective Incident Management System the above will hold true even for extremely complex incidents or multiple incidents (related to one another or not) occurring simultaneously in a given geographic region. These types of incidents can tax the resources of all agencies in a given geographic area and perhaps even across the whole of the United States. Without the effective use of an Incident Management System, chaos will reign for a longer period of time and control might never be regained.

OPERATING REQUIREMENTS

The following are operating requirements for the effectiveness of any Incident Management System:

- Single jurisdiction responsibility with single agency involvement.
  Car fire (Fire Department response)
  Heart attack (EMS Service response)
  Breaking and entering (Police Department response)

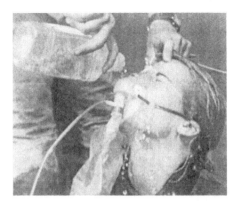

Figure 4.1 Single resource incident

- Single jurisdiction responsibility with multi-agency involvement
  Dwelling fire with entrapment (Fire, Police, and EMS response)
  Bank robbery with a person shot (Police and EMS response)
  Building collapse with entrapment (Fire, Police, EMS, and Public Works response)

Figure 4.2 Single jurisdiction incident

- Multi-jurisdictional responsibility with multi-agency involvement.
  Severe coastal storm with multiple city/county flooding (local, regional, and/or
  state-level responses with local city, county, and state jurisdiction)

Figure 4.3 Major urban flood

- Major bombing (local, state, and federal response and both state and federal jurisdictions)

Figure 4.4 Oklahoma City bombing

- Terrorist attack (nationwide response as well as a multitude of jurisdictions)

Figure 4.5 9/11 events

Following are the system requirements for effective operation:

- The system's organizational structure must be able to adapt to any emergency to which public safety agencies would be expected to respond.
- The system must be applicable, adaptable, and acceptable to all user agencies.
- The system must be able to expand in a rapid manner from an initial response into a major incident.
- The system must be able to reduce its size just as readily to meet the organizational needs of the incident as the situation decreases in scope and complexity.
- The system must have common elements in organization, terminology, and procedure, which will allow maximum effective application.
- The system should be implemented in a manner that causes the least possible disruption to existing procedures and operational systems.
- The system must be effective in fulfilling all the above requirements and be simple enough to ensure ease of understanding.
- The system must have the ability to adapt over time so that it is always current.
- The system must be integrated into the overall framework and concept of the National Incident Management System (NIMS) used by the federal government for management of all emergency incidents worldwide.

Incident Management System Commonalities

In order for any Incident Management System to fulfill the above operating requirements, several commonalities must be effectively in place during the design, implementation, and use of the system. Constant updating of certain elements must be built into the system to assure that it remains effective as both operational methods and technologies of the user agencies change.

The common elements of an effective Incident Management System are:

- Common terminology
- Modular organization
- Integrated communications
- Consolidated incident action plans
- Manageable span of control
- Predesignated incident faculties
- Comprehensive resource management

Common Terminology

One of most frustrating aspects about responding to an incident where multiple agencies are trying to work together for a seemingly common goal of saving lives and

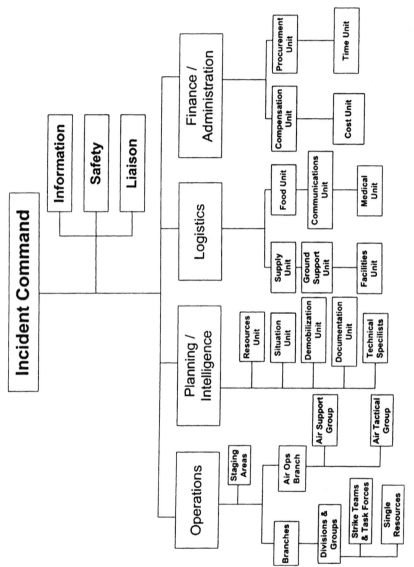

Figure 4.6 Full ICS command chart

protecting property is the fact that we all speak our own industry lingo with the only seeming commonality is that we speak English. Each of the major public-safety disciplines has its own "language" of sorts. Other agencies still use codes and other abbreviations. An example is the 10-Code system that many law-enforcement agencies still use even in the age of "clear text" speech across the airwaves. This system dates back to the days of tube-set radios. Further compounding these factors are the multitudes of "jargon dialects" that often tend to be regional in nature and even sub-regional or local at times.

The reasons for this variation are generally societal in nature and have to do with the makeup of the emergency services community and its geographic roots. Many of the terms tend to be holdovers from years ago, such as the use of the term "fire plug" in the northeastern United States to refer to a fire hydrant.

The above only represents a fraction of the overall problem. The real "minefields" of terminology occur when terms are used in radically differing manners by different services. This language can make a life-and-death environment even more hazardous by creating misunderstandings that cause individuals or agencies to act in a manner inconsistent with safety and counter to the tactical objectives of the incident. At best, this could lead to needlessly endangering lives, and at worst, a Line of Duty Death (LODD).

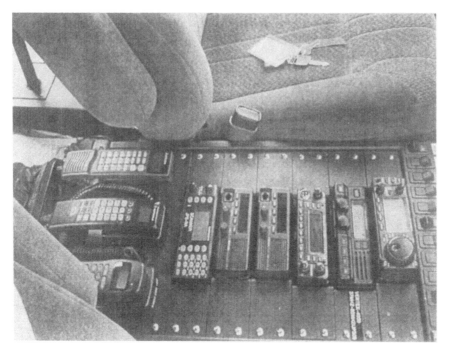

Figure 4.7 Modern-day mobile radio setup

One anecdotal example of this is a response to a shooting call with victims suffering gunshot wounds. Of course, a police and EMS response was indicated and commenced simultaneously. In most operating systems, the EMS component of the response would intentionally lag behind law enforcement and wait for an "all clear" signal from the police to ensure that they were not placed in any jeopardy from the perpetrator while doing their duty, as they generally have no defensive capability of their own. In this particular incident, a police officer told the responding EMS crew that the scene was "secure". The EMS crew took that to mean that the perpetrator was no longer a threat in the area. However, that was not the case. In the "police speak" of the area, "secure" referred to the surrounding of the perimeter so as to the prohibit anyone from entering or exiting without police knowledge of their presence. The perpetrator in this scenario was still in the area and still armed and dangerous. The EMS crew made the assumption that "secure" meant "safe," which was not the case. Luckily, the police had direct radio contact with the EMS unit and soon ordered it to a safe position. It was not until after the perpetrator was apprehended and the victims were transported off scene that the true nature of the "miscommunication" was revealed. As result of the incident, the agencies involved made adjustments to both operational methods and training of their respective personnel.

It is far beyond the scope of any Incident Management System to change such cultural realities. The only verbal communication problems that can be adjusted by the use of an Incident Management System are those related to the terminology used in direct relation to that system. Moreover, all Incident Management Systems must subscribe to a common terminology so that interoperability at all levels is assured no matter what system is in use at a local level.

Organizational Functions

The General and Command Staff designations refer to a standard set of major positions and functional units that have been pre-designated for the Incident Management System.

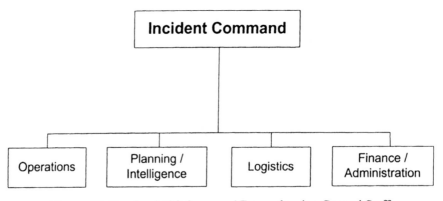

Figure 4.8 Top-level ICS Command Tree—showing General Staff

In many systems or for specific applications, subunits are also pre-designated, but the inherent flexibility of the Incident Management System allows for the creation of functional units that can be named "on the fly" as needed. The placement of these units in the incident management tree is dictated by their function in maintaining the integrity of the IMS.

This structure enables the Command position to concentrate on strategic objectives and allows the General Staff position to fulfill the tactical needs of the incident so that the strategic goals are met. On September 11, 2001, those strategic events were of national importance and at times drove the tactical realities of the day.

The Command position may consist of a single-agency structure or a unified multi-agency structure consisting of multiple management representatives and a variety of support personnel. The structure of this element is based on the type, scope, and complexity of the incident. By default the first-arriving response team functions in the Incident Commander's position until it can be relieved of this position.

Three additional functions, safety, information, and liaison, are the Command Staff positions and are essential for the successful completion of any incident response. The functions must be fulfilled in most if not all incidents. If they are not staffed, the responsibility falls back to the Incident Commander.

The Safety Officer's functions are to develop and recommend measures for assuring the personal safety of all incident personnel and to assess and/or anticipate hazardous and unsafe situations. Only one Safety Officer is assigned for a specific incident; this person becomes the Incident Safety Officer and has direct reporting responsibility to the Incident Commander. The Safety Officer may have assistants as necessary. These assistants may represent individual agencies or jurisdictions involved in the incident. Safety assistants may have specific responsibilities such as air operations, hazardous materials, or other specialty teams. The Safety Officer should have an understanding of all tactical operations in respect to the incident. The officer need not be a certified responder in every technical area represented in the incident scene. This would be impossible on a major large-scale incident such as the Oklahoma City bombing or

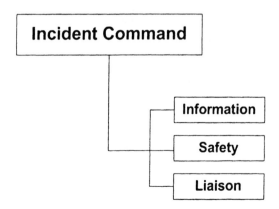

Figure 4.9 Top-level Command Tree—showing Command Staff)

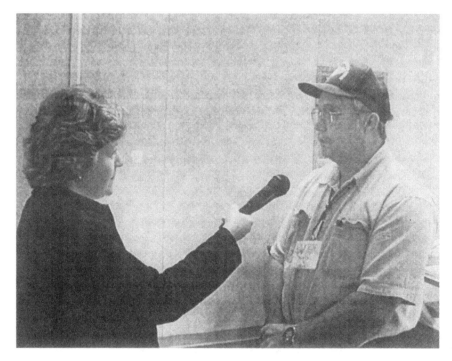

Figure 4.10 Information Officer in service with media representative

the 2001 World Trade Center incident. In order to be effective, the Safety Officer and assistants *must* be empowered to stop any action that he or she deems unsafe without going through Command. This cease of action *must* be reported to Command at once, as it may have a rippling effect on overall incident operations and tactical objectives.

The Information Officer is responsible for developing and releasing information about the incident to the news media, to incident personnel, and to other appropriate agencies and organizations. Only one Information Officer will be assigned for each incident, including multi-jurisdiction incidents. The Information Officer may have multiple assistants as necessary. Those assistants may represent assisting agencies or jurisdictions. The Information Officer's functions are not limited to the release of information to the press. He or she may also assist all members of the Command and General Staff in obtaining information needed by staff members to enable them to complete their respective functions in their role in the overall incident. In the case of operations conducted under Unified Command, a Joint Information Center (JIC) is likely to be needed to coordinate all information released to the media about the incident. This is especially vital in incidents with active crime scenes as in the case of suspected terrorism cases

Incidents that are multi-jurisdictional or with several agencies involved may require a Liaison Officer. The Liaison Officer is the contact for the personnel assigned to the incident by assisting or cooperating agencies. These are personnel other than those on direct tactical assignments or those involved in a Unified Command. The Liaison Officer will be the point of contact for all responding agencies that arrive on the scene of an incident without prior specific tactical tasking provided to them by Command.

General Staff

The four highest organizational functions of the Incident Management System are operations, planning, logistics, and administration/finance. They are referred to as the General Staff.

The operations function consists of all field operational elements encompassing the tactical aspects of the incident. The Operations Section is supported by the members of the Command Staff. The Operations Section Chief's position of the Incident Management System is filled in as soon as practical when operational units arrive on the incident scene. Often this position is assumed by the first arriving tactical element as his or her upper level chain of command arrives on the scene of the incident. Prime focus should be placed on the span of control concerns in respect to all tactical operations. The leadership element of the Operations Section is referred to as the Operations Section Chief.

The Logistics Section consists of specialists in medical, supply, ground support, facilities, food, communications, transportation, and security. Their task is to assure that those in the operations section as well as all others involved in the incident are supplied with all they need. Early recognition of the need for a separate logistics function and section

Figure 4.11 Liaison agency equipment

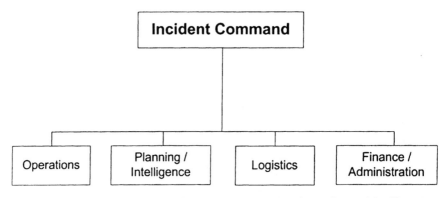

Figure 4.12 Top-level ICS Command Tree showing—General Staff

can reduce time and money spent on an incident. The leadership element is referred to as the Logistics Section Chief.

The Planning/Intelligence section is responsible for collecting and evaluating incident situation information, preparing situation status reports, displaying situation information, maintaining status of resources, developing an Incident Action Plan, and preparing incident-related documentation. This is done under the direction of the Planning/Intelligence Section Chief.

Finance/Administration consists of specialists in accounting, claims administration, and financial management services. In the case of small-scale incidents this is generally a function of the day-to-day budget process. However, in large scale or complex incidents it becomes very important in order to assure claims recovery from any number of sources, not limited to governmental cost recovery of disaster monies administered by the United States Department of Homeland Security. Generally this function is accomplished by using four standard units: a time unit, a procurement unit, a compensation/claims unit, and a cost unit. The leadership element is referred to as the Finance/Administration Section Chief.

Should an Incident Commander elect not to fill any of the General Staff positions, he or she by default assumes all the responsibilities for the tasks that fall under those unfilled positions. It is unlikely that this would occur with a complex incident. These General and Command Staff positions are likely to be filled in as rapidly as personnel qualified to handle the tasks and functions of each position arrive on the incident scene. Personnel assuming these roles are generally chosen for their knowledge, skills, and abilities as they pertain to the tasks at hand on a particular incident. Their proper selection is vital to the success in mitigating the incident.

All senior agency staff should be familiar with the capabilities and limitations of their junior or support staff. Senior staff should also constantly be looking for ways in which to improve the capabilities of these junior staffers from their entrance into their respective services. Further, senior staff should always foster a desire in those subordinate to them to aspire to higher levels. It is the role of every public safety leader to foster the development of future staff in their respective fields.

Not every person in every organization will understand all the roles and responsibilities of each operating position in the Incident Management System. It is not realistic to expect that every responder be prepared to fill every role, but in a marathon operation it will become necessary at times for people to operate outside their scope of knowledge in a position that needs to be filled. The tools found in the appendices of this book can assist people tasked with such roles in an actual situation.

Modular Organization

The organizational structures of any Incident Management System are developed in a modular fashion and are based on the type of incident and its needs as determined by the Incident Commander. As described earlier in this chapter the common functions of Command, Operations, Planning/Intelligence, Logistics, and Finance/Administration are utilized on all incidents no matter what their scope. As the needs of the incident grow, the system evolves to meet those needs. Within each of the five separate functional areas, several branches may be established.

Organizational Flexibility

The IMS organization adheres to a "form follows function" philosophy. In other words, the organization at any given time should reflect only what is required to meet planned tactical objectives. The size of the current organization and that of the next operational period are determined through the incident action planning process.

The specific organizational structure is established for any given incident and is based upon the management needs of that incident. If one individual can manage all major functional areas simultaneously, no further organization is necessary. If one or more of the functional areas require independent management, an individual will be named to be responsible for that particular area. As long as the functional needs are being met, the number of managers is not relevant.

Within the Incident Management System the first functional management assignments will be made by the Incident Commander and will normally be in the order of Operations, Logistics, Planning/Intelligence, and finally Finance/Administration. While this is a general rule in emergency operations, it is not a foregone conclusion. In the case of preplanned events or a slowly evolving emergency response, such as a hurricane, the system may deviate considerably from this model. The needs of the incident and the tactical objectives are the only elements that dictate how an incident management structure evolves.

The Incident Commander will generally establish a Section Chief. This Section Chief serves as the Officer-in-Charge (OIC) for a given functional area. His purpose is to supervise the specific functional area in all ways and to keep the Incident Commander abreast of the situation and the needs of the incident as they pertain to the specific functional area. Each Section Chief can further delegate management authority within his or her area as necessary, thereby maintaining a more efficient span of control for that element. These branches, as delegated authorities are called, are then assigned a Branch

Leader who will further delegate individual tasks within the branch as needed. Each activated element must have a person in charge. In some cases a single supervisor may initially be in charge of more than one element. Elements that have been activated and are clearly no longer needed should be deactivated to decrease organizational size and free up command level resources for more incident-pertinent issues. Again, the strategic and tactical needs of the incident and the foreseeable operational periods are the dictators of how this structure grows and shrinks as the incident progresses. The strategic and tactical needs of the incident alone will dictate the necessity for any given level of management. In some incidents the final management chart will be fairly flat and linear in nature, whereas in other more complex incidents, a very detailed incident management chart will evolve. Only the needs of the incident should dictate the management structure that is used at any point in any particular incident.

Integrated Communications

In the years since September 11, 2001 and the terrorist events of that day, the terms "interoperability" and "integrated communications" have become buzzwords in the emergency management and response business. Many commercial and technical solutions are being worked on by both public and private agencies to solve the problems of communications among public safety agencies. Many are based solely on technological solutions, but by themselves they will not solve all the communications problems that exist.

Integrated communications and the support systems needed to maintain them require considerable advance planning, which will coordinate tactical and support resources through the use of an incident-based communications center and of fixed emergency operations centers at several levels of government.

An awareness of available communications systems and frequencies, combined with an understanding of incident requirements, will enable the Communications Unit Leader to develop an effective plan for each operational period. This may include the use of several commonly used "NETS" or communication networks such as:

- Command net: links supervisory personnel from the Incident Commander down to and including division and group supervisors
- Tactical nets: agency, department, geographical area, or function connections established for branches or divisions, and groups depending upon hardware and frequency availability and specific incident needs
- Support nets: handle logistics traffic and resource status changes during larger incidents
- Ground-to-air nets: coordinate ground-to-air traffic
- Air-to-air nets: coordinate aircraft assigned to an incident

One major consideration that has implications for all emergency response forces is the use of "codes" during radio and other forms of communications. Wherever possible all radio communications between response agencies and their organizational elements at an incident should be accomplished in clear text. The use of clear text will do much to alleviate the misunderstandings described earlier in this chapter that have occurred in the past at emergency incidents.

Consolidated Action Plan

It is essential that every incident or event be managed according to a plan. In the Incident Management System the basic plan is called the Incident Action Plan (IAP). On day-to-day incidents and more complex incidents of relatively short duration, most Incident Action Plans will be written for a specific operational period such as twelve hours. While event action planning is slightly different in nature, similar principles will apply.

For simple incidents of short duration, the Incident Action Plan will be developed by the Incident Commander and communicated to subordinates verbally. Often it does not require much formal planning. Any written incident documentation such as an ICS 201 form will serve as the legal and historical record of such an Incident Action Plan.

Formal written Incident Action Plans are likely to be needed when the following incident conditions are met:

- Two or more jurisdictions are involved.
- The incident continues into another operational period.
- A number of organizational elements have been activated.
- It is required by agency policy.

Figure 4.13 IMS command vehicle

The ultimate decision to prepare a written Incident Action Plan rests with the Incident Commander. The use of a written plan provides the Incident Commander with:

- A clear statement of objectives and actions
- A basis for measuring work and cost effectiveness
- A basis for measuring work progress
- Accountability

One fairly easy method of preparing a written Incident Action Plan is to utilize a form such as the one developed by the National Wildfire Coordinating Group.

Incident Action Plans should be prepared for specific time periods called operational periods. They can be of various lengths, although they should normally be no longer than a full day. Generally a twelve-hour operational period is used.

An operational period will be affected by:

- Length of time available/needed to achieve tactical objectives
- Availability of fresh resources
- Future involvement of additional jurisdictions and/or agencies
- Environmental considerations such as remaining daylight or weather conditions
- Safety considerations

In order to develop an effective Incident Action Plan, the Incident Commander must do the following:

- Understand the situation
- Establish objectives and strategy
- Develop tactical direction and assignments
- Prepare the plan
- Implement the plan
- Evaluate the plan

The time required to develop a plan will vary based on the type of incident and the agencies involved.

Plans, informal and formal, are the glue that holds the Incident Management System together, enabling it to work more efficiently. The time spent to plan both the operational tactics and the strategic goals of an incident will greatly save time in the implementation of both.

Manageable Span of Control

A basic tenet of an effective Incident Management System is the concept of span of control. Span of control relates to how many organizational elements (resources) can be directly managed by one person. Maintaining an effective span of control is particularly important in emergency incidents, as personnel safety and accountability have top priority. Safety factors, as well as sound management planning, will dictate span-of-control considerations.

An important consideration is to anticipate and prepare for change. This is especially true during a rapid buildup of staff when management becomes difficult because of too many reporting elements. Planning is critical at this point to avoid unconsidered ordering of resources and consequent loss of effective span of control.

An effective span of control may vary from three to seven people under one manager. The rule of thumb for span of control in emergency incident management is one supervisor to four subordinates. If the number of reporting elements exceeds or falls short of this optimum, expansion or consolidation of the organization may be necessary. Maintaining an adequate span of control throughout the entire Incident Management System organization is very important. The more hazardous the operations that a group will be undertaking, the more people are needed to safely and effectively oversee units engaged in hazardous operations.

Pre-designated Incident Facilities

Several kinds of facilities may be established in and around the incident area. The determination of these facilities and their locations will be based upon the requirements of the incident and space availability as well as basic geography. The facilities will be established at the direction and discretion of the Incident Commander or his or her subordinates.

These facilities are commonly used at all but the smallest emergency incidents. Some are commonly used at specific types of incidents such as wildfires and/or HAZMAT incidents. The most common facilities are the Incident Command post, Staging Area, Rehabilitation Area, Base Camp, Helibase, Helispot, Triage Area, and Treatment Areas.

Comprehensive Resource Management

When dealing with resource management issues in terms of the Incident Management System, there are four main areas of consideration that are essential:

- Planning
- Organizing
- Directing
- Controlling

Planning is the process of evaluating the current situation, determining objectives, selecting a proper strategy, deciding acceptable outcomes, and determining which resources should be used to achieve those objectives and outcomes in the most efficient and cost-effective manner while maintaining safety for all responders.

Organizing is the next step of the management process after planning. The Incident Commander brings essential personnel and equipment resources together into a formalized relationship. A major part of the organizational process of the Incident Management System is the Incident Action Plan, discussed above.

Directing is the process of guiding and supervising the efforts of resources to attain the specified control objectives. Direction takes place at both a tactical and a strategic level and at all levels of the incident command process. A very important part of directing resources, particularly in the high-stress environment of an incident, is providing proper motivation, leadership, and delegation of authority in ways that enhance the organizational structure of the Incident Management System as well as ensuring the safety of those personnel operating on an incident scene.

Controlling involves evaluating the performance of the entire organization and applying the necessary corrections to make sure that progress is constantly directed toward accomplishing the established objectives.

Managing resources safely and effectively is the most important consideration at any incident regardless of size and complexity. The incident resource management process includes several interactive activities that take place simultaneously or separately and that help to achieve the goal of maintaining as safe an operational environment as possible:

- Establishing resource needs
- Resource ordering
- Check-in process
- Resource use
- Resource demobilization

Resource managers must be constantly aware that the decisions they make regarding the use of personnel and equipment resources will not only affect the timely and satisfactory conclusion of the incident but also may have significant cost implications.

Management by Objectives

Within Incident Management Systems the concept of management by objectives covers four essential steps. These steps take place during every incident regardless of size or complexity:

- Understand agency policy and direction
- Establish incident objectives
- Select appropriate strategy
- Perform tactical direction (applying tactics appropriate to the strategy, assigning the right resources, and monitoring performance)

Unified Command

Unified command is a management process that allows all agencies that have jurisdictional or functional responsibility for the incident to jointly develop a common set of incident objectives and strategies. This is accomplished without losing or giving up agency authority, responsibility, or accountability. Under unified command the following always apply:

- The incident will function under a single, coordinated Incident Action Plan.
- One Operations Section Chief will have responsibility for implementing the Incident Action Plan within that section.
- One Incident Command Post will be established (per Incident).

Unity and Chain of Command

In an effective Incident Management System the concept of unity of command means that every individual has a designated supervisor. A chain of command is an orderly line of authority and clear succession within the ranks of the organization, with lower levels subordinate to and connected to higher levels.

In the lingo of the fire service, "Ya got to know who your boss is and who he's working for too."

It is likely that, in 95 percent of incidents, the organizational structure for operations will consist of:

- Command
- Single resources

However, as incidents expand, the chain of command is established through an organizational structure that can consist of several layers as needed:

- Command
- Sections
- Branches
- Divisions/groups
- Units
- Resources

Establishment and Transfer of Command

Command at an incident is initially established by the highest-ranking authority at the scene that has jurisdiction for the incident.

Transfer of Command for that incident may take place for the following reasons:

- A more qualified person assumes command.
- The incident situation changes over time so that a jurisdictional or agency change in command is legally required or it makes good management sense to make a transfer of command.
- Normal turnover of personnel on long or extended incidents

Common Terminology

If an Incident Management System is to be effective at both smaller scales that are local in nature as well as multi-jurisdictional state or national incidents. common terminology must be applied to:

- Organizational elements
- Position titles
- Resources
- Facilities

With organizational elements there should be a consistent pattern for designating each level of the organization (e.g., sections, branches, etc.).

Those charged with management or leadership responsibility in the Incident Management System are referred to by position title, such as Chief Officer, Director, Supervisor,and Leader. This is done to place the most qualified personnel in organizational positions on multi-agency incidents without incurring the confusion caused by various multi-agency rank designations. It also provides a standardized method for ordering personnel to fill positions, and it eliminates the problems caused by the complexity of terms used to describe operational titles by the various responder disciplines.

Common designations are assigned to various kinds of resources. Many resources may also be classified by type, which indicates their capabilities (e.g., types of helicopters, patrol units, engines). For example, in an Incident Management System, a vehicle that is used in fire suppression is called an engine. Recognizing that there are a variety of engines, a type classification is given based on tank capacity, pumping capability, staffing, and other factors.

Several procedures used within the scope of an Incident Management System to ensure personnel accountability:

- Check-in is mandatory for all personnel upon arrival at an incident.
- Unity of command ensures that everybody has only one supervisor.
- Resource status unit maintains status of all assigned resources.
- Division/Group assignment lists identify resources with active assignments in the operations section.
- Unit logs provide a record of personnel assigned and major events in all ICS organizational elements.

CONCLUSION

The common components of the Incident Management System make the system work all the time, assuming that each component is applied on a universal basis by all responders who are working on any given incident. The components are needed in order to

Figure 4.14 IMS training session

Figure 4.15 IMS simulation in use

ensure that the system will fulfill the promise of "bringing control to chaos." The key to the implementation of an effective Incident Management System is the training of the personnel at all levels and in all organizations expected to be a part of the response to an emergency incident.

Chapter 5

MAJOR FUNCTIONS OF THE INCIDENT MANAGEMENT SYSTEM

Applying the theory and concepts of Incident Management to the real world is much like running a business or even a war. Life and death can hang in the balance quite literally in all three cases if taken to the extreme or if mistakes are made.

As we have noted previously, the five functional areas of an Incident Management System are:

- Command
- Operations
- Planning/Intelligence
- Logistics
- Finance/Administration

Simply put, Command has the overall authority and ultimately the overall responsibility for an incident. Operations is responsible for developing and directing the tactical actions designed to support and complete the incident objectives as specified by the Incident Commander. Planning/Intelligence is responsible for collecting, evaluating, disseminating, and using incident data and intelligence. Logistics creates a support and supply chain and Finance/Administration is responsible for cost accounting and procurement of supplies and services as well as cost documentation and final reporting.

COMMAND

As discussed in previous chapters, Command may be led by a single agency such as police, fire, or EMS, with the other agencies in supporting roles or by a unified

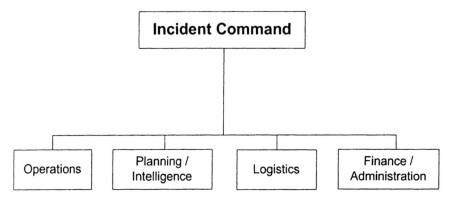

Figure 5.1 Top Level Incident Command System chart

structure consisting of multiple agency management, external agency representatives, and a variety of support personnel.

The chosen structure of the Command element used is based on the type, scope, and complexity of the incident and should be driven by the needs of the incident as opposed to the needs of any specific agency. By default, elements of the first arriving units of the responding agencies assume the Command position but may later be replaced.

Generally, the individual leading this function is referred to as the Incident Commander. Each incident will have only one Incident Commander unless a Unified Command structure is established. Often a single-agency Command is used at the onset of an incident and then transitions to a Unified Command as dictated by the needs of the incident.

An Incident Commander is accountable to the Agency Administrator(s) in charge of the agency in control of the incident. It is the Agency Administrator's responsibility to orient, counsel, and instruct the Incident Commander on agency specific management objectives and expected accomplishments. In some situations or agencies, a lower-ranking but more qualified person (for that incident) may be designated as the Incident Commander. The Incident Commander will perform the major ICS organizational functions until he or she determines that one or more should be delegated. This may occur at the beginning of the incident or later as needs determine.

Agency Administrator

An Agency Administrator is responsible for the management of resources in his or her jurisdiction or organizational unit (agency) and ultimately is held accountable for the overall results of that agency or jurisdiction. Protection from risk and hazard is a major part of that function's responsibility. Thus, the Agency Administrator has a responsibility to establish standards for expected performances and to hold the Incident Commander accountable. Standards and objectives are established through an agency strategic plan. That plan will guide agency personnel in day-to-day operations and in crisis situations. On large-scale incidents where the agency has a significant public exposure, it is not uncommon for Agency Administrators to respond to the scene for a face-to-face briefing from an Incident Commander.

Unified Command

In Unified Command individuals from respective agencies or departments who may be Incident Commander "types" work together to develop and achieve the stated incident objectives. Unified Command is an incident management process that allows all agencies with jurisdictional or functional responsibility for the incident to jointly develop a common set of incident objectives and strategies. This is accomplished without losing or giving up any jurisdictional or agency authority, responsibility, or accountability. It allows each agency having a legitimate responsibility at an incident to be part of the Command function.

Incident Commander Responsibilities

The Incident Commander has the following overall responsibilities for any incident or event, large or small. The Incident Commander must provide for the effective and safe execution of each of the five functional areas by doing the following:

- Assessing the situation and and/or obtaining a briefing from the prior Incident Commander
- Determining incident strategy and incident objectives
- Establishing immediate incident priorities
- Establishing an Incident Command Post
- Establishing an appropriate incident management organization
- Ensuring that planning meetings are scheduled as required
- Authorizing and approving the implementation of an Incident Action Plan

Figure 5.2 Agency Administrator being briefed by IC

- Ensuring that adequate safety measures are in place
- Coordinating activities for all Command and General Staff
- Coordinating with key people and officials from all concerned agencies and organizations
- Approving requests for additional resources or for the release of resources
- Informing Agency Administrators of incident status
- Approving the use of students, volunteers, and auxiliary personnel
- Authorizing release of information to the news media
- Ordering the demobilization of the incident
- Developing and overseeing the creation of any post-incident after action reports

Although some of the functional sections may be left unfilled, there will always be an Incident Commander. On small incidents he or she may perform all of the functional operations. On large incidents the Incident Commander may delegate the authority for managing certain functions. Let's discuss each of these responsibilities in turn.

Assessing the Situation

It is critical for Command to do an assessment of the situation and the surrounding environment. Sometimes this is referred to as "sizing up" the incident but it is more often referred to as the size-up process. Size-up is a mental evaluation of the situation. It includes the life hazard of occupants and responders, the location of the incident (where and how big), building construction, occupancy (contents) of affected buildings and exposures, height and area of affected buildings and exposures, exposures specifically within incident buildings and surrounding an area. Weather (current and forecasted), resources (personnel and equipment), water or other incident resources (available and needed), and finally time (duration and time of incident) are considered. Some incidents may not require all these elements, but in general the idea of a size-up is to rapidly determine the problems and issues immediately facing the Incident Commander in a given incident.

If Command is being transferred from one Incident Commander to another, it also critical that there be a face-to-face transition to determine that the intent of the incident objectives and strategy for the incident are understood by the incoming Incident Commander. This transfer of command is done to improve the quality of the command organization. Usually, the person in charge of the first arriving operational unit at the scene of an incident assumes the Incident Commander role. This will normally be someone who is designated by the organization or authority having jurisdiction. That person will remain in charge until formally relieved or until transfer of command is accomplished.

Determining Incident Objectives and Strategy

The determination of incident objectives and strategy is based on a continuous size-up process. The situational assessment may cause the objectives to be changed and updated. This is what is meant by Management-by-Objectives. Command must understand agency policy and direction. This may come from the Agency Adminis-trator, board of directors, city council, commissioners' court, or other governmental body. With the agency policy and direction, Command then establishes the respec-tive incident objectives. The objectives are used to select the most appropriate strat-egy to mitigate the incident.

Establishing Immediate Priorities

With objectives and strategy in place, we then can establish immediate priorities, which in turn determine the tactical direction, or the appropriate tactics based on resource allocation and performance monitoring. If additional resources are needed, they are ordered and obtained.

An Incident Commander's first priority should always be the safety of people involved in the incident, the responders, other emergency workers, and bystanders. The second priority should be incident stabilization. Stabilization time is normally tied directly to incident complexity.

In stabilizing the incident situation, the following "musts" are essential for the Incident Commander: to ensure life safety, stay in command, and manage resources efficiently and cost effectively

Establishing an Incident Command Post

Command must ensure that an Incident Command Post is established. Once it is in place, it is essential that its location and effectiveness are constantly evaluated.

Initially, the Incident Command Post will be wherever the Incident Commander is physically located. As the incident grows, it is important for the Incident Commander to establish a fixed location for the Incident Command Post and to work from that location.

The Incident Command Post provides a central coordination point from which the Incident Commander will normally operate. Depending on the incident, Command, Planning/Intelligence, and other members of the general staff may be operating in other locations. However, they will attend planning meetings and be in close contact with the Incident Commander.

The Incident Command Post can be any type of facility that is available and appro-priate e.g., vehicle, trailer, tent, an open area, or a room in a building. The Incident Command Post may be located at the Incident Base if that facility has been estab-lished. Once established, the Incident Command Post should not be moved unless absolutely necessary.

Establishing an Appropriate Incident Management Organization

Command must also evaluate whether an appropriate organization is in place. If not, Command must ensure that the right personnel are obtained. The Incident Commander must determine if span of control has approached or soon will approach practical limits. The appropriate span of control ensures that personnel safety factors are taken

Figure 5.3 Incident Command Post vehicle

into account. The organization at any given time should reflect only what is required to meet planned tactical objectives for the incident and the operational period.

One of the primary duties of the Incident Commander is overseeing the management organization. The organization needs to be large enough to do the job at hand, yet resource use must be cost-effective. Anticipated expansion or diminishment of the incident will call for corresponding changes to the organization. The Incident Commander is responsible for delegating authority as appropriate to meet the need.

Ensuring that Planning Meetings are Scheduled as Required

As part of the command function, the Incident Commander ensures that Planning Meetings are scheduled. Regularly scheduled and held planning meetings are essential to achieving the incident objectives. On many incidents, time factors do not allow for prolonged planning and operations must be conducted "on the fly" in more complex incidents planning will be crucial to the success of the response operations. On the other hand, lack of planning can be disastrous. Therefore, it is important to know and use an effective planning process. This includes having an agenda that is followed. Proactive planning is essential in considering future needs of the incident.

Approving and Authorizing the Implementation of an Incident Action Plan

At the Planning Meeting, an Incident Action Plan is developed. The Incident Commander is responsible for approving and authorizing the implementation of the Incident

Action Plan. The planning process will be covered in further detail in the planning section and subsequent chapters.

Incident Management Systems offer great flexibility in the use of Incident Action Plans. The plans may be oral or written. It should be noted that even with "oral" plans, the Incident Commander should document the objectives and directions that he has given to subordinates as directed. Hence, even with oral plans, there is a written plan, even if it is generated after the incident is completed in the form of an after-action report.

Written plans should be provided for all multi-jurisdictional or multi-agency incidents or when the incident will continue for more than one operational period. In this instance, we refer to written plans that are printed and provided to all supervisory personnel.

Ensuring that Adequate Safety Measures Are in Place

In developing the incident objectives, strategy, tactics, and plans, the Incident Commander ensures that adequate safety measures are in place. In most incidents, the Incident Commander will appoint a Safety Officer as part of the Command Staff. As a part of safety measures, a determination is also made of any environmental issues particular to the incident and all incident personnel must be made aware of such situations.

Life safety at the scene of an incident is always the top priority. If the incident is complex or if the Incident Commander is not a tactical expert in all the hazards present, a Safety Officer should be assigned. Note that under federal law, hazardous-materials incidents require the assignment of a Safety Officer who is competent in hazardous materials at a minimum training level of HAZMAT Technician pursuant to Occupational Safety and Health Administration regulations.

Coordinating Activity for Command and General Staff

As a manager, the Incident Commander coordinates the activities of his or her staff and monitors the progress of work being accomplished in support of the objectives. The Incident Commander reviews and modifies incident objectives and adjusts the Incident Action Plan as necessary. As a result of this coordination and the evaluation of span of control, the Incident Commander may mobilize or demobilize as needed personnel for the General and or Command Staff.

Coordinating with Key People and Officials from Concerned Agencies and Organizations

The Incident Commander also has the responsibility to coordinate with key people and officials from all concerned agencies and organizations. This includes local, state, and federal elected officials and may also include environmental groups and other atypical organizations. These individuals have a responsibility to their constituents to make sure their best interests are being looked after and the Incident Commander has a responsibility to assist them in meeting their needs based on the incident and its objectives.

Approving Requests for Additional Resources or for the Release of Resources

The Incident Commander approves the request for and release of resources by taking into account organizational and span-of-control issues.

On small incidents, the IC will personally determine additional resources needed and order them. As incidents grow in size and complexity, the ordering responsibility for required resources will shift to the Logistics Section Chief and the Supply Unit if those elements of the organization have been established. The Incident Commander will provide the parameters upon which these orders may be placed.

Keeping Agency Administrators Informed of Incident Status

The Incident Commander must keep the Agency Administrator(s) informed of incident status, since normally Agency Administrator(s) will not be present at the scene and they must keep their superiors or their constituents informed as well. Local, city, county, regional, state, and national interests may be involved depending on the scale and complexity of the incident.

Approving the Use of Students, Volunteers, and Auxiliary Personnel

The use of students, volunteers, and auxiliary personnel must be approved by the Incident Commander. Since there is a significant potential liability issue connected with the use of these resources, there needs to be close scrutiny of their use. Incident Commanders must be prepared for unqualified personnel volunteering to assist on incidents, especially those that have become high-profile. It may necessary to create a group in the Operations or Logistics Section to manage the use of such personnel.

Authorizing Release of Information to the News Media

Today, the news media has become an important element in any incident or event. To that extent, the Incident Commander needs to authorize the release of information to the media. It is important that the facts are given, but only so long as confidential information is not involved.

Agencies will have different policies and procedures relating to the handling of public information. What is presented here is general information as to how an Information Officer functions in the Incident Management System.

One significant change within recent years is the increased capability and desire of the media to obtain immediate access to information. The sophistication of modern news-gathering methods and equipment makes it very important that all incidents have procedures in place for managing the release of information to the media as well as responding appropriately to media inquiries.

It is not unusual that on some incidents the media may have recent and accurate information that is not yet available to the Incident Commander through internal lines of communication. In some cases media coverage may inadvertently affect priorities. An example of this is a wide area search, regardless of the weather, for individuals, missing children, or criminal suspects. In these cases the media is the key element in getting the proper information to the general public on a large scale in a timely manner.

Ordering the Demobilization of the Incident

Once resources become surplus and are deemed no longer needed for incident mitigation the Incident Commander orders their demobilization as well as that of the incident itself when objectives are accomplished. This demobilization will generally be coordinated by the planning section together with all command and general staff.

Developing and Overseeing the Creation of Any Post-Incident After Action Reports Regarding the Incident

The creation of some type of post-incident report will be needed in almost every response to an emergency due to the nature of the work that emergency services agencies are involved in. This may be as simple as a one-page report or as complex as the reports generated by the World Trade Center Commission and other such bodies seated after significant disaster situations. This process can become very complex and can literally take years to complete. Incident Commanders and the other

Figure 5.4 Public Information Officer briefing the media

members of the General and Command Staff will be expected to assist in the creation of these reports.

Characteristics of an Effective Incident Commander

With all the responsibilities that an Incident Commander is expected to perform, it goes without saying that it takes a very special person to be effective. He or she is normally the most visible person on the incident scene. Following are just some of the characteristics associated with an effective Incident Commander:

- Command presence
- Understands ICS
- A proven manager
- Puts safety first
- Proactive
- Decisive
- Objective
- Calm
- Quick-thinking
- Good communicator
- Adaptable and flexible
- Realistic about personal limitations
- Politically astute

Chain of Command Basics

The Incident Commander may have one or more Deputies. The only requirement regarding the use of Deputies, whether at the Incident Commander, Section, Division, or Branch level, is that they must be qualified to assume the position that they are subordinate to. The Deputy Incident Commander(s) may perform specific tasks as requested by the Incident Commander.

He or she may be called upon to perform the function in a relief capacity e.g., to take over for one or more operational periods. (In this case the Deputy Incident Commander will assume the primary Incident Commander role during that period.) He or she may also represent an assisting agency that may share jurisdiction or have jurisdiction in the future.

Command Staff

The Incident Commander will perform the functions of the Command Staff, Information, Safety, and Liaison unless and until he or she determines that one or more of these functions should be delegated. In some cases, there may be an Intelligence Officer as part of the Command Staff or this may be delegated to the Planning/Intelligence Section Chief depending on the exact structure of the Incident Management System in place. This option is becoming more and more popular in practice.

The difference between the Command Staff and the General Staff is that the former is considered to be the implementer of the plan and the latter is part of the Incident Commander's Staff. The General Staff has line authority, and the Command Staff is generally said to have an advisory authority or capacity. It should be remembered that both report to and work directly for the Incident Commander.

As stated, the Command Staff is appointed to release the Incident Commander from specific functions. Command Staff responsibilities will be performed by the Incident Commander unless the responsibility is delegated to a staff member.

Under the Chain of Command, assistants are other subordinate Command Staff positions, particularly those of Information Officer and Safety Officer. As with deputies, they too should be fully qualified for the positions they hold; however, they need not have the level of technical capability, qualifications, and responsibility characteristic of primary positions.

In Command Staff positions, there are some common rules. Only one person will be designated for each of the Command Staff positions. Command Staff positions should not be combined. Command Staff positions may be filled by persons from other agencies or jurisdictions. There are no Deputy positions at the Command Staff level. Each of the positions may have one or more Assistants as necessary. Assistants are recommended for larger incidents. Assistants can be designated from other jurisdictions or agencies as appropriate. Each member of the Command Staff reports directly to the Incident Commander. Command Staff members may interact with any position within the ICS for purposes of information exchange.

Information

The information function is normally set up when there is a high-visibility or sensitive incident or when media demands may obstruct the Incident Commander's effectiveness.

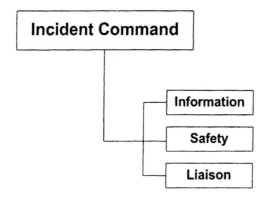

Figure 5.5 Top ICS Chart Showing Command Staff

The primary position is of the Information Officer is to serve as the Public Information Officer. The Public Information Officer is the central point for dissemination of information to the news media and to other agencies and organizations. Only one Information Officer will be named to an incident, including multi-jurisdictional incidents. In the case of extremely complex incidents a Joint Information Center will likely be formed where all information to be disseminated to the public will be vetted prior to release to assure that both accurate information is released and that any sensitive or classified information will be treated appropriately.

The media will often try to acquire information from its own sources. It is critical that Command Staff provide the information as opposed to unauthorized sources. Having a Public Information Officer reduces the risk of multiple sources of information being released. It also provides the ability to alert or warn the public if necessary. Agencies will have different policies and procedures relating to the handling of public information. What is presented here is general information as to how an Information Officer functions in the Incident Management System.

The Information Officer is responsible for developing and releasing information about the incident to the news media, to incident personnel, and to other appropriate agencies and organizations. The Information Officer may have assistants as necessary. These Assistant Information Officers may also represent assisting agencies or jurisdictions.

Major responsibilities of the Information Officer are to determine from the Incident Commander if there are any limits on Information releases. They develop material for use in media briefings and obtain the Incident Commander's approval of such materials prior release. They are responsible for informing the media and conducting media briefings. As a part of that function, they may arrange for tours and other interviews or briefings that may be required. The Information Officer function may also obtain information from the media that may be useful to incident planning. It may come from the community in the vicinity of the incident. Additionally, they maintain current information summaries and/or displays of the incident and provide information on the status of the incident to assigned personnel. This could include the development and maintenance of bulletin boards as well as a websites.

The Information Officer may also provide advice and counsel on information matters to the Incident Commander. He or she is essentially the eyes and ears of the incident command team.

When a Joint Information Center is set up, the Public Information Officer will coordinate its operation. This is all a part of shaping the public perception of the incident and the agency. It also gives the Information Officer the ability to disseminate information regarding the incident, especially emerging issues and opportunities for positive coverage. The Information Officer would also participate in any Incident Management Team planning meetings and briefings.

In setting up a Joint Information Center or any media area, it is important that it be separate from the Incident Command Post but still close enough to access information. If multiple agencies are involved, this would be formally referred to as a Joint Information Center. It would be a facility established to coordinate all incident-related public information activities. It is the central point of contact for all news media at

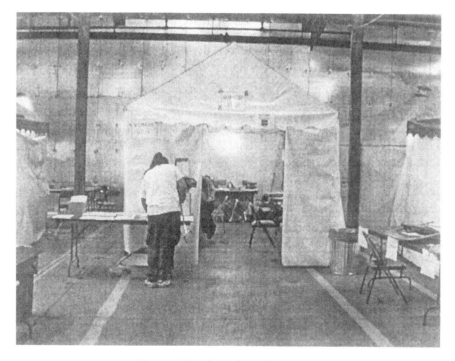

Figure 5.6 Joint Information Center

the scene of the incident. Information officials from all participating agencies should meet at the Joint Information Center.

Safety Officer

The Safety Officer's functions are to develop and recommend measures for assuring personnel safety and to assess and/or anticipate hazardous and unsafe situations.

Only one Safety Officer will be assigned for each incident and will report directly to the Incident Commander. The Safety Officer may have Assistants as necessary; they may represent assisting agencies or jurisdictions. Safety Assistants may have specific responsibilities such as air operations, hazardous materials incidents, Division, Camp assignments, or building construction. The assistants would be assigned to various parts of the incident depending on its complexity. A Line or Team Safety Officer is an assistant unique to a portion of the incident that warrants special safety considerations as deemed necessary by the Safety Officer in charge.

The Safety Officer participates in the Planning Meeting and any briefings. As a part of the Planning Meeting as well as throughout an Operational Period, the Safety Officer identifies hazardous situations associated with the incident. Additionally, the Safety Officer will review the Incident Action Plan for safety implications and determine possible ways to reduce the risks and/or hazards. Once a tactical plan is implemented, the

Safety Officer may exercise emergency authority to directly stop unsafe acts if personnel are in imminent, life-threatening situations. If an accident occurs, the Safety Officer initiates the initial investigation.

The Safety Officer is responsible for reviewing and approving the incident Medical Plan, which is normally developed by the Medical Unit in the Logistics Section. The Safety Officer assists in developing the Incident Action Plan, determines what risks and hazards are present within the incident area, develops a mitigation plan, and prepares and sends a safety message as part of the IAP. When working with hazardous materials, the Safety Officer would also review and approve the Hazardous Materials Site Safety & Control Plan (ICS Form 208-HM) as required.

When the Demobilization Plan is developed, the Safety Officer has a critical role. He is responsible for the evaluation of personnel prior to release and becomes important in demobilization. The Incident Commander may also request that contingency plans, such as an Evacuation Plan, be developed. The Safety Officer would become involved in those as well. The Safety Officer will be involved in any such plans and should be consulted in the creation of any such plan.

Liaison Officer

Incidents that are multi-jurisdictional in nature or that have several agencies involved may require a Liaison Officer, since the Incident Commander may not have time for individual coordination of the multitudes of agencies that will respond to the incident. The Liaison Officer is the point of contact at the incident for personnel from assisting or cooperating agencies. Only one Liaison Officer will be assigned for each incident, including those operating under Unified Command. As with the Information and Safety Officers, on very large incidents the use of Assistants may be required. The Assistants may also represent assisting agencies or jurisdictions. Any agency participating in the incident by directly contributing tactical resources to the responsible agency or jurisdiction is known as an assisting agency. Thus, fire, police, or public-works equipment sent to another jurisdiction would be considered assisting agency resources. An agency that supports the incident or supplies assistance other than tactical resources would be considered a cooperating agency. Examples include the American Red Cross, Salvation Army, or a utility company. On some law-enforcement incidents, a fire agency may not send fire equipment but may supply representatives for coordination purposes. In this case, the fire agency would be considered a cooperating agency.

The Liaison Officer is a contact point for agency representatives, and he or she maintains a list of assisting and cooperating agencies and their representatives. He or she assists in establishing and coordinating interagency contacts as well as keeping supporting agencies aware of incident status. The Liaison Officer monitors incident operations to identify current or potential inter-organizational problems. He or she also participates in planning meetings, provides current resource status, including limitations and capability of assisting agency resources.

When the Demobilization Plan is developed, the Liaison Officer provides agency-specific demobilization information and requirements. Some agencies may have union agreements, which may be different from other agencies or departments; thus the Liaison Officer would ensure that those agreements are addressed in the planning and implementation phases of the incident.

Agency Representatives

While the Incident Commander may designate other special staff positions, the agency representative is the final position that we will discuss as part of the Command Staff. Agency representatives report to the Liaison Officer; if there is none, they report directly to the Incident Commander.

An outside agency may send a representative to work with the incident management staff to coordinate among agencies or jurisdictions. It is important to note that that representatives need to have decision-making authority for their respective agencies. A lack of such authority can delay implementing actions on an incident and ultimately have a negative impact.

Some agencies may have a designated representative and also a more senior person involved in another assignment on the same incident. The respective roles need to be clarified. Agency representatives should not be used in a Unified Command structure unless they are qualified by their agency as an Incident Commander.

Figure 5.7 Agency Representative

Agency representative responsibilities include ensuring that all agency resources used are properly recorded at the incident. They should obtain briefing information from the Liaison Officer or Incident Commander. They should inform assisting or cooperating agency personnel that the position for that particular agency has been filled. Representatives should attend briefings and planning meetings as required by the Liaison Officer or Incident Commander. They should provide input on the use of agency resources unless technical specialists are assigned from that agency. One very important role of the agency representative is to ensure the well-being of agency personnel assigned to the incident and to advise the Liaison Officer of any special agency needs or requirements. They should report to their home-agency dispatch or headquarters on a prearranged schedule, ensuring that all agency personnel and equipment are properly accounted for and released prior to departure. They should finally ensure that all required agency forms, reports, and documents are complete and attend a debriefing session with the Liaison Officer or Incident Commander prior to departure.

Intelligence Officer

In a terrorist or law-enforcement incident, an Intelligence Officer may be assigned. The Intelligence Officer is responsible for the management of internal information, intelligence, and operational security requirements. These may include information security and operational security activities, as well as the complex task of ensuring that sensitive information of all types (e.g., classified information, law-enforcement information, proprietary information, or export-controlled information) is handled in a way that not only safeguards the information but also ensures that it is received by those who need access to it to effectively and safely perform their missions.

General Staff

As noted earlier, the General Staff is the executor or implementer of the Incident Action Plan. General guidelines related to General Staff positions dictate that only one person be designated to lead each General Staff position. General Staff positions may be filled by qualified persons from any agency or jurisdiction. Members of the General Staff report directly to and only to the Incident Commander. If a General Staff position is not activated, the Incident Commander will retain responsibility for that functional activity. Deputy positions may be established for each of the General Staff positions. Deputies are individuals fully qualified to fill the primary position. Deputies may be designated from other jurisdictions or agencies as appropriate. Doing so is a good way to bring about greater interagency coordination.

General Staff members may exchange information with any person within the organization. Direction takes place through the chain of command. This is an important concept in the Incident Management System. General Staff positions should not be combined. It is better to initially create separate functions and if necessary for a short time to place one person in charge of them, allowing, the subsequent transfer of responsibility to be made more easily.

Operations Section

The Operations Section is responsible for directing and coordinating all tactical actions to meet incident objectives. The Operations Section can be managed by the Incident Commander; however, in most cases the Incident Commander will designate an Operations Section Chief to direct and coordinate it. The Operations Section Chief ensures the safety of section resources, personnel, and equipment. The Operations Section Chief assists the Incident Commander in developing response goals and objectives for the incident. He or she is responsible for implementing the Incident Action Plan. The Operations Section Chief determines what tactical actions will be necessary to accomplish the Incident Commander's objectives and strategy. This will include determining what resources will be needed. Once the tasks are accomplished, the Operations Section Chief determines what resources are surplus to his needs. The Operations Section Chief keeps the Incident Commander informed of the situation and resource status within the operation. Only one person is assigned as Operations Section Chief; he or she reports directly to the Incident Commander. Frequently, the Operations Section Chief will have been originally assigned as the initial attack Incident Commander.

The Operations Section is responsible for carrying out the response activities described in the Incident Action Plan. The Operations Section Chief coordinates Operations Section activities and has primary responsibility for receiving, implementing, and executing the plan. The Operations Section Chief also determines the required resources and organizational structure within the Operations Section. As it is the Operations Sec-

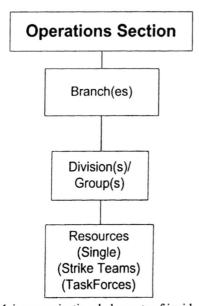

Figure 5.8 Major organizational elements of incident operations

tion Chief's responsibility to request additional resources to support tactical operations, he or she also approves release of resources from active assignment.

The Operations Section Chief's main responsibilities are to direct and coordinate all operations, ensuring the safety of all incident personnel. He or she may assist the Incident Commander in developing response goals and objectives for the incident. It may be necessary to make or approve expedient changes to the operations portion of the Incident Action Plan. If this occurs, the Operations Section Chief should inform the Incident Commander of those changes.

The Operation Section Chief ensures that interaction is taking place with other agencies. It is critical that close contact is maintained with subordinate positions to assure safe tactical operations. It is imperative that the Safety Officer and Operations Section Chief work in close coordination.

Operations at an incident or event can be set up in a variety of ways depending upon the kind of incident, the agencies involved, and the objectives and strategy. The Operations Section will expand or contract based upon the existing and projected needs of the incident. Initially, the Operations Section usually consists of those few resources first assigned to an incident. (These resources will initially report directly to the Incident Commander.) As additional resources are committed and the incident becomes more complex, an Operations Section may be established. The Operations Section develops from the bottom up by first establishing divisions, then groups, and, if necessary, branches. Also, the Operations Section may have staging areas and, in some cases, an air assets organization.

We will briefly examine a number of organization types and four methods of organization.

Geographic Divisions

A common method of organizing tactical operations at an incident is for the Incident Commander to first establish two or more Divisions. Divisions always refer to geographically defined areas, such as the area around a stadium, the inside or floors of a building, or an open area. Initially, divisions may be established to define the incident and may or may not include the designation of separate Division Supervisors.

If there are two floors in a building, for example, Division 1 may be designated the first floor, and Division 2 the second floor. The Incident Commander or Operations Section Chief may not have named separate supervisors. When the resources assigned within a Division exceed or will soon exceed the recommended span-of-control guidelines of one to five, Division Supervisors should be designated. Divisions not under the direct management of the Incident Commander or Operations Section Chief are managed by these Division Supervisors.

Another example might be to designate the front of a building Division A, the left side Division B, the rear Division C, and the right side Division D; the interior could be referred to as Division E or more commonly "interior." The designated divisions are assigned by starting at the front and going clockwise. If this is your agency's guideline, then responders will recognize these designations on any incident.

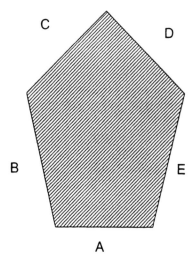

Figure 5.9 Use of geographic divisions designators(A to E)

The difference in the examples is that a vertical geographic area division designations utilize numbers and a horizontal area the designations use alpha characters.

Functional Groups

Another common method of organizing operations at an incident is to establish Functional Groups. Just as the name implies, this form of organization deals not with geographic areas but with activities. Examples of Functional Groups include medical, search and rescue, perimeter security, and maritime salvage groups. Functional Groups, like Divisions, are managed by supervisors, referred to as Group Supervisors.

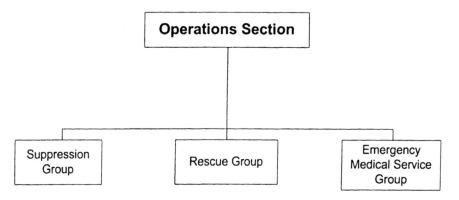

Figure 5.10 Use of Functional Groups

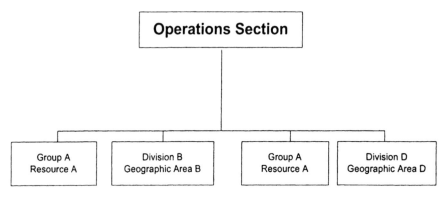

Figure 5.11 Combined Divisions and Groups

Combined Divisions and Groups

A third method is the use of combined geographic Divisions and Functional Groups. This approach is commonly used when a functional activity operates across divisional lines. For example, a specialized Canine Search Group would be used wherever required and moved geographically as needed in an earthquake incident. In any organization in which combined Divisions and Groups are used, it is important that the respective supervisors establish and maintain close communications and coordination. Each will have equal authority; neither supervisor will be subordinate to the other.

Branches

A fourth method of Operations Section organization is to establish a Branch. In a Branch structure Branches may be either geographic or functional. Geographic Branches may

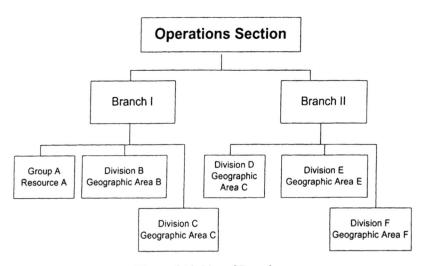

Figure 5.12 Use of Branches

be established due to span-of-control considerations when more than five Divisions are established. Functional Branches may also be established to manage various operations functions. Geographic and Functional Branches may be used together on an incident.

Branches will be managed by a Branch Director. Branch Directors may have Deputy positions as required. In multi-agency incidents the use of Deputy Branch Directors from assisting agencies can be of great benefit to ensure and enhance interagency coordination.

In addition to the Operations Section positions discussed so far, there are two additional important organizational elements that should be discussed: Staging Areas, and Air Operations.

Staging Areas

Staging Areas are facilities within the Incident Management System. They are locations set up at an incident where resources can be placed while awaiting tactical assignment. Once a Staging Area has been designated and named, a Staging Area Manager will be assigned. The Staging Area Manager reports to the Operations Section Chief or to the Incident Commander in the absence of the Operations Section Chief.

It should be noted that in Staging Areas all resources are assigned and should be ready for deployment. Staging Areas should not be used to locate out-of-service resources or for logistics functions, such as Rehab. They may be relocated as necessary.

In some applications, Staging Areas may be functional in nature as opposed to general in nature. Branches may have separate Staging Areas also. For example, a

Figure 5.13 Staging area

Medical Branch may have an Ambulance Staging Area assigned to the Branch or a Law Enforcement Branch may stage their vehicles separately for security reasons.

Air Operations Branch

Some types of incidents may require use of aviation resources to provide tactical or logistical support. On smaller incidents, aviation resources will be limited in number and will report directly to the Incident Commander or to the Operations Section Chief once that position has been established. On larger incidents, it may be desirable to acti-

Figure 5.14 Air operations

vate a separate Air Operations Branch to coordinate the use of aviation resources. The Air Operations Branch will then be established at the Branch level, reporting directly to the Operations Section Chief via the Air Operations Branch Director.

The Air Operations Branch Director can establish two functional groups. The Air Tactical Group coordinates all airborne activity. The Air Support Group provides all ground-based support to aviation resources.

Resource Organization

Initially, at any incident, individual resources that are assigned will report directly to the Incident Commander. As the incident grows in size or complexity, individual resources may be organized and employed in a number of ways to facilitate incident management. These resources are designated as single resources, task forces, or strike teams.

Single Resources may be employed on an individual basis. This is typically the case in the context of the initial response to the incident. During sustained operations, situations will typically arise that call for the use of a single helicopter, vehicle, or piece of mobile equipment.

Task Forces are any combination of resources put together to accomplish a specific mission. They have a designated leader and operate with common communications. The combining of resources into task forces allows for several key resource elements to be managed under one individual's supervision, thus aiding in span of control.

Strike Teams are a set number of resources of the same kind and type operating under a designated leader with common communications between them. They represent a known capability and are highly effective management units.

Planning/Intelligence Section

The Planning/Intelligence Section, if established by the Incident Commander, will assume responsibility for several important functions. The Planning/Intelligence Section is responsible for the collection, evaluation, and dissemination of incident information, preparing situation status reports, displaying status information, maintaining status of resources, conducting planning meetings, and preparing required incident-related documentation. This section's responsibilities can also include creation of the Incident Action Plan, which defines the response activities and resource utilization for a specified time, as well as development of the Incident Demobilization Plan. The section also provides a primary location for technical specialists assigned to an incident. This is done under the leadership of the Planning/Intelligence Section Chief. A Deputy Planning/Intelligence Section Chief may be assigned as required.

The Planning/Intelligence Section also consists of personnel focused on longer-term issues, such as natural-resource damage assessment, waste management and response strategy development, and agency or company image and resource management. This function is utilized in all incidents regardless of the duration. In smaller events, the Incident Commander is responsible for planning, but when the incident is of larger scale, the Incident Commander establishes the Planning/Intelligence Section. The Planning/

Intelligence Section Chief will normally come from the jurisdiction with primary incident responsibility and may have one or more Deputies from other participating jurisdictions.

Planning Section Chief

As a member of the Incident Commander's General Staff, the Planning/Intelligence Section Chief oversees all incident-related data gathering and analysis regarding operations and assigned resources, develops alternatives for tactical operations, and prepares the Incident Action Plan for each Operational Period. The Planning/Intelligence Section Chief determines the current status of the section's activities, evaluates the need to activate or deactivate units within the planning section, evaluates the need to replace, demobilize, or order personnel for the section, evaluates the current strategic plan against the incident objectives and the complexity analysis, gathers data on alternate strategies and contingency plans, evaluates the need for technical specialists, and evaluates the planning section's job performance.

The Planning/Intelligence Section Chief's job is more involved with data evaluation, while a Unit Leader's job is more concerned with information gathering. He or she collects and manages all relevant operational data, provides input to the Incident Commander and Operations Section Chief for use in preparing the Incident Action Plan, supervises preparation of the plan, conducts and facilitates planning meetings, reassigns personnel already on site to Incident Management System organizational positions as appropriate, establishes information requirements and reporting schedules for resources and situation units, and determines the need for specialized resources to support the incident.

The Planning/Intelligence Section Chief assembles and disassembles Task Forces and Strike Teams not assigned tactical operations, establishes specialized data-collection systems as necessary (e.g., for weather), assembles information on alternative strategies, provides periodic predictions on incident potential, reports any significant changes in incident status, compiles and displays incident status information, oversees preparation of the Incident Demobilization Plan, and incorporates traffic, medical, communications, and other supporting material into the Incident Action Plan.

One of the most important functions of the Planning/Intelligence Section is to look beyond the current and next Operational Periods and anticipate potential problems or events in a long-term operation.

The Planning/Intelligence Section may be organized into four and possibly five unit-level positions. These positions include the Resource Unit, Situation Unit, Documentation Unit, Demobilization Unit, and possibly an Intelligence Unit.

Resource Unit
The Resource Unit is responsible for all resource-reporting activity and for maintaining the status on all resources assigned to the incident. Physical resources consist of personnel, teams, facilities, supplies, and major items of equipment available for assignment to or employment during incidents. With a large and complex incident, the

Resource Unit will grow to support the incident operations. The leadership within the Resource Unit is the Resource Unit Leader. He or she will manage unit operations and delegate tasks. The Resource Unit Leader may have Status Check-in Recorders to assist in their operations. The Resources Unit makes certain that all assigned personnel and other resources have checked in and reported at the incident.

This unit should have a system for keeping track of the current location and status of all assigned resources and should maintain a master list of all resources committed to incident operations. Ensuring the completion of a Check-in List (ICS Form 211, See Appendix B) is major task of the Resource Unit Leader. A Check-in List may be developed to obtain and record the essential information that will be needed to complete the Incident Action Plan, Master Resource List, and in time the Incident Demobilization Plan. This can be accomplished utilizing Resource Status Cards (Also known as T-cards or ICS Form 219, see Appendix B) or other methods of displaying resource status. These systems provide for tracking incident-assigned single resources, strike teams, task forces, and overhead personnel and a way of tracking resources' condition for safety purposes, rehab, and release of resources.

These systems also assist in the preparation of resource summaries such as the Incident Status Summary (ICS Form 209, See Appendix B) and the Operational Worksheet (ICS Form 215, See Appendix B) used during planning meetings.

The Resource Unit Leader also collaborates with the Operation Section Chief in preparing resource assignments, reporting locations, control operations, and radio communications information to produce an assignment list (ICS Form 204, See Appendix B) for the Incident Action Plan.

Additional forms prepared include the organization assignment list (ICS Form 203, See Appendix B) and the organization chart (ICS Form 207, See Appendix B). The organization chart, along with resource allocation and deployment, is a part of the Incident Command Post display that the Resource Unit prepares and updates as necessary.

A Check-in/Status Recorder reports to the Resources Unit Leader and assists with the accounting of all incident-assigned resources. Once the Resource Unit Leader establishes a check-in function at incident locations, the Recorder will document check-in information on a list and forward that information to the Resource Unit if located at a remote location. The Recorder will transmit check-in information to the Resource Unit on a regular prearranged schedule or as needed.. The Recorder will also establish communications with the Incident Communication Center and Ground Support Unit and post signs to direct arriving resources to incident check-in location(s).

In keeping up with the status of resources, normally three status conditions are utilized. Resources that are received for and are supporting incident operations are considered to be "assigned." Resources assigned to an incident but not actively engaged in supporting operations are considered "available," in staging, ready to be activated. The third status condition is "out-of-service." Out-of-service resources are personnel, teams, equipment, or facilities that have been assigned to an incident but are unable to function for mechanical, rehabilitation, or personnel reasons. The individual who changes the status of a resource, such as equipment location, is responsible for promptly informing the Resources Unit.

Situation Unit

The Situation Unit collects, processes, and organizes ongoing situation information; prepares situation summaries; and develops projections and forecasts of future events related to the incident. The Situation Unit also prepares maps and gathers and disseminates information and intelligence for use in the IAP. This unit may also require the expertise of Technical Specialists and operations and information security specialists. The leadership of the Situation Unit is the Situation Unit Leader.

The Situation Unit Leader ensures that collection and analysis of incident data are started as soon as possible. The Situation Unit Leader will prepare, post, or disseminate resource and situation status information as required, including special requests. He or she also prepares periodic predictions or as requested. The Situation Unit Leader provides photographic services and maps as needed for the incident. In order to keep agency administrators and others informed, an Incident Status Summary (ICS Form 209, See Appendix B) is prepared in collaboration with the Resource Unit Leader.

This summary is a responsibility of the Situation Unit Leader and can take different forms, depending on the needs of the Incident Commander, agency directives, and local, state, and federal government. Items reported in the summary include incident information such as the name, size, time it occurred, cause, location, resources, critical resource needs, expected containment and control, perimeter, progress of tactical operations, weather data, and injuries and fatalities.

Map products that the Situation Unit might develop may include incident maps that depict the perimeter, control lines, Branch and Division boundaries, facility and reporting locations, and internal and external traffic plans. Progression maps as well as future projections maps might also be developed. Geographic Information Systems often referred to as GIS, can greatly enhance these products.

Typical subordinate positions within the Situation Unit are the Field Observer, Display Processor or GIS Specialist, and Weather Observer. In wildfire incidents, a Fire Behavior Analyst and Incident Meteorologist might be a part of the Situation Unit. In other incidents, they might be considered as Technical Specialists.

Field Observer

The Field Observer is responsible for collecting situation information from personal observations at the incident and providing this information to the Situation Unit Leader. These responsibilities include determining on-the-ground perimeters of the incident, location of problem areas, actual weather conditions, hazards including escape routes and safe areas, and progress of operations resources. They verify actual locations on Helispots and Division and Branch boundaries.

The Field Observer reports information to the Situation Unit Leader by established procedure, including immediate notification of any condition observed that may cause danger and safety hazard to personnel. The Field Observer also gathers intelligence that will lead to accurate predictions. Field Observer products in a wildland situation might include information such as improved properties, vegetation types, and hazards in advance of the incident, water sources, and various other mapped data, other types of situations will of course generate more incident related Field Observer products.

Display Processor/Geographic Information System Technician

The Display Processor or Geographic Information System Technician is responsible for the display of incident status information obtained from Field Observers, resource status reports, aerial and orthographic photographs, and infrared data. Based on guidance from the Situation Unit Leader, the Display Processor or Geographic Information System Technician determines map requirements for Incident Action Plans, time limits for completion, and communications means with Field Observers or other field operations personnel. He or she might also assist the Situation Unit Leader in analyzing and evaluating field reports and develop various other products and displays as determined and needed by the other sections.

Weather Observer

The Weather Observer is responsible for collecting current incident weather information and providing the information to an assigned Meteorologist, Fire Behavior Specialist, or Situation Unit Leader. If a Weather Observer is assigned, he or she may determine weather data collection methods to be used, priorities for collection, specific types of information required, frequency of reports, method of reporting, and sources of equipment. The Weather Observer obtains weather data collection equipment through appropriate channels. Once the equipment is set up, the data is recorded and reported as weather observations at assigned locations on a schedule.

Documentation Unit

The Documentation Unit maintains all incident-related documentation, prepares the Incident Action Plan, and provides duplication services to incident personnel. Incident-related documentation includes a complete record of the major steps taken to resolve the incident. The documentation process includes filing, maintaining, and storing incident files for legal, analytical, and historical purposes. Documentation is part of the Planning/Intelligence Section primarily because this unit prepares the Incident Action Plan and maintains documents and records developed as part of the overall Incident Action Plan and planning function. The leadership in the Documentation Unit is the Documentation Unit Leader.

The Documentation Unit Leader ensures that the incident files are properly organized, duplication service provided, and requests processed. The Documentation Unit collects materials from all sections, reviews records for accuracy and completeness, and informs appropriate units of errors or omissions. One critical element of an incident today is on-site duplication capability. The Documentation Unit leader responsibilities include the maintenance of the equipment on an emergency and on-site basis.

Demobilization Unit

On large, complex incidents, the Demobilization Unit will assist in ensuring that an orderly, safe, and cost-effective movement of personnel will be made when they are no longer required at the incident. This unit should begin its work early in the incident in order to create rosters of personnel and resources and obtain any missing information as check-in proceeds.

The Demobilization Unit develops an Incident Demobilization Plan that includes specific instructions for all personnel and resources that require demobilization. Many locally provided resources do not require specific demobilization instructions; but it is the responsibility of the Incident Commander that they return to their home station in safe and orderly fashion. Once the Incident Demobilization Plan has been approved, the Demobilization Unit ensures that it is distributed both at the incident and elsewhere as necessary. The Demobilization Unit Leader is the supervisor of the unit.

In addition to the development of the Incident Demobilization Plan, specific tasks associated with the Demobilization Unit include: verifying resource condition, whether resources are reassignable or non-reassignable; determining travel method; identifying home units; assigning transportation vehicles; and coordinating with local or expanded dispatch. This unit may review incident resource records to determine the likely size and Demobilization Plan.

The Demobilization Unit Leader ensures that all sections and units understand their specific demobilization responsibilities and supervises execution of the Incident Demobilization Plan. As with any subordinate unit, he or she keeps the Planning/Intelligence Section Chief updated on demobilization progress.

Intelligence Unit

An Intelligence Unit may be established in the Planning/Intelligence Section on an incident with a need for tactical intelligence, and where a law-enforcement entity is not a member of the Unified Command. The analysis and sharing of information and intelligence are important elements of an Incident Management System. In this context, intelligence includes not only national security or other types of classified information but also other operational information, such as risk assessments, medical intelligence, weather information, geospatial data, structural designs, toxic contaminant levels, and utilities and public works data that may come from a variety of different sources.

Traditionally, information and intelligence functions are located in the Planning/Intelligence Section. The Intelligence Unit may develop, conduct, and manage various information and security plans as directed by the Planning/Intelligence Section Chief or the Incident Commander.

Technical Specialists

The Incident Management System is designed to function in a wide variety of incident scenarios, some of which may require the use of Technical Specialists. Technical

Figure 5.15 State Police/PD Operations at the WTC September 11, 2001

Specialists are advisors with special knowledge, skills and abilities. Technical Specialists will initially report to the Planning/Intelligence Section Chief. Such specialists may serve anywhere within the organization, including the Command Staff and work within that Section but may be reassigned to another part of the organization. No minimum qualifications are prescribed, as Technical Specialists normally perform the same duties during an incident that they perform in their everyday jobs, for which they are typically certified in their fields or professions.

Technical Specialists assigned to the Planning/Intelligence Section may report directly to the Planning/Intelligence Section Chief or to any function in an existing unit. They may also form a separate unit within the Planning/Intelligence Section, depending upon the requirements of the incident and the needs of the Section Chief. Technical Specialists may be assigned to other parts of the organization (e.g., to the Operations Section to assist with tactical matters or to the Finance/Administration Section to assist with fiscal matters). For example, a legal specialist or legal counsel may be assigned directly to the Command Staff to advise the Incident Commander on legal matters such as emergency proclamations, legality of evacuation orders, and legal rights and restrictions pertaining to media access. Generally, if the expertise is needed for only a short period of time and normally involves only one individual, that individual should be assigned to the Situation Unit. If the expertise will be required on a long-term basis and may require several personnel, it is advisable to establish a separate Technical Unit in the Planning/Intelligence Section.

The incident itself will primarily dictate the necessity for Technical Specialists. Examples of specialists are meteorologist, environmental-impact specialist, flood-control specialist, water-use specialist, explosives specialist, structural engineer, firefighter, medical/healthcare specialist, medical intelligence specialist, pharmaceutical specialist, veterinarian, agricultural specialist, toxicologist, radiation-health physicist, intelligence specialist, infectious-disease specialist, chemical or radiological decontamination specialist, law-enforcement specialist, attorney, industrial hygienist, transportation specialist, scientific support coordinator, fire-behavior specialist, environmental specialist, resource-use specialist, and training specialist.

Fire Behavior Specialist

The Fire Behavior Specialist is a position found most commonly only in the wildland event. He or she is primarily responsible for establishing a weather data collection system and developing required fire behavior predictions based on fire history, fuel, and weather and topography information. Tasks associated with a Fire Behavior Specialist would be to establish weather data requirements, verify dispatch of a meteorologist, order a mobile weather station and direct the setup on arrival, inform the meteorologist of weather data requirements, and forward weather data to incident personnel. Other tasks might be to collect, review, and compile fire history data, exposed fuel data, and information about topography and fire barriers; provide weather information and other pertinent information to the Situation Unit Leader for inclusion in the incident status summary (ICS Form 209, See Appendix B); and prepare fire behavior prediction information at periodic intervals or upon request.

Environmental Specialist

The Environmental Specialist is another position likely to be used only in a wildland event. He or she collects and validates environmental information within the incident area by reviewing pre-attack land use and management plans, determines environmental restrictions within the incident area, and develops suggested priorities for preservation of the environment. He or she may provide environmental analysis information, as requested, and collect and transmit required records and logs to the Documentation Unit at the end of each Operational Period.

Resource Use Specialist

The Resource Use Specialist may respond to requests for information about limitations and capabilities of resources and collect and transmit records and logs to the Documentation Unit at the end of each Operational Period.

Training Specialist

A Training Specialist may inform the Planning/Intelligence Section Chief of planned use of trainees, review trainee assignments and modify if appropriate, coordinate the assignments of trainees to incident positions with the Resources Unit, brief trainees and trainers on training assignments and objectives, coordinate use of unassigned trainees, and make follow-up contacts on the job to provide assistance and advice for trainees to meet training objectives as appropriate and with approval of unit leaders. Additionally, he or she would ensure that trainees receive performance evaluations, monitor operational procedures and evaluate training needs, respond to requests for information concerning training activities, and maintain and report training specialist records and logs to the Documentation Unit at the end of each operational period.

Technical Units

A specific example of the need to establish a distinct Technical Unit within the General Staff is the requirement to coordinate and manage large volumes of environmental sampling and analytical data from multiple sources in the context of certain complex incidents, particularly those involving biological, chemical, and radiation hazards. To meet this requirement, an Environmental Unit may be established within the Planning Section to facilitate inter-agency environmental data management, monitoring, sampling, analysis, and assessment. The Environmental Unit would prepare environmental data for the Situation Unit and work in close coordination with other units and sections within the Incident Management System structure to enable effective decision support to the Incident Commander or Unified Command. Technical specialists assigned to the Environmental Unit might include a scientific support coordinator and sampling, response technologies, weather forecasting, resources at risk, cleanup assessment, and disposal specialists. Sample tasks accomplished by the Environmental Unit would include the following: identifying sensitive areas and recommending response priorities; developing a plan for collecting, transporting, and analyzing samples; providing input on wildlife protection strategies; determining the extent and effects of site contamination; developing site cleanup and hazardous-material disposal plans; and identifying the need for and obtaining permits and other authorizations.

Logistics Section

The Logistics Section is without a doubt an essential part of the Incident Management System. Early recognition of the need for a separate logistics function may reduce time and money spent on an incident. The Logistics Section meets all support needs for the incident except aircraft, including ordering resources through appropriate procurement authorities from off-incident locations. Logistics also provides facilities, transportation, supplies, equipment maintenance and fueling, food service, communications, and medical services for incident personnel. All off-incident resources (for example, those that are enroute or returning from deployment) are a responsibility of Logistics.

Logistics Section organization

The Logistics Section is responsible for providing facilities, services, and support material to meet all incident or event needs. This is accomplished under the direction of the Logistics Section Chief, a member of the General Staff. A Deputy Logistics Section Chief may also be assigned. The Logistics Section Chief has responsibility for six principal activities at an incident:

- Communications
- Medical support
- Food service
- Supply
- Facilities
- Ground support, transportation, fuel, and maintenance

Each of these activities is assigned as a Unit within the Logistics Section. It is important to remember that the Units' functions, except for the Supply Unit, are geared to supporting personnel and resources directly assigned to the incident.

The Logistics Section Chief participates in the development and implementation of the Incident Action Plan and activates and supervises the Branches and units within the Logistics Section based on known and anticipated needs. He or she estimates future service and support requirements and may establish separate units for one or more of the logistics support or service activities. On large incidents when all six Logistics Section Units are activated or there are many facilities and large amounts of equipment, it may be desirable or necessary to establish a two-branch structure. This will reduce the span of control for the Logistics Section Chief.

These two Branches are designated as the Service Branch and the Support Branch. It should be emphasized that this division of organization is available if necessary. Each branch will be assigned a Branch Supervisor.

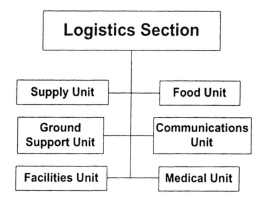

Figure 5.16 Logistics Section: two-branch structure

It should be note that several of the Logistics Section Units may be designed to perform certain detailed activities which may not be appropriate for some kinds of incidents. Only those units that perform functions that are needed for the incident at hand should be activated by the Logistic Section Chief.

The Logistics Section Chief participates in the development and implementation of the Incident Action Plan and activates and supervises the branches and units within the section. He or she also assists in the development and implementation of the demobilization plan.

Service Branch

The Service Branch Director, when activated, is under the supervision of the Logistics Section Chief and is responsible for the management of all service activities at the incident. The Branch Director supervises the operations of the Communications, Medical, and Food Units.

Communications Unit

The Communication Unit is responsible for developing plans for the effective use of incident communications equipment and facilities, installing and testing communications equipment, supervision of the Incident Communications Center, distribution of communications equipment to incident personnel, and maintenance and repair of communications equipment. Collectively these activities are known as the Communications Plan, The plan is documented using a Communication Plan Form (ICS Form 205, See Appendix B)

The Communications Unit's major responsibility is effective communications planning for the incident and the managers working the command elements of the incident, especially in the context of a multi-agency incident. It is critical to determine required radio nets, establish interagency frequency assignments, and ensure interoperability and the optimized use of all assigned communications capabilities.

The leadership for the Communication Unit is the Communication Unit Leader. The Communications Unit Leader will attend all Incident Planning Meetings to ensure that the communication systems available for the incident can support tactical operations planned for the next Operational Period. This may include setting up telephone and public-address systems and establishing appropriate communications distribution centers within the base.

The Communications Unit is responsible for planning the use of radio frequencies; establishing networks for command, tactical, support, and air units; setting up on-site telephone and public-address equipment; and providing any required off-incident communication links. Codes of any type (numeric or alpha based) should not be used for radio communication. The use of clear text speech based on common terminology that avoids misunderstanding in complex and noisy situations reduces the chance for error.

Advance communications planning is required to ensure that an appropriate communications system is available to support incident operations requirements. This planning includes the development of frequency inventories, frequency-use agreements, and interagency radio caches.

Radio networks or "nets" for large incidents will normally be organized as follows:

- Command Net
- Tactical Net
- Support Net
- Ground-to-Air Net
- Air-to-Air Net

The functional area of each Net is as follows:

- The Command Net links together incident command, command staff, section chiefs, branch directors, and division and group supervisors.
- The tactical nets are established to connect agencies, departments, geographical areas, or specific functional units. There may be multiple tactical nets. The number of nets is dependent on how large and complex the operation is.
- The support net may be established primarily to handle changes in resource status but also to logistical requests and other non-tactical functions.
- The ground-to-air net is used to coordinate ground-to-air traffic; either a specific tactical frequency may be designated, or regular tactical nets may be used.
- Air-to-air nets will normally be predesignated and assigned for use at the incident. These are generally established in conjunction with local aviation authorities.

Subordinate elements within the Communications Unit are the Incident Communication Technician and Incident Communications Center Manager.

The Incident Communication Technician is responsible for installation, maintenance, and tracking of communication equipment. Specific tasks include assisting in designing communication systems to meet operational needs, installing and testing communication equipment, cloning or programming radios, and repairing and/or replacing communication equipment.

The Incident Communications Center Manager is responsible for receiving and transmitting radio and telephone messages among and between personnel and providing dispatch services at the incident. Specific tasks include determining frequencies in use, nets established or to be established, and location of repeaters; setting up a communication center; and receiving and transmitting messages internally and externally.

Medical Unit

The Medical Unit is responsible for the effective and efficient provision of medical services to incident-assigned personnel. The Medical Unit is responsible for the development of the Medical Plan (ICS Form 206, See appendix B) and obtaining first aid, medical treatment, and transportation for injured or ill incident personnel. The Medical Plan will include procedures for handling any major medical emergency involving incident personnel. This unit also develops the Emergency Medical Transportation Plan (ground and/or air) to provide transportation for injured incident personnel. The Medical Unit continually provides medical care, including vaccinations, vector control, occupational health, prophylaxis, and mental-health services for incident personnel.

The Medical Unit ensures tracking of patient movement from origin to care facility to final disposition. The unit also prepares medical reports and assists in processing all paperwork related to injuries or deaths of incident-assigned personnel and coordinates personnel and mortuary affairs for fatalities when necessary. It is the responsibility of the Medical Unit to set up and maintain Rehabilitations Areas (rehab) facilities within the incident area.

The leadership for the Medical Unit is the Medical Unit Leader, who develops the Medical Plan, which in turn forms part of the overall incident Action Plan. The Medical Plan should provide specific information on medical-assistance capabilities at incident locations, potential hazardous areas or conditions, and off-incident facilities and procedures for handling complex medical emergencies. The Medical Unit will also assist the Finance/Administration Section with the administrative requirements related to injury compensation, including written authorizations, billing forms, witness statements, administrative medical documents, and reimbursement as required. The Medical Unit will ensure patient privacy to the fullest extent possible.

As noted in the Command Staff, the Safety Officer will work with the Medical Unit Leader and approve the Medical Plan. Subordinate elements within the Medical Unit include Emergency Medical Technicians and Paramedics as staff.

Food Unit

The Food Unit is responsible for determining and supplying the feeding and potable-water requirements at all incident facilities and for active resources within the

Operations Section. The unit may prepare menus and food, provide them through catering services or use some combination of both methods. This may include maintaining food-service areas to include food security and safety concerns. Efficient food service is important, especially during any extended incident. The leadership for the Food Unit is the Food Unit Leader, who is responsible for the general maintenance of the food-service areas.

The Food Unit Leader must be able to anticipate incident needs, both in terms of the number of people who will need to be fed and whether there may be special food requirements based on the type, location, or complexity of the incident. The Food Unit must supply food needs for the entire incident, including all remote locations (such as camps and staging areas), as well as to operations personnel unable leave operational assignments.

The Food Unit must interact closely with the Planning/Intelligence Section, to determine the number of personnel that must be fed as well as the Facilities Unit, to arrange food-service areas; and the Supply Unit, to order food and other supplies related to feeding services; the Ground Support Unit, to obtain ground transportation, either for the food service personnel and or the food and supplies themselves; and the Air Operations Branch Director, to obtain air transportation where needed. Careful planning and monitoring are required to ensure food safety before and during food- service operations, including the assignment, as indicated, of public health professionals with expertise in environmental health and food safety.

It should be noted that while the feeding of victims (when present) is a critical operational activity, which will be incorporated into the Incident Action Plan and coordinated via the incident Management System. The actual feeding activities for the victims or other on incident related external personnel will normally be conducted by appropriate non-government organization staff, such as the American Red Cross, Salvation Army, or other similar entities. It is generally not the mission of the Food Unit to undertake such activities.

Support Branch

The Support Branch Director supervises the operations of the Supply, Facilities, and Ground Support Units.

Supply Unit

The Supply Unit is primarily responsible for the ordering of personnel, equipment, and supplies; receiving and storing all supplies for the incident; maintaining an inventory of supplies; and servicing nonexpendable supplies and equipment. In the Incident Management System, all resource orders are placed through the Logistics Section's Supply Unit. This includes all tactical and support resources (including personnel) and all expendable and non-expendable supplies required for incident support. The unit also handles tool operations, which include storing, disbursing, and servicing of all tools and portable, nonexpendable equipment. If the Supply Unit has not been established, the responsibility for ordering rests with the Logistics Section Chief.

The standard elements of the Supply Unit are an Ordering Manager, a Receiving Manager, and Distribution Manager. The Ordering Manager is responsible for placing all orders for supplies and equipment for the incident. Specific tasks include establishing ordering procedures, establishing name and telephone numbers of agency(s) personnel receiving orders, setting up a filing system, obtaining names of incident personnel who have ordering authority, consolidating orders when possible, and keeping the Receiving and Distribution Managers informed of orders placed.

The Receiving and Distribution Managers are responsible for receiving and distribution of all supplies and equipment (other than primary resources) and the service and repair of tools and equipment. Specific tasks include establishing procedures for the operating supply area, setting up a filing system for receiving and distribution of supplies and equipment, maintaining an inventory of supplies and equipment, developing security requirements for the supply area, and establishing procedures for receiving supplies and equipment.

It is not necessary to have both a Receiving and Distribution Manager unless the magnitude of ordering is such that both positions are needed. One Manager accomplishes both functions of receiving and distribution in smaller less complex incidents.

Facilities Unit

Facilities referred to in the Incident Management System include the Incident Command Post, Base, Camp, Staging Areas, Helibase, and Helispot. The Facilities Unit is responsible for the layout and operation of the Base, Camp(s), and Incident Command Post. The leadership element is the Facilities Unit Leader. The Facilities Unit manages the Base and Camp(s) operations. Each Base/Camp may be assigned a Manager, if required and he or she reports to the Facilities unit Leader. The Facilities Unit Leader sets up and maintains all facilities necessary to support the incident.

The Facilities Unit also provides and sets up necessary personnel support facilities, including areas for food and water service, sleeping, sanitation and showers, and staging. These facilities may be fixed buildings, office trailers, travel trailers, rental trucks, or tents. The Facilities Unit also orders, through supply, such additional support items as portable toilets, shower facilities, and lighting units. The Facilities Unit Leader is responsible for providing security support for the facilities as required.

Like victim food services, sheltering for victims is a critical operational activity, which will be incorporated into the Incident Action Plan and coordinated via the Incident Management System but the actual sheltering activities themselves again will normally be conducted by appropriate non-government organization staff, such as the American Red Cross, Salvation Army, or other similar entities.

Ground Support Unit

The Ground Support Unit is responsible for transportation of personnel, supplies, food, and equipment; fueling, service, maintenance, and repair of vehicles and other ground support equipment; support of out-of-service resources; and developing and implementing the incident transportation plan. The Ground Support Unit also maintains a transportation pool for major incidents. This pool consists of vehicles (e.g., staff cars, buses, pick-up trucks and specialty vehicles) that are suitable for transporting personnel. The unit will perform pre- and post- inspections of incident and other ground support equipment.

The leadership of the Ground Support Unit is the Ground Support Unit Leader. He or she provides up-to-date information on the location and status of transportation vehicles to the Resources Unit and supervises a staff of equipment managers, equipment operators, and drivers. These individuals maintain and repair primary tactical equipment, vehicles, and mobile ground-support equipment and document usage time for all ground equipment (including contract equipment) assigned to the incident. The unit supplies fuel for all mobile equipment.

Subordinate elements of the Ground Support Unit may include Equipment Managers who provide service, repair and fuel for all apparatus and equipment and provide transportation and support vehicle services. The Equipment Manager also maintains records of equipment use and service provided and conducts the pre- and post- incident inspection of equipment.

Finance/Administration Section

Just as a budget is critical to a business, a Finance/Administration Section is essential to the Incident Management System. The Finance/Administration Section is responsible for on-site financial and administrative management. The section keeps track of incident-related costs and personnel and equipment records and administers procurement contracts associated with the incident or event. When there is a possibility of reimbursement, whether it is for an agency or individual, tracking incident costs is critical. Unless this is carefully recorded and justified, reimbursement is difficult if not impossible. In addition to monitoring multiple sources of funds, the Finance/Administration Section Chief must track and report to the Incident Commander the financial "burn rate" as the incident progresses. This allows the Incident Commander to forecast the need for additional funds before operations are affected negatively. This is particularly important if significant operational assets are under contract from the private sector. The Finance/Administration Section Chief may also need to monitor cost expenditures to ensure that statutory rules are met. Close coordination with the Planning/Intelligence and Logistics Sections Chiefs is also essential to ensure that operational records can be reconciled with financial documents. Note that in some cases only one specific function may be required (e.g., cost analysis), which a Technical Specialist in the Planning/Intelligence Section could provide.

The Finance/Administration Section may not be activated on all incidents. The Incident Commander will retain responsibility for all finance-related activities until the Finance/Administration Unit or Section has been activated.

The Finance/Administration Section is especially important when the incident is of a magnitude that may result in a governmental disaster declaration. Each of the functional areas can be expanded into additional organizational units with further delegation of authority. These units may also be contracted as the incident deescalates. In past incidents involving the Federal Emergency Management Agency and a presidential disaster declaration, state and local governments must contribute up to 25 percent of disaster costs. In the event of a federally-declared disaster, reimbursement for expenses associated with the response will only occur when they have been tracked and documented.

The Finance/Administration Section Chief is responsible for managing all financial aspects of an incident, providing financial and cost-analysis information as requested, ensuring that compensation and claims functions are being addressed, gathering pertinent information from briefings with responsible agencies, developing an operating plan for the Finance/Administration Section; filling section supply and support needs, meeting with assisting and cooperating agency representatives as needed, maintaining daily contact with agency administrative headquarters on finance matters, ensuring that all personnel time records are accurately completed and transmitted to home agencies according to policy, providing financial input for demobilization planning, ensuring that all obligation documents initiated at the incident are properly prepared and completed, and briefing agency administrative personnel on all incident-related financial issues requiring attention or follow-up. When necessary the Finance Section Chief will determine the need to set up and operate an incident commissary.

The Finance/Administration Section Chief reports directly to the Incident Commander and is responsible for communicating issues, concerns, and problems related to incident business management. He or she notifies the Incident Commander as soon as possible of critical issues. As a primary member of the General Staff, the Finance/Administration Section Chief interacts daily with all members of the Command and General staff during scheduled briefings and meetings and as necessary to receive and provide current information.

The Finance/Administration Section Chief is responsible for ensuring that information is exchanged between the Finance/Administration Section and appropriate incident personnel. The Finance/Administration Chief may not actually exchange the information but rather Finance/Administration Section Unit Leaders may communicate directly with members of Command and General Staff. The Finance Section Chief determines the level and method of communication appropriate to the complexity of the incident.

Interacting with the Planning/Intelligence Section, the Finance/Administration Section Chief provides cost information for the Incident Status Report (ICS Form 209, See Appendix B). The Finance/Administration Section Chief provides a review and updates cost information for the situational analysis. The Finance/Administration Section obtains resource status information from the Resource Unit.

With the Operations Section, the Finance/Administration Chief provides information on expensive or underutilized equipment. The Finance Section obtains aircraft costs from the Operations Section or the Air Branches (if established) and exchanges information regarding costs and appropriate expenditure of funds.

With the Logistics Section, the Finance Section Chief obtains information on facility, support, and property costs. The Finance/Administration Section Chief provides information on contract costs. By interacting with the Information Officer the Finance/Administration Section Chief provides factual information on incident cost estimates and projections to members of the general public, the media and other external entities as needed and determined by the Incident Commander or by law.

The Finance/Administration Section Chief establishes time submission requirements and procedures. The primary incident agency advises the Incident Commander of the rest and recuperation policy, and the Incident Commander implements the policy through the General and Command Staff of the Incident Management team, generally through the Finance/Administration Section Chief.

The Finance/Administration Section Chief advises other Section Chiefs, as well as the Incident Commander, of personnel who have excess hours on shift and are not meeting the established work/rest guidelines. The Finance/Administration Section Chief provides information and answers questions relating to incident time and pay issues, such as meal breaks, hazard pay, standby orders, or spot change tour of duty.

The Finance/Administration Section Chief also exchanges information with other Section Chiefs regarding equipment under the management of that Section, which might include copy and facsimile machines, portable toilets, generators, buses, heavy equipment, land-use agreements, facility agreements, and contract claims. He or she establishes equipment time submission procedures and requirements and answers questions relating to contracts and agreements pursuant to contract requirements and agency guidelines.

The Finance/Administration Section Chief exchanges information with other Section Chiefs regarding injury compensation claims, employee claims, tort claims, and government property claims. He or she will along with the Safety and Security Officers and other involved personnel participate in the investigation and documentation of accidents, injuries, and claims. The Finance Section assists in the review, approval, and implementation of the demobilization plan. Throughout the incident, the Finance/Administration Chief identifies section resources that are available for release and provides this information to the Planning/Intelligence Section. The Finance/Administration Chief also identifies underutilized equipment that may be released or reassigned and notifies the section responsible for management of the equipment.

Communication with Agency Administrative Representatives begins upon arrival at the incident and continues throughout the duration of the incident. The Finance/Administration Section Chief ensures daily contact to provide updates on issues, concerns, and the progress of the section. The Finance/Administration Section participates in a final closeout with the administrative representative, separate from the Incident Management Team closeout, and provides an opportunity to review the incident finance package.

As with other General Staff positions, the Finance/Administration Section Chief may assign a Deputy Finance/Administration Section Chief to assist in managing the Section. This Deputy must be fully qualified as well and may be from another agency.

There are four Units that may be established in the Finance/Administration Section. These are the Time Unit, Procurement Unit, Compensation/Claims Unit, and Cost Unit. Primary positions in the Units include: Time Unit Leader, Procurement Unit Leader, Compensation/Claims Unit Leader, and Cost Unit Leader.

Time Unit

The Time Unit ensures that times for all personnel on an incident or event are recorded. It is the responsibility of the Time Unit Leader to ensure accurate recording of daily personnel time, and compliance with all agency time-recording policies. He or she communicates requirements to subordinate staff, establishes a record-keeping system, and coordinates on agency-specific time-recording requirements. He or she may require the assistance of personnel familiar with the relevant policies of any affected agencies.

The Time Unit also ensures that the Logistics Section records or captures equipment usage time through the Ground Support Unit for ground equipment and the Air Operations Support Group for aircraft. These records must be verified, checked for accuracy, and posted according to existing policies. Excess hours worked must also be determined, for which separate logs must be maintained.

Procurement Unit

The Procurement Unit processes administrative paperwork associated with equipment rental and supply contracts and is responsible for equipment time reporting. The Procurement Unit Leader handles financial matters pertaining to vendor contracts, leases, and fiscal agreements; maintains equipment time records; establishes local sources for equipment and supplies; and manages all equipment rental agreements and all rental and supply billing invoices. In some agencies, the Supply Unit in the Logistics Section will be responsible for certain procurement activities. The Procurement Unit will also work closely with local cost authorities.

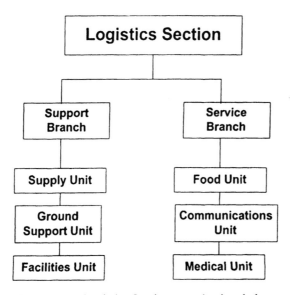

Figure 5.17 Logistics Section organizational chart

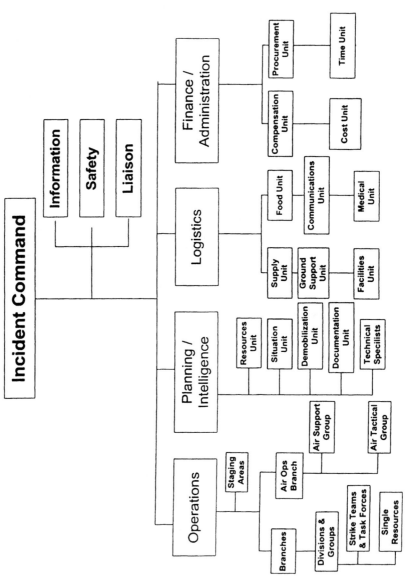

Figure 5.18 Full ICS chart

Compensations/Claims Unit

The Compensation/Claims Unit combines two important functions. The Compensation Unit is responsible for seeing that all documentation related to worker pay is correctly completed. The Compensation Unit also maintains files of injuries and/or illnesses associated with the incident. The Claims Unit handles investigation of all claims involving damaged property associated with or involved in the incident. The Compensation/ Claims Unit Leader provides documentation of continuation of pay/time loss/restricted duty to the Time Unit as well as personnel medical costs. He or she also coordinates with the incident agency to obtain medical treatment options, agency-provided medical care, and claims investigation. Since the Medical Unit may also perform many of these tasks as well, close coordination between the Medical Unit and the Compensation and Claims Units is essential. The Compensation and Claims Unit maintains logs on claims, obtains witness statements, and documents investigations and agency follow-up requirements. On larger incidents they may be split out as separate but interdependent Units whereas on smaller incidents they may function as as single Unit.

It should be highly stressed that with some incidents this unit may be extremely important. Establishing this unit and assigning personnel to monitor this function could result in substantial cost savings. If claims are made without complete documentation or with delays in follow-up, an agency may not be able to defend against them. Activating the Compensation/Claims Unit is basically good insurance for everyone involved in the incident. The files prepared and processed through this element may serve as valid and acceptable documentation for any legal claims pertaining to the incident that may occur at a later date.

CONCLUSION

While the aforementioned system and processes may seem overly complex at first to a reader with limited field experience in any type of emergency incident management, the exact opposite is true. The systems described above have evolved over time and have been proven to bring calm to the chaos that is an emergency regardless of its nature and scope. The systems described above have been in place for many years in one form or another, yet they are still evolving and will likely continue to do so for many years to come, for as soon as we think we "get it" both people and nature seem to conspire against us to make it harder, hence we adapt, for this is the only option available to us in this business we call emergency response.

Chapter 6

THE PLANNING PROCESS IN INCIDENT MANAGEMENT SYSTEMS

In incident management, much discussion has been given to strategy and tactics, but in actuality they are simply a part of the planning process. In the Incident Management System, considerable emphasis is also placed on developing effective Incident Action Plans. A planning process has been developed as a part of the Incident Management System to assist incident planners in the development of this plan in an orderly and systematic manner. It is important to understand that the planning process starts prior to an incident. It starts with the policy and direction received from an administrator. We generally call this individual an Agency Administrator.

The Agency Administrator is the executive or Chief Executive Officer, or designee of the agency or political subdivision that has responsibility for the incident. The title can also be given to an executive from the private sector. "Executive" and "Agency Administrator" are synonymous terms in this text.

Agency Administrators are responsible for the management of resources in their jurisdictions or organizational units and are held accountable for the overall results. Protection from risk and hazards is a part of that responsibility. Thus, the Agency Administrator has a responsibility to establish standards for expected performance and to hold the Incident Commander accountable.

The Executive will establish the policy and provide guidelines on priorities, objectives, and constraints to a qualified Incident Commander. In many agencies this is as a matter of policy through a written delegation of authority to the designated Incident Commander.

In Incident Management System, the Executive establishes policy and direction and allocates authority to the Incident Commander. Generally, the Executive is not at the scene of the incident, but he or she must have the ability to communicate and meet with

Figure 6.1 Texas Governor Rick Perry at an Incident Command Post exercise

the Incident Commander as necessary. The Executive could be stationed at a jurisdictional or regional Emergency Operations Center (EOC)

The establishment of objectives is part of the incident action planning process. It uses the classic Management-by-Objectives concept first outlined by Peter Drucker in 1954 in his book *The Practice of Management*. Drucker felt that managers (here, Incident Commanders) should not be so involved in the day-to-day activities that they forget about their main purpose. He also felt that all managers and supervisors should participate in the planning process in order to improve the implementation of the plan. This also includes the performance evaluation of the plan.

The Incident Commander utilizes the policies, priorities, and directions from the agency he or she is supporting and develops the incident objectives. These objectives must be specific, measurable, attainable, realistic, action-oriented, flexible, and often time-sensitive, which is particularly true in the case of emergency response.

The objectives are utilized to determine the strategy that will be utilized for an incident. The strategy is used by the Operations Section Chief to determine the tactics and tasks that will accomplish the objectives and achieve the goals set by the Agency Administrator or other executive.

These objectives, strategy, and tactics in turn will dictate the management organization that needs to be put in place. The Operations, Logistics, and Planning/Intelligence Sections will play a major role in this overall process.

In every incident, no matter how small or large, some kind of planning is essential. The better the planning, the more efficient the operation will be. The Incident Management System uses a planning process that will accommodate a small, simple incident or a large, complex incident. For small and simple incidents, the first arriving official will become the Incident Commander and will accomplish the planning process in an intuitive manner. On large and complex incidents, the Incident Commander will start in an intuitive manner and expand through a systematic and formal planning process. The planning process used in the Incident Management System begins with the objectives, which again come from agency policy, priorities, and direction. Incident Management Systems emphasize that the mental and verbal procedures used in the early stages of the crisis should be rapidly replaced by the more formal and systematic planning process. Learning this formal process sets a mental pattern that allows for more accurate application of the principles when intuitive planning is necessary.

The steps outlined in this chapter will allow for the development of an Incident Action Plan in a minimum amount of time. Incidents vary in their kind, complexity, size, and requirements for detailed and written plans. The planning process described in this chapter has been utilized successfully on many different incidents of all types and scopes.

Not all incidents require detailed written plans. Recognizing this, the following planning process provides a series of basic steps that are generally appropriate for use in any incident situation. The determination of the need for written plans and attachments is based on the requirements of the incident and the judgment of the Incident Commander.

The process follows the management-by-objectives sequence, uses various worksheets and forms, and allows for both functional and geographic response to objectives and actions.

The process starts with documentation of the Incident Commander's objectives. The organization required to meet the objectives is designed to utilize multi-agency resources according to span-of-control, unit-integrity, and functional guidelines. Support, services, and communications requirements are obtained and assigned. Branch, Division, and Unit assignments are detailed. Financial considerations are defined and agreements are documented. The developed plan is returned to the Incident Commander for approval.

Meetings, Huddles, Plays

The Incident Management System may require numerous meetings, or huddles, as some managers would regard them. There will be major differences in the planning process in the case of a preplanned event in contrast to the emergency event. In the case of the preplanned event, one has the luxury of time to plan the Planning Meetings and to make sure that all required personnel attend those meetings. In the case of the happening event or emergency situation, the first meeting we will talk about is the initial 9-1-1 call in which a Telecommunicator or operator receives a request for assistance. The Call Taker gets as much information about the incident as possible and then passes this information on to a responder, a law enforcement officer, firefighter, paramedic, or public-works technician. This could be considered a verbal delegation of authority to a responder, which is where we find our first indication of incident objectives-basically, why a response is necessary. The individual arriving first on the scene is by default the initial Incident Commander. At this point there are no additional Command and General Staff; however, the planning process has begun in the mind of the Incident Commander.

The initial Incident Commander arrives at the scene of the crisis and requests additional resources. He or she will begin to develop and refine incident objectives. Based on these objectives and the subsequent strategies and tactics, the planning process continues. Because of the complexity of the incident, the initial Incident Commander may recognize that additional management resources and perhaps an organized management team are necessary. This is communicated to an Agency Administrator, who may be a Police Chief, Fire Chief, Hospital Administrator, or City Manager.

An organized management team is ordered and dispatched to the incident. The team members may come from local resources or supporting agencies. Two meetings should occur at this point, which may be held in conjunction or separately. The team meets with the Agency Administrator and with the initial Incident Commander. They may be the same individual. This gathering is designated as the transfer-of-command meeting or initial briefing.

During the transfer-of-command meeting, the ICS 201 form used for the incident briefing is prepared by the initial Incident Commander; it provides the incoming Incident Commander or members of the Unified Command with the basic information regarding the incident situation and resources allotted to the incident. Most importantly, it provides the Incident Action Plan for the initial response; they will remain in force and continue to develop until the response ends or the Planning/Intelligence Section Chief generates the incident's formal Incident Action Plan. It is appropriate for personnel who will be assigned as Command and General Staff positions to attend this briefing.

Figure 6.2 9-1-1 Call Taker

Figure 6.3 Fire Chief at an incident

Figure 6.4 Transfer of command at the field level

Discussion and decisions made in this meeting would detail the parameters under which the Incident Commander and the team are to work, including jurisdictional priorities and objectives, present jurisdictional limitations, concerns, and restrictions.

When multiple agencies are involved and the incident is going to run under a Unified Command, the next meeting will be the Unified Command Meeting. This meeting is held during or prior to the first operational period when the initial Incident Commander is transferring command to the final Incident Commander. At the Unified Command Meeting, every jurisdictional or functional agency's Incident Commander level representative assembles prior to the first operational period Planning Meeting in a Command Meeting to be briefed on the Incident Action Plan.

The Command Meeting provides the responsible agency officials with an opportunity to discuss and concur on important issues prior to joint incident action planning. The agenda for the meeting should include stating jurisdictional or agency priorities and objectives; presenting jurisdictional limitations, concerns, and restrictions; developing a collective set of incident objectives; and establishing and agreeing on acceptable priorities. Additionally, the agenda should include adopting an overall strategy or strategies to accomplish objectives; agreeing on the basic organizational structure; designating the best-qualified Operations Section Chief; agreeing on General Staff personnel designations and on planning, intelligence, logistical, and finance agreements and procedures; agreeing on the resource ordering process to be followed; agreeing on cost-sharing procedures; agreeing on informational matters; and perhaps most important designating one agency official to act as the Unified Command spokesperson.

Figure 6.5 Agency briefing

Figure 6.6 Unified Command training

There are some inherent Command Meeting requirements. The meeting should include only agency Incident Commander level representatives. The meeting should be brief, and important points should be documented. Prior to the meeting, the respective responsible officials should review the purposes and agenda items described above and be prepared to discuss them.

Objectives may be widely different depending on incident character, agency roles, and other factors. It is extremely important to understand that these separate and perhaps diverse objectives do not have to be forced into a consensus package. Unified planning is not a "committee" process that must somehow resolve all differences in agency objectives before any action can take place. It is, however, a "team" process, and that promotes open sharing of objectives and priorities. Through the process, the team formulates collective (which is significantly different than common) directions to address the needs of the entire incident.

The process follows the management-by-objectives sequence, uses the same worksheets and forms, and allows for both functional and geographic response authorities to combine objectives and actions. Once objectives and priorities are documented, the process continues as it would for single-agency involvement, except that all agencies are equally included.

Following the Command Meeting, a Command Staff Meeting may be conducted. The purpose of this meeting is to coordinate functions, responsibilities, and objectives at the Command Staff level.

The next meeting is a Strategy/Tactics Meeting where the blueprint for tactical deployment during the next Operational Period is created. Objectives and strategies are reviewed with Incident Commander and the Operations Section Chief. The Incident Commander and/or Unified Command will review the current Incident Action Plan and situation status information as provided by the Planning/Intelligence Section to access work progress against the current incident objectives. The Planning/Intelligence and Operations Section Chiefs will jointly develop primary and alternate strategies to meet objectives for consideration at the next Planning Meeting.

During the Strategy/Tactics Meeting, the Planning/Intelligence Section Chief or Situation Unit Leader briefs the pertinent strategies and any other issues, such as environmental sensitivities of the incident area. The Incident Commander presents objectives for the next operational period. These objectives are clearly stated and attainable with the resources available, yet flexible enough to allow the Operations Section Chief, in concert with the rest of the Operations Section staff, to choose tactics. During the meeting, the Planning/Intelligence Section Chief prepares an initial draft ICS Form 202 for use during the Planning Meeting. Primary and alternate strategies are developed with input from the Operations Section Chief and the Safety Officer. The Operations Section Chief and Planning/Intelligence Section Chiefs develop a draft of the Operational Planning Worksheet (ICS Form 215) to identify resources necessary for tactics that are agreed upon. Resources and other support needs are developed and ordered in conjunction with the Logistics Section Chief. The Safety Officer develops the Incident Safety Analysis using ICS Form 215A in conjunction with the ICS Form 215. This process can be completed by the Operations Section Chief and the Resource Unit Leader after the meeting adjourns, with Safety Officer preparing a poster-size Form 215 for the planning meeting.

The Planning Meeting appropriately follows the Strategy/Tactics Meeting. In this meeting incident objectives, strategies, and tactics are developed and resource needs for the next operational period are identified. Here the fine tuning of objectives and priorities takes place; problems are identified and solved; work assignments are defined; and responsibilities are compiled and completed on a ICS Form 215 (Operations Planning Worksheet). As noted, the strategy and tactics meeting is a preparation for this Meeting. Objectives (ICS Form 202) for the next Operational Period, large sketch maps or charts clearly dated and time-stamped, poster-size Operations Planning Worksheets, current resource inventory, and current situation status displays prepared by the Planning/Intelligence Section should be displayed in this meeting. After the meeting, the ICS Form 215 is used by the Logistics Section Chief to prepare the off-incident tactical and logistical resource orders, and by the Planning/Intelligence Section Chief to develop the Incident Action Plan assignment lists. This meeting is generally attended by all Command and General Staff, Technical Specialists, Agency Administrators and Representatives, as well as politicians and other concerned ranking officials as needed and directed by the Incident Commander or the Unified Command.

Following the Planning Meeting, participants immediately prepare their assignments for the Incident Action Plan to meet the Planning/Intelligence Section Chief's deadline for assembling their Incident Action Plan components. The deadline will be early enough to permit timely Incident Command/Unified Command approval and duplication of sufficient copies for the Operations Briefing and for supervisory and team personnel. After the Incident Action Plan is prepared, it is immediately submitted to the Incident Commander/Unified Command for approval.

After the Incident Action Plan is completed, an Operations Briefing is conducted. At this briefing, the Planning/Intelligence Section Chief will review the Incident Commander/Unified Command objectives and make any changes to the Incident Action Plan accordingly. The Situation Unit Leader will provide weather conditions and the forecast for the Operational Period. The Operations Section Chief will review current response actions and the last shift's accomplishments and make Division/Group and Air Operations assignments. The Safety Officer will provide a Safety Message. The Logistics Section Chief will report any transportation, communications, and supply updates. The Finance/Administration Section Chief will provide a financial report and state other issues pertinent to the group. The Public Information Officer will provide a media report and review any issues relating to the media. The Liaison Officer will report on any contributing organization or agency concerns. Finally, the Incident Command/Unified Command will give their Incident Action Plan approval and motivational remarks.

The Operations Briefing conveys the Incident Action Plan for the oncoming shift to the response organization. After this meeting, field supervisors who are rotating off should be debriefed by their oncoming resources and by the Operations Section Chief in order to further confirm or adjust the course of the new operational period's Incident Action Plan. Shifts in tactics may be made by the Operations Section Chief and or his subordinate Supervisors. Similarly, a Supervisor may reallocate resources within that Division to adapt to changing conditions. It is critical that when there are shifts from the Incident Action Plan that they are communicated back to the Incident Command Post during the Operational Period.

Other meetings that may occur during an incident may include business management, agency representative, Command and General Staff meetings, and press conferences on an as needed basis and or as directed by Incident Command/Unified Command.

In the Business Management Meeting team members develop and update the operating plan for finance and logistics support. This agenda could include:

- Finance requirements (to include any criteria imposed by contributing organizations)
- Business operating plan for resource procurement and incident funding
- Cost analysis
- Financial summary data

The Business Management Meeting is attended by the Finance/Administration Section Chief, who generally facilitates, the Cost Unit Leader, the Logistics Section Chief, the Supply Unit Leader, and the Documentation Unit Leader and is generally held prior to the planning meeting.

The Agency Representative Meeting is held to update Agency Representatives and ensure their support of the Incident Action Plan. It is most appropriately held after the Planning Meeting in order to announce plans for the next Operational Period allowing for changes in case the Incident Action Plan's goals are unattainable by a team. It is facilitated by the Liaison Officer and should be attended by all agency representatives or their designee.

Press Conferences must be planned and facilitated. The Information Officer is responsible for coordinating them. Their purpose is to brief media and the public that the incident is being handled competently and appropriately. Normally, it is conducted by the Incident Commander or the Unified Command structure, with the assistance of response organization members required addressing any particular issues specific to their agency or operations.

Some Incident Commanders will request a Command and General Staff Meeting in conjunction with a meal, as everyone needs to eat, and this provides an opportunity for the Command and General Staff to gather under more informal and relaxing conditions to share and update each other on developing issues.

The Incident Action Plan

The Incident Action Plan contains objectives reflecting the overall incident strategy, specific tactical actions, and supporting information for the next Operational Period. It is important that all incidents have some form of an Incident Action Plan. The plan is developed around a specified duration of time called an Operational Period. It states the objectives to be achieved and describes the strategy, tactics, resources, and support required to achieve the objectives within the specified time frame. Generally, the length of the Operational Period is determined by the length of time required to achieve the objectives.

The Incident Action Plan may be oral or written. Small incidents with only a few assigned resources may have a very simple oral plan, which may not be written out until the reporting phase of the incident if at al. As incidents become larger or require multi-agency involvement, the action plan should be written. Incident Action plans will vary in content and form depending upon the kind and size of the incident. The Incident Management System provides for the use of a systematic planning process, along with forms and formats for developing the Incident Action Plan. The general guidelines for use of a written versus a verbal action plan are when:

- Two or more jurisdictions are involved.
- A number of organizational elements have been activated.
- The incident continues into another planning or operational period.
- It is required by agency policy.

The Incident Action Plan provides all agencies with a clear set of objectives, actions, and assignments. It also provides the organizational structure and the Communications Plan required to manage the incident effectively under Unified Command. Written Incident Action Plans have four main elements that should be included:

- Statement of objectives: what is expected to be achieved; objectives must be measurable, attainable, and flexible
- Organization: which elements of the Incident Management System organization will be in place for the next Operational Period

- Tactics and assignments: tactics and control operations, including what resources will be assigned; resource assignments are often made by a Division or Group
- Supporting material: examples could include a map of the incident, a Communications Plan, Medical Plan, Traffic Plan, weather data, special precautions, and Safety Messages

Management by Objectives

Whether oral or written, the Incident Action Plan process relies on the Management-by-Objectives framework and the use of the Incident Command Management System forms to aid response:

- Core organizational functions are assured
- Policy, objectives, and priorities are set by Command to meet the objectives designed by the Operations and Planning/Intelligence Section
- Support and service needs, including communications and medical requirements, are clearly identified from the beginning of the incident, typically by the Logistics Section
- Financial parameters, abilities, and constraints are considered by an activated
- Finance/Administration Section
- A review of the initial work is carried out; all participants in the process examine the tentative plan for completeness, feasibility, and capability to meet objectives; results of the review are used to revise or strengthen the plan

During multi-agency incidents organized under Unified Command, the Incident Action Plans should always be written. Incident Action Planning Meetings will use the results of the Command Meeting to decide on tactical operations for the next Operational Period, establishing resource requirements and determining resource availability and sources, making resource assignments, establishing the unified Operations Section organization, and establishing combined Planning/Intelligence, Logistics, and Finance/Administration Sections as needed.

The end result of the planning process will be an Incident Action Plan that addresses multi-jurisdiction or multi-agency priorities and provides tactical operations and resource assignments for the unified effort.

Incident Command System Forms

The Incident Command System has utilized numerous forms to aid the planning process. The individuals who developed the Incident Command System spent years in creating forms that are still in use today. It was important for those individuals to follow the Management-by-Objectives concept. The process should answer the questions:

- "What do we need to know to accomplish the objectives?"
- "What do we need to do to accomplish the objectives?"
- "What resources do we need to accomplish the objectives?"

This process needs to be relatively easy to complete and must be realistic for incident personnel in emergency conditions.

The one recurring problem with forms is that while almost everyone involved in the process agrees on what information needs to be captured on a given form and even on who needs to maintain and update that information on a given form, the actual forms that have evolved over the last several decades are at times similar yet different. Similar in that they all manage to capture the same relevant data but different in content, format, and flow.

This planning process has been around for over thirty years. The forms that are a part of the planning process are those that become a part of the Incident Action Plan and those that support and document the process.

Specific forms that become a part of the Incident Action Plan and are an output of the planning process are the Incident Objectives, Organization Assignment List, Operational Assignment List, Communication Plan, Medical Plan, Air Operations Summary, Safety Plan or message, and incident area map. Others forms such as T-Cards, etc. may be added as needed.

Other forms that support and document the planning process include the Operational Planning Worksheet, Radio Requirements Worksheet, Incident Safety Analysis Worksheet, Organization Chart, Unit Logs, and Demobilization Checkout. Others provide an accountability tracking system, which is critical to the planning process. These include the Check-in List, Support Vehicle Inventory, Status Cards (often called T-cards), and Resource Order Forms.

Figure 6.7 T-cards in use

Many forms that share the same name and even number designator are often slightly different to the point of confusion. The forms referenced in this chapter are based on the National Wildfire Coordinating Group set. In Appendix B examples of the most commonly used forms from other major "players" in the "Incident Command World" are provided as well.

ICS 201 Incident Briefing Form

Although there are different versions of the Incident Briefing Form also known as the ICS 201, it is generally a four-part form. Part 1 is a sketch map of the incident. It is valuable to have the ability to record decisions made in the early parts of an incident. The description of traffic plans as well as the location of incident facilities may be reflected on this map. Part 2 is a summary of current objectives and control actions. Part 3 is an organizational chart from the early phases of the incident. Part 4 provides a summary of resources on location and those that have been ordered. The whole form becomes a way of documentation for simple incidents as well as a briefing document for succeeding Incident Commanders.

Incident Objectives Form

The Incident Objectives Form also known as the ICS 202, describes the basic incident strategy and control objectives for use during each Operational Period. It lets everyone know why they are there and what they are supposed to accomplish. It documents the Incident Commander's objectives as well as provides weather forecasts for the operational period and a general safety message. The ICS 202 is completed by the Planning/Intelligence Section Chief and approved by the Incident Commander in conjunction with the Planning Meeting

Organization Assignment Form/Organization Chart

The Organization Assignment Form/Organization Chart also known as the ICS 203 and/or ICS 207 form, chart provides information on the response organization and personnel staffing. The list is prepared and maintained by the Resource Unit under the direction of the Planning/Intelligence Section Chief. An Organization Assignment List or Organization Chart may be completed any time the number of personnel assigned to the incident increases or decreases or a change in assignment occurs. The forms basically tell who works for whom and whom you work for. With each Operational Period, the changes made on these forms keep a record of the resources. The ICS 203 is a part of the Incident Action Plan, while the ICS 207 is not.

Assignment List

The Assignment List also known as the ICS 204 is used to inform the Operations Section of the resources and the tasks assigned to a give Division or Group. It will specify the resources assigned, whether they are a Strike Team, Task Force, or Single Resource. The tasks are described as well as special instructions and safety issues. The Communication Plan with respect to that operational element is also described on it. It will let the assigned resources know whom they are being supervised by. This particular form indicates the primary results of the planning process.

INCIDENT BRIEFING	1. INCIDENT NAME	2. DATE PREPARED	3. TIME PREPARED
	4. MAP SKETCH		
ICS 201 (12/93) NFES 1325 PAGE 1	5. PREPARED BY (NAME AND POSITION)		

Figure 6.8 NWCG ICS-201

6. SUMMARY OF CURRENT ACTIONS

ICS 201 (12/93) NFES 1325	PAGE 2	

Figure 6.8 NWCG ICS-201 *(continued)*

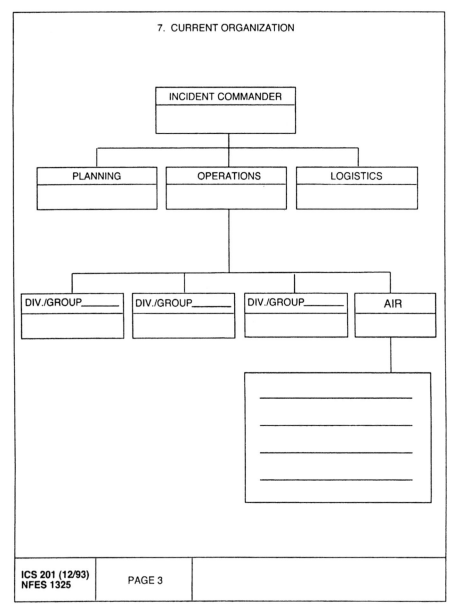

Figure 6.8 NWCG ICS-201 *(continued)*

8. RESOURCES SUMMARY				
RESOURCES ORDERED	RESOURCES IDENTIFICATION	ETA	ON SCENE √	LOCATION/ASSIGNMENT

Figure 6.8 NWCG ICS-201 *(continued)*

INCIDENT OBJECTIVES	1. INCIDENT NAME	2. DATE PREPARED	3. TIME PREPARED

4. OPERATIONAL PERIOD (DATE/TIME)

5. GENERAL CONTROL OBJECTIVES FOR THE INCIDENT (INCLUDE ALTERNATIVES)

6. WEATHER FORECAST FOR OPERATIONAL PERIOD

7. GENERAL SAFETY MESSAGE

8. ATTACHMENTS (✔ IF ATTACHED)

☐ ORGANIZATION LIST (ICS 203)   ☐ MEDICAL PLAN (ICS 206)   ☐ _____
☐ ASSIGNMENT LIST (ICS 204)   ☐ INCIDENT MAP   ☐ _____
☐ COMMUNICATIONS PLAN (ICS 205)   ☐ TRAFFIC PLAN   ☐ _____

9. PREPARED BY (PLANNING SECTION CHIEF)   10. APPROVED BY (INCIDENT COMMANDER)

Figure 6.9 NWCG ICS-202

ORGANIZATION ASSIGNMENT LIST	1. INCIDENT NAME	2. DATE PREPARED	3. TIME PREPARED

POSITION NAME	4. OPERATIONAL PERIOD (DATE/TIME)

5. INCIDENT COMMANDER AND STAFF
INCIDENT COMMANDER
DEPUTY
SAFTEY OFFICER
INFORMATION OFFICER
LIAISON OFFICER

6. AGENCY REPRESENTATIVES
AGENCY   NAME

7. PLANNING SECTION
CHIEF
DEPUTY
RESOURCES UNIT
SITUATION UNIT
DOCUMENTATION UNIT
DEMOBILIZATION UNIT
TECHNICAL SPECIALISTS

8. LOGISTICS SECTION
CHIEF
DEPUTY

a SUPPORT BRANCH
DIRECTOR
SUPPLY UNIT
FACILITIES UNIT
GROUND SUPPORT UNIT

b. SERVICE BRANCH
DIRECTOR
COMMUNICATIONS UNIT
MEDICAL UNIT
FOOD UNIT

9. OPERATIONS SECTION
CHIEF
DEPUTY
a. BRANCH I- DIVISION/GROUPS
BRANCH DIRECTOR
DEPUTY
DIVISION/GROUP
DIVISION/GROUP
DIVISION/GROUP
DIVISION/GROUP
DIVISION/GROUP
b. BRANCH II- DIVISION/GROUPS
BRANCH DIRECTOR
DEPUTY
DIVISION/GROUP
DIVISION/GROUP
DIVISION/GROUP
DIVISION/GROUP
DIVISION/GROUP
c. BRANCH III- DIVISION/GROUPS
BRANCH DIRECTOR
DEPUTY
DIVISION/GROUP
DIVISION/GROUP
DIVISION/GROUP
DIVISION/GROUP
DIVISION/GROUP

d. AIR OPERATIONS BRANCH
AIR OPERATIONS BR. DIR.
AIR TACTICAL GROUP SUP
AIR SUPPORT GROUP SUP.
HELICOPTER COORDINATOR
AIR TANKER/FIXED WING CRD.

10. FINANCE/ADMINISTRATION SECTION
CHIEF
DEPUTY
TIME UNIT
PROCUREMENT UNIT
COMPENSATION/CLAIMS UNIT
COST UNIT

PREPARED BY(RESOURCES UNIT)

Figure 6.10 NWCG ICS-203

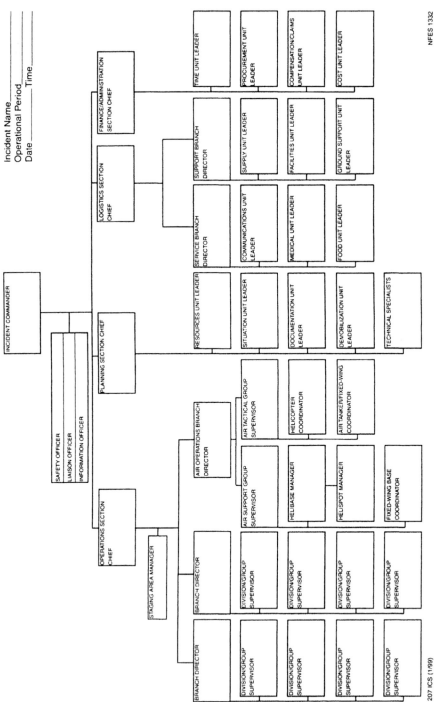

Figure 6.11 NWCG ICS-207

1. BRANCH	2. DIVISION/GROUP	**ASSIGNMENT LIST**					
3. INCIDENT NAME		4. OPERATIONAL PERIOD DATE _____ TIME _____					

5. OPERATIONAL PERSONNEL

OPERATIONS CHIEF _____   DIVISION/GROUP SUPERVISOR _____

BRANCH DIRECTOR _____   AIR TACTICAL GROUP SUPERVISOR _____

6. RESOURCES ASSIGNED THIS PERIOD

STRIKE TEAM/TASK FORCE/ RESOURCE DESIGNATOR	EMT	LEADER	NUMBER PERSONS	TRANS. NEEDED	PICKUP PT./TIME	DROP OFF PT./TIME

7. CONTROL OPERATIONS

8. SPECIAL INSTRUCTIONS

9. DIVISION/GROUP COMMUNICATIONS SUMMARY

FUNCTION		FREQ.	SYSTEM	CHAN.	FUNCTION		FREQ.	SYSTEM	CHAN.
COMMAND	LOCAL				SUPPORT	LOCAL			
	REPEAT					REPEAT			
DIV./GROUP TACTICAL					GROUND TO AIR				

PREPARED BY (RESOURCE UNIT LEADER)	APPROVED BY (PLANNING SECT. CH.)	DATE	TIME

204 ICS (1/99)   NFES 1328

Figure 6.12 NWCG ICS-204

Incident Communications Plan

The Incident Communication Plan, also known as the ICS 205, provides information on assignments for all communications equipment for each Operational Period. The plan is a summary of information related to all aspects of communications as they relate to the incident. The information on frequency assignments can also be placed on the appropriate assignment list (ICS 204) as well. This form is prepared by the Communication Unit Leader in the Logistics Section. This document can be critical in bringing order out of chaos at an incident. The Communication Unit Leader may also use the Radio Requirements Worksheet also known as the ICS 216 and the Frequency Assignment Worksheet also known as the ICS 217 in the development of the Incident Communication Plan.

Medical Plan

The Medical Plan, also known as the ICS 206, provides information on incident medical-aid stations, medical transportation services, hospitals, and emergency medical procedures. While the ICS 206 is intended for incident assigned personnel, it can also assist in medical operations conducted in the Operations Section. The Medical Plan is prepared by the Medical Unit Leader and reviewed by the Safety Officer.

Site Safety Plan

The Site Safety Plan, also known as ICS 208, is a detailed safety plan normally utilized during hazardous-material incidents. Most other incidents will utilize a general Safety Message form. The primary intent is to inform incident personnel of the risks and hazards that they may be confronted with and be made aware of mitigating actions. It is prepared by the Safety Officer.

Incident Status Summary

The Incident Status Summary form, also known as ICS 209, summarizes information for team members as well as various external agencies. It also provides information for the Public Information Officer for preparation of media releases and briefings to elected officials. The form is prepared by the Situation Unit Leader with input from the Resource Unit Leader in plans and the Cost Unit Leader in the Finance/Administration Section. The schedule for submission is usually dictated by federal or state authorities based on the needs, scope and complexity of the incident.

Check-in List

The Check-in List, also known as the ICS 211 can be one of the most valuable forms on an incident. It shows personnel and equipment arriving at or departing from the incident. These resources can check in or out at various locations set up by the Plans/Intelligence Section. Checking in or out consists of reporting specific information, which is recorded on the form. It is initiated at a number of locations including Staging Areas, Security Posts, Bases, Camps, Helibases, and the Incident Command Post. Staging Area Managers, Base/Camp Managers, or Status Check-in Recorders record the information and relay it to the Resources Unit as soon as possible.

INCIDENT RADIO COMMUNICATIONS PLAN	1. INCIDENT NAME	2. DATE/TIME PREPARED	3. OPERATIONAL PERIOD DATE/TIME

4. BASE RADIO CHANNEL UTILIZATION

SYSTEM/CACHE	CHANNEL	FUNCTION	FREQUENCY/TONE	ASSIGNMENT	REMARKS

5. PREPARED BY (COMMUNICATIONS UNIT)

205 ICS (9/66) NFES 1330

Figure 6.13 NWCG ICS-205

MEDICAL PLAN	1. INCIDENT NAME	2. DATE PREPARED	3. TIME PREPARED	4. OPERATIONAL PERIOD

5. INCIDENT MEDICAL AID STATIONS				
MEDICAL AID STATIONS	LOCATION		PARAMEDICS	
			YES	NO

6. TRANSPORTATION				
A. AMBULANCE SERVICES				
NAME	ADDRESS	PHONE	PARAMEDICS	
			YES	NO

B. INCIDENT AMBULANCES			
NAME	LOCATION	PARAMEDICS	
		YES	NO

7. HOSPITALS								
NAME	ADDRESS	TRAVEL TIME		PHONE	HELIPAD		BURN CENTER	
		AIR	GRND		YES	NO	YES	NO

8. MEDICAL EMERGENCY PROCEDURES

206 ICS 8/78	9. PREPARED BY (MEDICAL UNIT LEADER)	10. REVIEWED BY (SAFETY OFFICER)

NFES 1331

Figure 6.14 NWCG ICS-206

SITE SAFETY AND CONTROL PLAN ICS 208 HM	1. Incident Name:	2. Date Prepared:	3. Operational Period: Time:

Section I. Site Information

4. Incident Location:

Section II. Organization

5. Incident Commander:	6. HM Group Supervisor:	7. Tech. Specialist - HM Reference:
8. Safety Officer:	9. Entry Leader:	10. Site Access Control Leader:
11. Asst. Safety Officer - HM:	12. Decontamination Leader:	13. Safe Refuge Area Mgr:
14. Environmental Health:	15.	16.

17. Entry Team: (Buddy System) Name:	PPE Level	18. Decontamination Element: Name:	PPE Level
Entry 1		Decon 1	
Entry 2		Decon 2	
Entry 3		Decon 3	
Entry 4		Decon 4	

Section III. Hazard/Risk Analysis

19. Material:	Container type	Qty.	Phys. State	pH	IDLH	F.P.	I.T.	V.P.	V.D.	S.G.	LEL	UEL

Comment:

Section IV. Hazard Monitoring

20. LEL Instrument(s):	21. O₂ Instrument(s):
22. Toxicity/PPM Instrument(s):	23. Radiological Instrument(s):

Comment:

Section V. Decontamination Procedures

24. Standard Decontamination Procedures:	YES:	NO:

Comment:

Section VI. Site Communications

25. Command Frequency:	26. Tactical Frequency:	27. Entry Frequency:

Section VII. Medical Assistance

28. Medical Monitoring:	YES:	NO:	29. Medical Treatment and Transport In-place:	YES:	NO:

Comment:

ICS 208 HM   Page 1   3/98

Figure 6.15 NWCG ICS-208

Incident Status Summary (ICS-209)

| 1: Date   2: Time   3: Initial | Update | Final   4: Incident Number   5: Incident Name |

6: Incident   7: Start Date   8:   9: Incident   10: IMT   11: State-
   Kind   Time   Cause   Commander   Type   Unit

12:   13: Latitude and   14: Short Location Description (in reference to nearest
County   Longitude   town):
   Lat:
   Long:

Current Situation

15:   16: %   17: Expected   18: Line to   19:   20: Declared
Size/Area   Contained or   Containment   Build   Costs   Controlled
Involved   MMA   Date: to Date   Date:
   Time: Time:

21: Injuries   22:   23:
this   Injuries   Fatalities   24: Structure Information
Reporting   to Date:
Period:

   #   #   #
   Type of Structure   Threatened   Damaged   Destroyed

25: Threat to Human Life/Safety: Residence
Evacuation(s) in progress ----
No evacuation(s) imminent --   Commercial
Potential future threat -------   Property
No likely threat --------------   Outbuilding/Other

26: Communities/Critical Infrastructure Threatened (in 12, 24, 48 and 72 hour time frames):

12 hours:

24 hours:

48 hours:

72 hours:

27: Critical Resource Needs (kind & amount, in priority order):
1.
2.
3.

28: Major problems and concerns (control problems, social/political/economic concerns or impacts, etc.) Relate critical resources needs identified above to the Incident Action Plan.

29: Resources threatened (kind(s) and value/significance):

Figure 6.16 NWCG ICS-209

CHECK-IN LIST

1. INCIDENT NAME

2. CHECK-IN LOCATION ☐ BASE ☐ CAMP ☐ STAGING AREA ☐ ICP RESOURCES ☐ HELIBASE

3. DATE/TIME

CHECK-IN INFORMATION

4. PERSONNEL (OVERHEAD) BY AGENCY & NAME -OR- LIST EQUIPMENT BY THE FOLLOWING FORMAT				5. ORDER/ REQUEST NUMBER	6. DATE/TIME CHECK-IN	7. LEADERS NAME	8. TOTAL NO. PERSONNEL	9. MANIFEST YES | NO	10. CREW WEIGHT INDIVIDUAL WEIGHT	11. HOME BASE	12. DEPARTURE POINT	13. METHOD OF TRAVEL	14. INCIDENT ASSIGNMENT	15. OTHER QUALIFICATION	16. SENT TO RESOURCES TIME/INT.
AGENCY	SINGLE T/F S/T	KIND	TYPE	I.D. NO./NAME												

17. PAGE ___ OF ___

18. PREPARED BY (NAME AND POSITION)

USE BACK FOR REMARKS OR COMMENTS

211 ICS (1/99)

NFES 1335

Figure 6.17 NWCG ICS-211

General Message Form
The General Message Form, also known as the ICS 213, is used by Incident Dispatchers to record incoming messages that cannot be orally transmitted to the intended recipients. These forms are used to order or document resource and supply orders or to pass other pertinent information along as needed.

Unit Log
The Unit Log, also known as the ICS 214, is used to record details of unit activity including specialized team activity. Unit Logs can provide a basic reference from which to extract information for inclusion in any post incident report. This form is initiated and maintained by all Command and General Staff members, Unit Leaders, and any other supervisors on the incident. Completed logs are forwarded to the Documentation Unit for the incident documentation file.

Operational Planning Worksheet
The Operational Planning Worksheet, also known as an ICS 215, is a form that may be used to communicate the decisions made concerning resource needs for the next Operational Period. The Operational Planning Worksheet is used by the Planning/Intelligence Section to complete the Assignment List (ICS 204) and by the Logistics Section for ordering resources for the incident. This form may be used as a source document for updating resource information on other incident command forms such as the Incident Status Summary (ICS 209). It is primarily initiated by the Operations Section Chief with input from the Planning/Intelligences and Logistics Sections. It is the basis of the Planning Meeting. It provides the required, have, and need-to-order resource numbers. If Logistics is unable to fill the "need" quantities, the plan may need to be adjusted.

Radio Requirements and Frequency Assignment Worksheets
The Radio Requirements Worksheet is also known as ICS 216 and the Frequency Assignment Worksheet is known as ICS are the initiators of the Incident Communications Plan (ICS 205) as described above. They are used primarily by the Communications Unit Leader to develop the Incident Communications Plan and are very dynamic documents as an incident begins and/or demobilizes.

Support Vehicle Inventory
The Support Vehicle Inventory, also known as the ICS 218, is prepared by the Ground Support Unit Leader in the Logistics Section to provide records and maintain availability information on support and service vehicles. It is a form for Finance/Administration and the Logistics Sections to account for what vehicles are on an incident and when they were where doing what task.

ICS 219 Resource Status Cards
Resource Status Cards, commonly known as T-cards and referred to as ICS 219 forms, are used by the Planning/Intelligence and Logistics Sections to record status and location information on resources, transportation, and support vehicles and personnel.

*U.S. GPO: 1009-793-975

GENERAL MESSAGE

TO: POSITION

FROM   POSITION

SUBJECT   DATE

MESSAGE:

DATE   TIME   SIGNATURE/POSITION

213 ICS 1/79
NFES 1336

PERSON RECEIVING GENERAL MESSAGE KEEP THIS COPY

SENDER REMOVE THIS COPY FOR YOUR FILES

Figure 6.18 NWCG ICS-213

UNIT LOG	1. INCIDENT NAME	2. DATE PREPARED	3. TIME PREPARED
4. UNIT NAME/DESIGNATORS.	5. UNIT LEADER (NAME AND POSITION)	6. OPERATIONAL PERIOD	

7. PERSONNEL ROSTER ASSIGNED		
NAME	ICS POSITION	HOME BASE

8. ACTIVITY LOG (CONTINUE ON REVERSE)	
TIME	MAJOR EVENTS

NFES 1337

Figure 6.19 NWCG ICS-214

OPERATIONAL PLANNING WORKSHEET	1. INCIDENT NAME	2. DATE PREPARED	3. OPERATIONAL PERIOD (DATE/TIME)
		TIME PREPARED	

4. DIVISION OR OTHER LOCATION	5. WORK ASSIGNMENTS	6. RESOURCE	RESOURCES BY TYPE (SHOW STRIKE TEAM AS ST)						7. REPORTING LOCATION	8. REQUESTED ARRIVAL TIME
		TYPE								
		REQ								
		HAVE								
		NEED								
		REQ								
		HAVE								
		NEED								
		REQ								
		HAVE								
		NEED								
		REQ								
		HAVE								
		NEED								
		REQ								
		HAVE								
		NEED								
		REQ								
		HAVE								
		NEED								
		REQ								
		HAVE								
		NEED								
		REQ								
		HAVE								
		NEED								

	9. TOTAL RESOURCES REQUIRED	SINGLE RESOURCES / STRIKE TEAMS							10. PREPARED BY (NAME AND POSITION)
	TOTAL RESOURCES ON HAND								
215 ICS 9-86	TOTAL RESOURCES NEEDED								

NFES 1338

Figure 6.20 NWCG ICS-215

RADIO REQUIREMENTS WORKSHEET						1. INCIDENT NAME			2. DATE	3. TIME	
4. BRANCH			5. AGENCY			6. OPERATIONAL PERIOD			7. TACTICAL FREQUENCY		
8. DIVISION/GROUP ____ AGENCY ____			DIVISION/ GROUP ____ AGENCY ____			DIVISION/ GROUP ____ AGENCY ____			DIVISION/ GROUP ____ AGENCY ____		
9. AGENCY	ID NO.	RADIO RQMTS	AGENCY	ID NO.	RADIO RQMTS	AGENCY	ID NO.	RADIO RQMTS	AGENCY	ID NO.	RADIO RQMTS
216 ICS 3-82			PAGE			5. PREPARED BY (COMMUNICATIONS UNIT)					

NFES 1339

*GPO 1985-0-593-005/14028

Figure 6.21 NWCG ICS-216

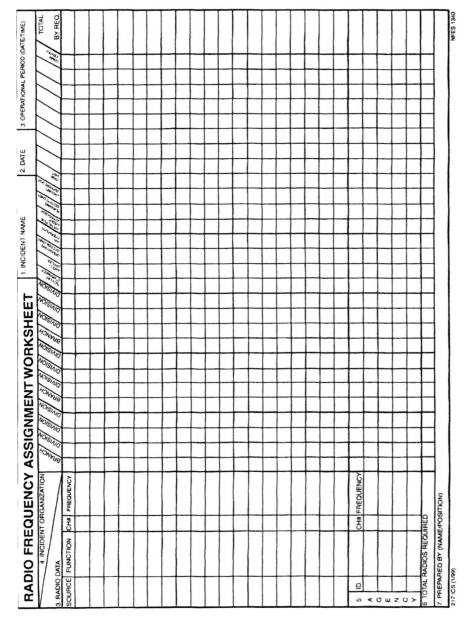

Figure 6.22 NWCG ICS-217

SUPPORT VEHICLE INVENTORY (USE SEPARATE SHEET FOR EACH VEHICLE CATEGORY)			1. INCIDENT NAME		2. DATE PREPARED	3. TIME PREPARED
VEHICLE INFORMATION						
a. TYPE	b. MAKE	c. CAPACITY/SIZE	d. AGENCY/OWNER	e. I.D. NO.	f. LOCATION	g. RELEASE TIME
218 ICS 8-78			PAGE		5. PREPARED BY (GROUND SUPPORT UNIT)	
NFES 1341						

Figure 6.23 NWCG ICS-218

The cards provide a visual display of status and location of resources assigned to the incident. Information on the cards is derived from several sources such as the ICS 201 and ICS 211 as well as information supplied by the Operations Section and the Incident Dispatchers.

Air Operations Summary

The Air Operations Summary also known as the ICS 220 provides information on air operations including the number, type, location, and specific assignments of helicopters and fixed-wing aircraft. It is completed by the Air Operations Branch Director or the Operations Section Chief. The form is also used by Finance/Administration in cost accounting for such assets.

The Incident Action Planning Process

In the Incident Command System, considerable emphasis is placed on developing effective Incident Action Plans. A planning process has been developed as a part of the system itself to give planners an orderly and systematic methodology. The steps outlined in this chapter will allow for the development of an Incident Action Plan in a minimum amount of time. Incidents vary in their kind, complexity, size, and requirements for detailed and written plans. The planning process described in this chapter is based on the development of Incident Action Plans to support major wildfire incidents but is applicable to any type of emergency response. Not all incidents require detailed written plans. Recognizing this, the following planning process provides a series of basic steps that are generally appropriate for use in any incident situation. The determination of the need for written Incident Action Plans and attachments is based on the requirements of the incident and the judgment of the Incident Commander.

General Responsibilities

The general responsibilities associated with the Planning Meeting and the development of the Incident Action Plan is described below. The Planning/Intelligence Section Chief should review these with the General Staff prior to the Planning Meeting.

The Incident Commander:

- Provides overall control of objectives and strategy
- Establishes the procedure for off-incident resource order
- Approves requests for off-incident action plan by signature
- Approves completed incident action plan by signature

The Planning/Intelligence Section Chief:

- Conducts the Planning Meeting and coordinates preparation of the Incident Action Plan

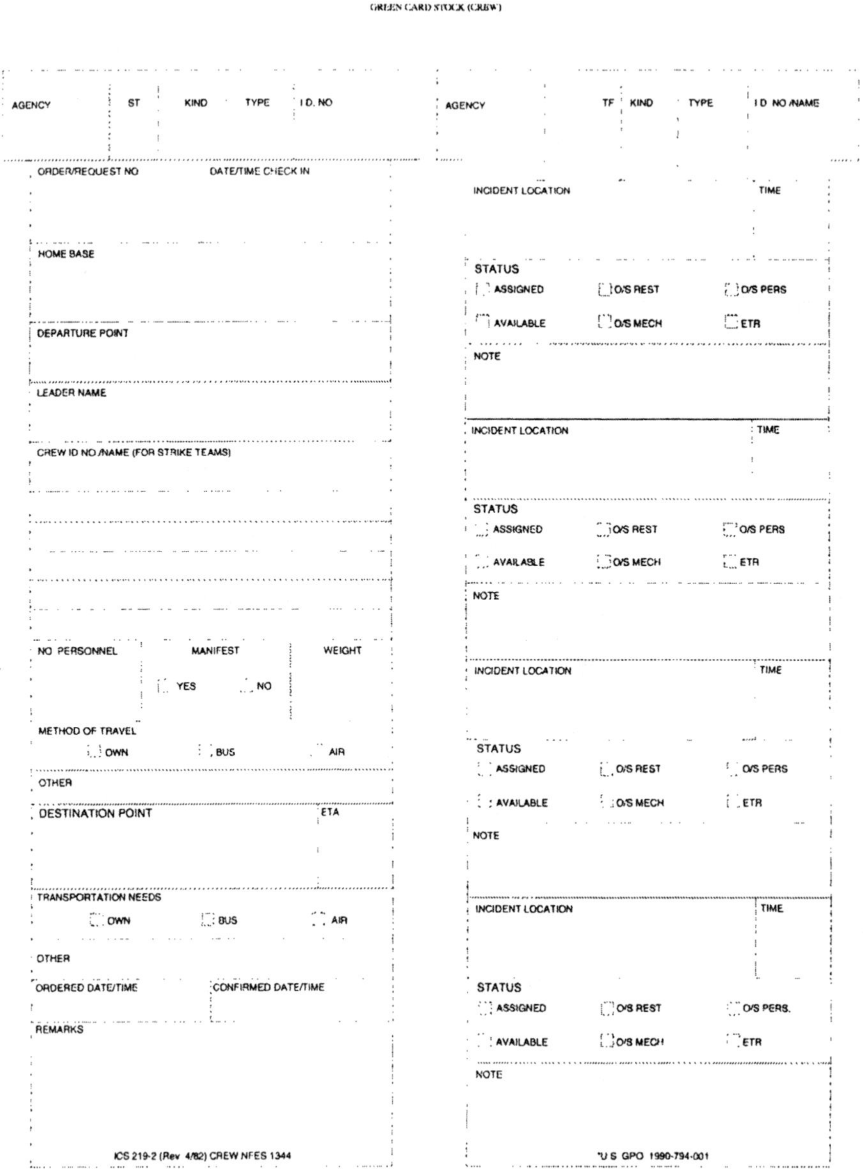

GREEN CARD STOCK (CREW)

AGENCY	ST	KIND	TYPE	I.D. NO		AGENCY	TF	KIND	TYPE	I.D. NO./NAME

ORDER/REQUEST NO   DATE/TIME CHECK IN

INCIDENT LOCATION   TIME

HOME BASE

STATUS
- [ ] ASSIGNED   [ ] O/S REST   [ ] O/S PERS
- [ ] AVAILABLE   [ ] O/S MECH   [ ] ETR

DEPARTURE POINT

NOTE

LEADER NAME

INCIDENT LOCATION   TIME

CREW ID NO./NAME (FOR STRIKE TEAMS)

STATUS
- [ ] ASSIGNED   [ ] O/S REST   [ ] O/S PERS
- [ ] AVAILABLE   [ ] O/S MECH   [ ] ETR

NOTE

NO. PERSONNEL   MANIFEST   WEIGHT

INCIDENT LOCATION   TIME

[ ] YES   [ ] NO

METHOD OF TRAVEL
[ ] OWN   [ ] BUS   [ ] AIR

STATUS
- [ ] ASSIGNED   [ ] O/S REST   [ ] O/S PERS

OTHER

- [ ] AVAILABLE   [ ] O/S MECH   [ ] ETR

DESTINATION POINT   ETA

NOTE

TRANSPORTATION NEEDS
[ ] OWN   [ ] BUS   [ ] AIR

INCIDENT LOCATION   TIME

OTHER

ORDERED DATE/TIME   CONFIRMED DATE/TIME

STATUS
- [ ] ASSIGNED   [ ] O/S REST   [ ] O/S PERS.

REMARKS

- [ ] AVAILABLE   [ ] O/S MECH   [ ] ETR

NOTE

ICS 219-2 (Rev. 4/82) CREW NFES 1344   *U.S. GPO 1990-794-001

Figure 6.24 NWCG ICS-219-2

BLUE CARD STOCK (HELICOPTER)

| AGENCY | ST | KIND | TYPE | I.D. NO | | AGENCY | TYPE | MANUFACTURER | I.D. NO |

ORDER/REQUEST NO.   DATE/TIME CHECK IN

INCIDENT LOCATION   TIME

HOME BASE

STATUS
  ASSIGNED   O/S REST   O/S PERS.
  AVAILABLE   O/S MECH   ETR

DEPARTURE POINT

NOTE

PILOT NAME

INCIDENT LOCATION   TIME

DESTINATION POINT   ETA

STATUS
  ASSIGNED   O/S REST   O/S PERS.
  AVAILABLE   O/S MECH   ETR

REMARKS

NOTE

INCIDENT LOCATION

INCIDENT LOCATION   TIME

STATUS
  ASSIGNED   O/S REST   O/S PERS.
  AVAILABLE   O/S MECH   ETR

STATUS
  ASSIGNED   O/S REST   O/S PERS.
  AVAILABLE   O/S MECH   ETR

NOTE

NOTE

INCIDENT LOCATION   TIME

INCIDENT LOCATION   TIME

STATUS
  ASSIGNED   O/S REST   O/S PERS
  AVAILABLE   O/S MECH   ETR

STATUS
  ASSIGNED   O/S REST   O/S PERS
  AVAILABLE   O/S MECH   ETR

NOTE

NOTE

ICS 219-4 (Rev 4/82) HELICOPTER NFES 1346   *U.S. GPO 1988-594-771   NFES 1346

Figure 6.25 NWCG ICS-219-4

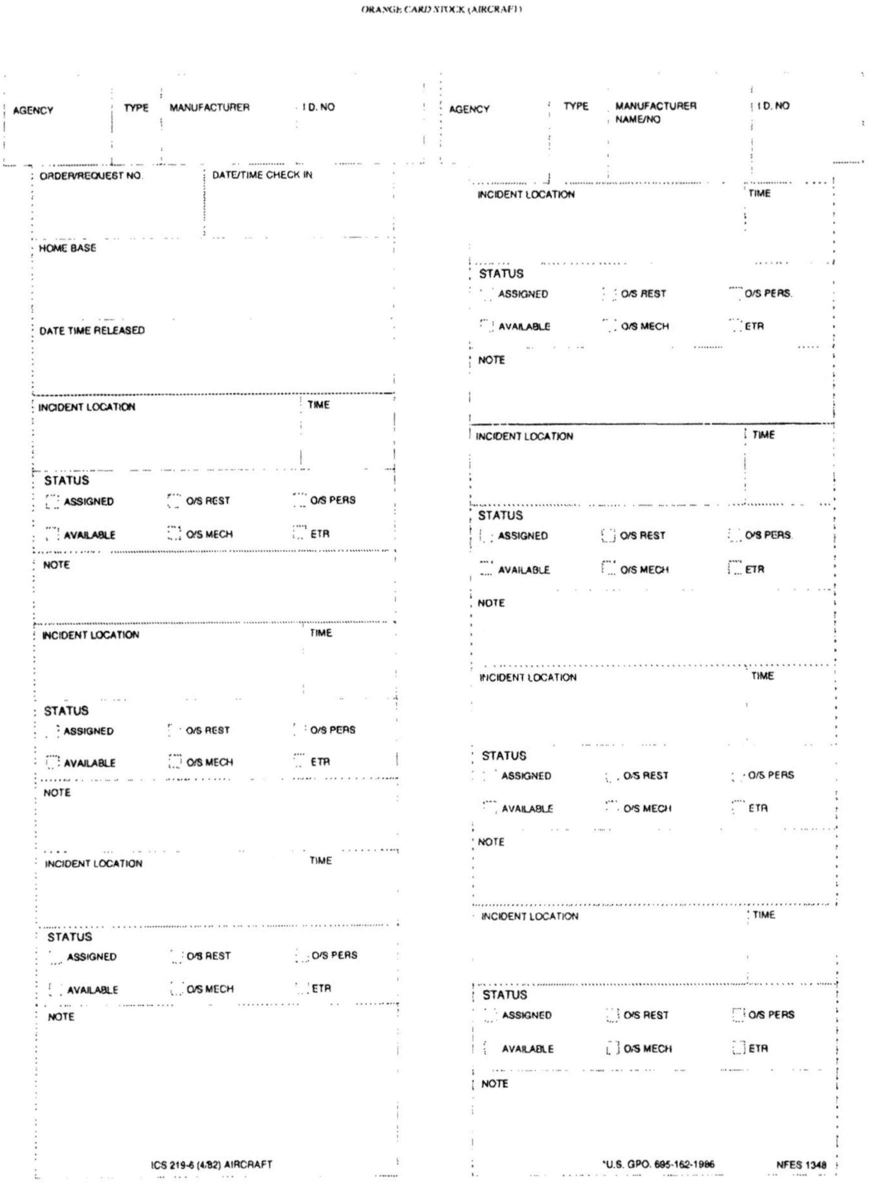

ORANGE CARD STOCK (AIRCRAFT)

AGENCY	TYPE	MANUFACTURER	I D. NO

ORDER/REQUEST NO.	DATE/TIME CHECK IN

HOME BASE

DATE TIME RELEASED

INCIDENT LOCATION	TIME

STATUS
- ASSIGNED
- O/S REST
- O/S PERS
- AVAILABLE
- O/S MECH
- ETR

NOTE

INCIDENT LOCATION	TIME

STATUS
- ASSIGNED
- O/S REST
- O/S PERS
- AVAILABLE
- O/S MECH
- ETR

NOTE

INCIDENT LOCATION	TIME

STATUS
- ASSIGNED
- O/S REST
- O/S PERS
- AVAILABLE
- O/S MECH
- ETR

NOTE

ICS 219-6 (4/82) AIRCRAFT

AGENCY	TYPE	MANUFACTURER NAME/NO	I D. NO

INCIDENT LOCATION	TIME

STATUS
- ASSIGNED
- O/S REST
- O/S PERS.
- AVAILABLE
- O/S MECH
- ETR

NOTE

INCIDENT LOCATION	TIME

STATUS
- ASSIGNED
- O/S REST
- O/S PERS
- AVAILABLE
- O/S MECH
- ETR

NOTE

INCIDENT LOCATION	TIME

STATUS
- ASSIGNED
- O/S REST
- O/S PERS
- AVAILABLE
- O/S MECH
- ETR

NOTE

INCIDENT LOCATION	TIME

STATUS
- ASSIGNED
- O/S REST
- O/S PERS
- AVAILABLE
- O/S MECH
- ETR

NOTE

*U.S. GPO. 695-162-1986   NFES 1348

Figure 6.26 NWCG ICS-219-6

YELLOW CARD STOCK (DOZERS)

AGENCY	ST	TF	KIND	I TYPE	I.D. NO.

ORDER/REQUEST NO	DATE/TIME CHECK IN

HOME BASE

DEPARTURE POINT

LEADER NAME

RESOURCE ID. NO.S/NAMES

DESTINATION POINT	ETA

REMARKS

INCIDENT LOCATION	TIME

STATUS
☐ ASSIGNED   ☐ O/S REST   ☐ O/S PERS
☐ AVAILABLE   ☐ O/S MECH   ☐ ETR

NOTE

ICS 219-7 (Rev. 4/82) DOZERS NFES 1349

AGENCY	ST	TF	KIND	I TYPE	I D NO.

INCIDENT LOCATION	TIME

STATUS
☐ ASSIGNED   ☐ O/S REST   ☐ O/S PERS
☐ AVAILABLE   ☐ O/S MECH   ☐ ETR

NOTE

INCIDENT LOCATION	TIME

STATUS
☐ ASSIGNED   ☐ O/S REST   ☐ O/S PERS
☐ AVAILABLE   ☐ O/S MECH   ☐ ETR

NOTE

INCIDENT LOCATION	TIME

STATUS
☐ ASSIGNED   ☐ O/S REST   ☐ O/S PERS
☐ AVAILABLE   ☐ O/S MECH   ☐ ETR

NOTE

INCIDENT LOCATION	TIME

STATUS
☐ ASSIGNED   ☐ O/S REST   ☐ O/S PERS.
☐ AVAILABLE   ☐ O/S MECH   ☐ ETR

NOTE

*U S GPO 1990-794-006

Figure 6.27 NWCG ICS-219-7

AIR OPERATIONS SUMMARY

PREPARED BY:	PREPARED DATE/TIME:

1. INCIDENT NAME | **2. OPERATIONAL PERIOD DATE:** | START TIME: | END TIME: | SUNRISE: | SUNSET:

3. REMARKS (Safety Notes, Hazards, Air Operations Special Equipment, etc.):

4. MEDEVAC A/C:

5. TFR:
Radius: _____ NM
Altitude: _____ 'MSL
Centerpoint: Lat: _____
Long: _____

6. PERSONNEL	Phone		7. FREQUENCIES	AM	FM	8. FIXED-WING #Available/Type/ Make-Model/ FAA N#/ Bases
AOBD:			AIR/AIR FW:			Airtankers
ATGS:			AIR/AIR RW:			
HLCO:			AIR/GROUND:			
ASGS:			COMMAND: (Simplex)			Leadplanes
HEBM:			COMMAND RPT	Rx:	Tx:	Base FAX#
ATB MGR:			DECK FREQ.:			ATGS Aircraft
			TOLC FREQ.:			
						Other

9. HELICOPTERS (Use Additional Sheets As Necessary)

FAA N#	TY	MAKE/MODEL	BASE	AVAIL	START	REMARKS	FAA N#	TY	MAKE/MODEL	BASE	AVAIL	START	REMARKS

220 ICS (2/99)

PAGE 1 OF 2

NFES 1351

Figure 6.28 NWCG ICS-220

The Finance/Administration Section Chief:

- Provides cost implications of control objectives as required
- Evaluates facilities being used to determine if any special arrangements are needed
- Ensures that the Incident Action Plan is within the finance limits established by the Incident Commander

The Operations Section Chief:

- Determines division work assignments and resource requirements

The Logistics Section Chief:

- Ensures that incident facilities are adequate
- Ensures that the resource ordering procedure is made known to appropriate agency dispatch center(s)
- Develops a transportation system to support operations needs
- Ensures that the section can logistically support the Incident Action Plan
- Places orders for resources

Preplanning Steps
The Planning/Intelligence Section Chief:

- If possible obtains completed Incident Briefing Form ICS 201) prior to the initial Planning Meeting
- Evaluates the current situation and decides if the current planning is adequate for the remainder of the Operational Period (until next Incident Action Plan takes effect)
- Advises the Incident Commander and Operations Section Chief of any suggested revisions to the current Incident Action Plan as necessary
- Establishes the planning cycle for the Incident Commander
- Determines Planning Meeting attendees with the Incident Commander
- Establishes the location and time for the Planning Meeting
- Ensures that planning boards and forms are available
- Notifies necessary support staff (recorders, etc.) of meetings and assignments
- Ensures that a current situation and resource briefing will be available for meetings
- Obtains an estimate of regional resource availability from agency dispatch for use in planning for the next Operational Period
- Obtains necessary agency policy and legal or fiscal constraints for use in the Planning Meeting

Conducting the Planning Meeting

For major incidents, attendees should include:

- Incident Commander
- Command Staff members
- General Staff members
- Resource Unit Leader
- Situation Unit Leader
- Air Operations Branch Director
- Communications Unit Leader
- Technical Specialists (as required)
- Agency Representatives (as required)
- Recorders (as required)

The Planning Meeting is normally conducted by the Planning/Intelligence Section Chief. The checklist that follows is intended to provide a basic sequence of steps to aid the Planning/Intelligence Section Chief in developing the Incident Action Plan. The Planning Checklist is intended to be used with the (Operational Planning Worksheet (ICS 215). Every incident must have an Incident Action Plan. However, NOT ALL INCIDENTS REQUIRE WRITTEN PLANS.

Figure 6.29 Planning meeting in progress

The need for written plans and attachments is based on the requirements of the incident and the decision of the Incident Commander. Table 6.1 shows a list of checklist items and who has the primary responsibility to maintain those items in preparation for the PlanningMeeting.

The Planning/Intelligence Section Chief and/or Resources and Situation Unit Leaders should provide an up-to-date briefing on the situation as it currently exists. Information for this briefing may come from any or all of the following sources:

- Initial Incident Commander
- Incident Briefing Form (ICS 201)
- Field observations
- Operations reports
- Incident behavior modeling (when appropriate)
- Regional resources and situation reports

TABLE 6.1   Checklist Items

CHECKLIST ITEM	PRIMARY RESPONSIBILITY
Briefing on situation and resource status	Planning/Intelligence Section Chief
Set control objectives	Incident Commander
Plot control lines and division boundaries	Operations Section Chief
Specify tactics for each Division/Group	Operations Section Chief
Specify resources needed by Division/Group	Operations Section Chief, Planning/Intelligence Section Chief
Specify facilities and reporting locations plot on map	Operations Section Chief, Planning/Intelligence Section Chief, Logistics Section Chief
Place resource and overhead personnel order	Logistics Section Chief
Consider Communications, Medical and Traffic Plan requirements	Planning/Intelligence Section Chief, Logistics Section Chief
Finalize, approve, and implement Incident Action Plan	Planning/Intelligence Section Chief, Incident Commander, Operations Section Chief
Brief on situation and resource status	Planning/Intelligence Section Chief, Resource Unit Leader, Situation Unit and Leader

Set Control Objectives
This important step is performed by the Incident Commander. The control objectives are not limited to any single Operational Period but rather consider the total incident situation. The Incident Commander will establish the general strategy to be used and state any major policy, legal, or fiscal constraints in accomplishing the objectives as well as appropriate contingency considerations.

Plot Control Lines and Division Boundaries on Map
This step is normally accomplished by the Operations Section Chief (for the next Operational Period) in conjunction with the Planning/Intelligence Section Chief, they will:

- Determine control line locations
- Establish Division/Branch boundaries for Geographical Divisions
- Determine the need for Functional Group assignments for the next Operational Period.

These decisions will be plotted on the map once they have been made.

Specify Tactics for Each Division
Once the Division geographical assignments have been determined, the Operations Section Chief will establish the specific work assignments to be used for each Division for the next Operational Period. Note that it may be necessary or desirable to establish a Functional Group in addition to Geographical Divisions. Tactics (work assignments) must be specific and must be within the boundaries set by the Incident Commander's general control objectives (strategies). These work assignments should be recorded on the Operational Planning Worksheet. At this time that the Operations Section Chief, Incident Commander, and Logistics Section Chief should consider the need for any alternative strategies or tactics and see that these are properly noted on the planning matrix.

Specify Resources Needed by Division
After specifying tactics for each Division, the Operations Section Chief, in conjunction with the Planning/Intelligence Section Chief, will determine the necessary resources by Division to accomplish the work assignments. The resource needs will be recorded on the Planning Worksheet. Resource needs should be considered according to the type of resource required to complete the assignment for example, "use a Type 2 Engine, rather than a Type 1 Engine for mop-up situations."

The Planning/Intelligence Section Chief should also ensure that the Air Operations Summary (ICS 220) is being developed by the Operations Section Chief or Air Operation Branch Director as appropriate. The Air Operations Summary Worksheet brings together in one place all tactical and logistical air assignments, with information on kinds and numbers of air resources required, reporting locations, and designation of resources assigned. Information is obtained from the Operational Planning Worksheet (ICS 215), and is used by the Planning/Intelligence, Operations, and Logistics Sections in establishing the Incident Air Operations Program for the next Operational Period.

FEMA-1391-DR, New York
Remote Sensing - 15 September
Debris Field and Building Damage Levels

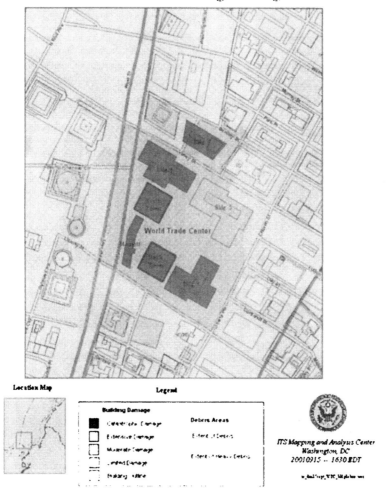

Figure 6.30 Map of incident boundaries

Specify Operations Facilities and Reporting Locations
The Operations Section Chief, in conjunction with the Planning/Intelligence and Logistics Section Chiefs, should designate and facilitate locations necessary to accomplish the work assignments, such as Staging Areas and Helispots. These designated locations should be charted on the Incident Area Map. Depending upon the situation, it may be appropriate to establish a Camp or Helibase location. Operations should also at this time indicate the reporting time requirements for the resources and any special resource

assignments. At the conclusion of this step, Operations Section personnel at the planning meeting may be released if desired.

Place Resource and Personnel Order

At this point in the process, the Planning/Intelligence Section Chief should perform a resource needs assessment based on the needs provided by the Operations Section Chief and resources data available from the Planning/Intelligence Section's Resources Unit. The Operational Planning Worksheet, when properly completed, will show resource requirements and availability to meet those requirements. By subtracting resources available from those required, any additional resource needs can be determined. From this assessment, a new resource order can be established and provided to the Incident Commander for approval before being processed through normal dispatch channels by the Logistics Section.

Consider Communications, Medical and Traffic Plan Requirements

The Incident Action Plan consists of the Incident Objectives (ICS 202), Organization Chart (ICS 203), Division Assignment List (ICS 204), and a map of the incident area. Larger incidents may require additional attachments, such as a separate Communications Plan (ICS 205), a Medical Plan (ICS 206), and possibly a Traffic Plan. The Planning/Intelligence Section Chief must determine the need for these attachments to be included as part of the written plan and ensure that these documents are prepared by the appropriate units. For major incidents, the Incident Action Plan and attachments will normally include all these items. They are prepared by specific individuals as shown in Table 6.2.

Prior to the completion of the Incident Action Plan, the Planning/Intelligence Section Chief should review the Division/Group tactical work assignments for any changes due to lack of resource availability. Recorders may then transfer Division assignment information including alternatives from the Planning Matrix Board or form ICS 215 onto the Division Assignment Lists (ICS-204).

Finalize, Approve, and Implement Incident Action Plan

The Planning/Intelligence Section Chief is responsible to ensure that the Incident Action Plan is completed, reviewed, and distributed. The sequence of steps to accomplish this is listed below:

- State the time that Incident Action Plan attachments are required to be completed
- Obtain Incident Action Plan attachments and review for completeness and approvals
- Determine numbers of copies of the Incident Action Plan required
- Arrange with documentation unit to reproduce the Incident Action Plan
- Review the Incident Action Plan to ensure it is up-to-date and complete prior to
- Operations Briefing and distribution of the Incident Action Plan

- Provide briefing on the Incident Action Plan as required
- Distribute on the Incident
- Action Plan prior to beginning of new Operational Period

How to Write a Good Objective

Writing an objective that is both specific and measurable can be an art form of sorts. Perhaps the best way to convey this art is to teach by example, but there are some basic rules that may be helpful.

Rule # 1 State your goal.

Define the objective by stating the actions that must be accomplished by the goal. "Prevent fire from reaching Sunnybrook Lake subdivision" is a good start. "Dig a fire line between the Sunnybrook Lake subdivision and the fire" is not a good way to begin as this statement describes how to accomplish the objective. We'll show you where a statement like this fits into your objective in Rule #4.

Rule # 2 Give details.

Provide enough detail to make the objective meaningful. "Control fire within 24 hours of initial attack" is a good start. "Put the fire out" is not.

Rule # 3 Can the results be measured?

When the fire is out, you should be able to answer "yes" or "no" to the question, "Did we accomplish this objective?" This is what we mean by measurability. "Preserve historic lodge structures in east end of park" is measurable. In other words, you can observe historic lodge structures after the fire and determine whether they were preserved or not. "Minimize damage to historic structures" is not really measurable. For one thing, people may differ on what is meant by minimal damage. At the very least, you should be able to measure the outcome against the objective in a meaningful way:

- "All but one of the lodge buildings remain undamaged"
- "Burned acreage exceeded the objective by 10 percent"
- "90 percent of the Spotted Owl nesting territory was preserved"

Rule # 4 How will the objectives be accomplished?

Where possible, include guidelines for how the objective is to be accomplished. "Keep firefighter injuries and accidents to zero" is an incomplete objective, but by adding the phrase "by reducing time on the line and in the air," the objective takes on additional meaning in the context of the current fire. Our example in Rule #1 would now read "Prevent fire from reaching Sunnybrook Lake subdivision by digging a fire line between the fire and the subdivision." This information is important because it gives the staff an opportunity to evaluate the fire response directly against the objective. For example, say the objective is "Preserve Pre-Columbian ruins on the south end of ridge." Before the Incident Management Team is deployed, the wind shifts and the fire moves in

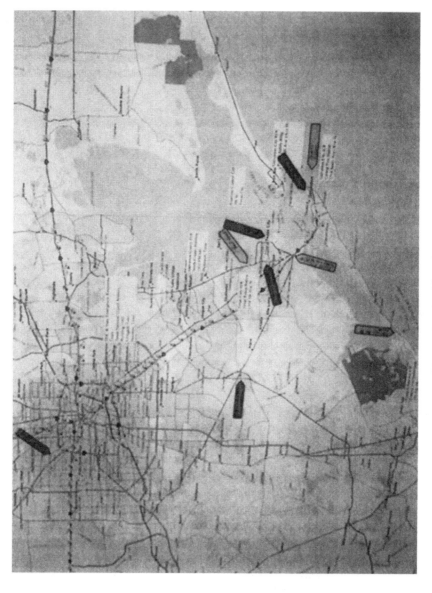

Figure 6.31 Incident Area map

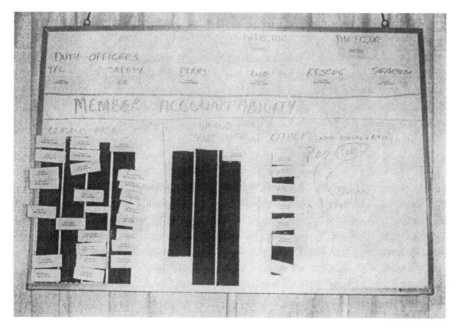

Figure 6.32 Resource board

TABLE 6.2   Specific Duties and Who Does Them

ITEM	WHO PREPARES
Incident Objectives (ICS 202)	Planning Section Chief
Organization List (ICS 203)	Resources Unit Leader
Division Assignments Lists (ICS 204)	Resources Unit Leader
Communications Plan (ICS 205)	Communications Unit Leader
Medical Plan (ICS 206)	Medical Unit Leader
Incident Map	Situation Unit Leader
Traffic Plan	Situation Unit Leader

an entirely different direction from its previous path, bypassing the ruins. The objective has been accomplished. To say that the selected strategy was effective in accomplishing the objective may be misleading, however. "Preserve Pre-Columbian ruins on the south end of ridge by directly attacking fire at its southern extremity" allows us to determine that the happy ending in this scenario was not due to the selection of a particular strategy.

Rule # 5 Is the objective reasonable?

Some goals are worthy in theory, but can't be met in practice. For example, the objective "Keep total loss and suppression costs under $1,000 by using local volunteers and free resources" sounds like a great idea, but it has no chance of being accomplished in responding to a 500-acre fire. Use caution to prevent developing strategies that are not viable. Avoid stating objectives that cannot be reasonably met.

Conclusion

Planning is an essential part of any efficient and successful Incident Management System. One old adage that speaks to the failure to plan is "Failing to plan is planning to fail." Another popular saying, "Prior Planning Prevents Poor Performance," is also known as the "5 P's of Planning."

Both statements are essentially true and seemingly more so when taken in the context of the control of emergency incidents. Simply put, one must become a good planner if one is going to become a good Emergency Manager, and a good Emergency Manager must be a good planner.

Chapter 7

THE LOGISTICS PROCESS

The logistics needs of a large, complex incident may require tens, hundreds, or even thousands of different types of personnel, heavy equipment, buses, tools and equipment, apparatus and vehicles, containment and absorbent materials for hazardous substances. The on-scene repair, storage, and location of the equipment must be completed and documented. Communications, medical support for response personnel, food, ordering and maintaining supplies, setting up facilities, and supporting and transporting personnel, as well as repair and maintenance of apparatus and equipment, must be accomplished to establish a successful outcome.

When the incident requires a number of additional emergency responses, outside agency, and possibly private contractor resources, there is a need to implement the Logistics Section Chief position. When an incident requires special apparatus such as a foam unit and additional foam, which are resources that are available through the fire service, there is a need for logistics. Requirements for booms, dump trucks, and front-end loaders from public works, state highways, or private contractors indicate a need for logistics. Acquiring any type of resource outside the original responding agencies should be handled by the Logistics Section of the Incident Command System. What each piece of equipment is, where it came from, how and when it was used, how and when it was maintained and stored, as well as when it was returned, will be documented under the Logistics Section.

Incident Command Post and Base facilities may have to be set up. Medical resources for response personnel (separate from those for victims) and Rehabilitation Areas (often referred to as REHAB) will be required. Effective Incident Communications and Traffic Plans will have to be developed. Food and drink as well as other supplies for response personnel may be needed. It would not be possible for the Incident Commander to do all these jobs to support incident operations himself. As part of the Incident Management System, the Incident Commander will by delegating authority establish a Logistics Section Chief to manage these vital operations.

Figure 7.1 Incident Command Post

Logistics is the service organization of the Incident Management System, and it should meet all the reasonable needs of internal and external customers. Each Unit Leader will manage his or her own units. Any problems with personnel, firefighters, or other team members will involve the Logistics Section Chief.

In this chapter, we will cover in detail the logistics process. The Logistics Section Chief position requires an individual with substantial knowledge and experience who is also an overall good manager and leader. The person selected should understand the availability of the resources needed and where to get them. The Logistics Section Chief will need to anticipate the lead time for resources to get to the scene and be put to work. Experience in the logistics function is a critical component of effectively managing the logistical needs of any incident.

When a large and complex incident occurs, which generally dictates extended Operational Periods, the Logistics Section Chief has many responsibilities and duties that must be accomplished. This is all a part of the logistics process of incident management.

The Logistics Section Chief must provide the Resources Unit Leader with plans of all of the incident units being activated, including names and locations of assigned personnel. There may be a significant operation requiring a large number of personnel, possibly as many as there are in the Operations Section to support the needs of the incident, strictly from the logistics perspective.

Figure 7.2 Base Camp

Depending on the span of control and the complexity of the incident, the Logistics Section Chief may need to assemble and brief Branch Directors and Unit Leaders. Large, complex incidents require a great deal of coordination and understanding of present and future logistical concerns to effectively support all phases and parts of the organization.

Continued concern for coordination by the Logistics Section Chief and the Branch Directors down to the Unit Leader level is mandatory.

As a part of the planning process, the Logistics Section Chief needs to participate in the preparation of the Incident Action Plan. It is important to Incident Support that the Logistics Section Chief be represented during the planning phase for incident operations. The Logistics Section Chief must know the plan in order to effectively support the proposed operations.

As a result of coordinating with the Operations and Planning/Intelligence Sections, the Logistics Section Chief identifies service and support requirements for planned and expected operations for the current and future needs of the incident. The Logistics Section Chief must focus on supporting logistical needs at a preplanning or Planning Meetings. He or she must provide input on the predicted availability and/or arrival of basic or specialized resources so that the plan is realistic from a time-frame standpoint. Any unmet needs must be brought to the attention of the Operations and Planning Intelligence Sections prior to the Operational Period Briefing.

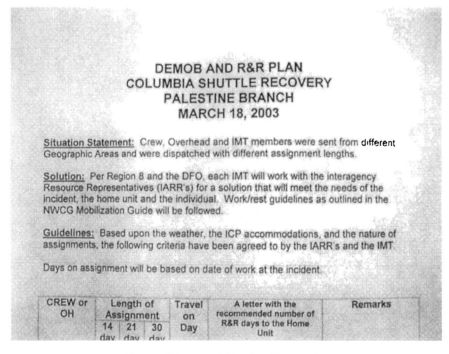

DEMOB AND R&R PLAN
COLUMBIA SHUTTLE RECOVERY
PALESTINE BRANCH
MARCH 18, 2003

Situation Statement: Crew, Overhead and IMT members were sent from different Geographic Areas and were dispatched with different assignment lengths.

Solution: Per Region 8 and the DFO, each IMT will work with the interagency Resource Representatives (IARR's) for a solution that will meet the needs of the incident, the home unit and the individual. Work/rest guidelines as outlined in the NWCG Mobilization Guide will be followed.

Guidelines: Based upon the weather, the ICP accommodations, and the nature of assignments, the following criteria have been agreed to by the IARR's and the IMT.

Days on assignment will be based on date of work at the incident.

CREW or OH	Length of Assignment			Travel on Day	A letter with the recommended number of R&R days to the Home Unit	Remarks
	14 day	21 day	30 day			

Figure 7.3 Demobilization Plan poster

As part of planning, the Logistics Section Chief also provides input to and review of the Incident Communications, Medical, and Traffic Plans. These plans will be developed by the Unit Leaders of the Logistics Section. Communication with the Planning/Intelligence Section is necessary so that the intent of the various plans can be smoothly implemented within the Incident Action Plan. It should be noted that the Safety Officer is the primary reviewer of the Medical Plan, but since the Medical Unit Leader works for the Logistics Section Chief, he or she should also be involved in this review as well.

The Logistics Section coordinates and processes requests for additional resources that are needed to control the incident. These resources will include personnel, equipment, and supplies. This is a very essential element of the logistics process. Large, complex incidents usually require the accumulation of numerous resources. These resources may be fire/EMS resources but also will include many resources from other assisting and supporting agencies. This lesson was learned very quickly in the aftermath of the World Trade Center events in 2001 when supplies not normally needed in massive quantities for fire and rescue operations (such as general duty leather work gloves) were quickly exhausted by local responders. The Logistics Section at the incident was tasked with finding replacements rapidly. Since then some major rethinking of some basic supply concerns has been done at the federal level.

The Logistics Section Chief reviews the Incident Action Plan and estimates the needs of each Section needs for the next Operational Period. Many large, complex incidents are of long duration and stretch across many Operational periods. The Logistics Section Chief and supporting Branches and Units will be directly involved in predicting and procuring the resources required for the whole of the Incident.

Keeping the Command and General Staff advised of current service and support capabilities of the Logistics Section is a responsibility of the Logistics Section Chief. The Logistics Section Chief should request from Command the necessary personnel to effectively meet the needs of the incident from the Logistics Sections perspective.

The Logistics Section Chief receives the Incident Demobilization Plan from the Planning/Intelligence Section Chief and recommends release of Unit level resources in conformity with that plan. As a large, complex incident scales down, there will be less need for a full-blown Logistics Section operation. Logistics personnel may be released as the workload becomes less, but sufficient personnel must be kept ready to complete all necessary tasks required to support the Logistics Sections operation until the incident is terminated completely.

Branch Directors

Based on complexity or span-of-control factors, Branch Directors determine the level of service required to support incident operations. The Branch Director must meet with the Logistics Section Chief to receive information on the level of services required for the units under their Branch. He or she then requests the necessary personnel to accomplish the service needs required at the incident of their Branch.

Branch Directors participate in Planning Meetings involving Logistics Section personnel. On large, complex incidents this affords the Service Branch Director the ability to provide input into the capability of the Branch, understand the needs of the incident, and request any additional personnel required for his branch. Information obtained at Logistics Planning Meeting must be passed on to Unit Leaders so they can adjust their specific operations accordingly.

Like the Logistics Section Chief, the Branch Directors should review the Incident Action Plan. It is important for them to have a working understanding of the Incident Action Plan so that effective service levels can be delivered by Branch personnel.

Branch Directors must inform the Logistics Chief of Branch activities. On large, complex incidents it is especially critical to effective service that the Logistics Section Chief be periodically apprised of Branch capability and level of performance.

Branch Directors determine unit personnel needs. For example, additional personnel may be needed effectively provide communications services for the incident.

Service Branch
The Service Branch is composed of the Communications, Medical, and Food Units.

Communications Unit
The Communications Unit Leader prepares and implements the Incident Radio Communications Plan (ICS Form 205). Most large, complex incidents require the segregation of various parts of the operation to a number of radio frequencies. This will make

Figure 7.4 Service Branch chart

it necessary to take into account the radio frequencies of other agencies and to devise a plan to provide those agencies with specific radios or otherwise ensure that they can be easily communicated with.

The Communications Unit Leader ensures that the communications and message centers are established. The Incident Communications Center is staffed by an Incident Dispatcher. On large, complex incidents, on-scene control of communications can prove useful to the overall flow of the incident.

Many written messages may start to flow at a large, complex incident. They are necessary for effective control and coordination of the incident and supporting operations. These messages need to be delivered to the addressees as soon as possible, so personnel staffing needs to be appropriate to the job.

The Communications Unit Leader establishes appropriate communications distribution and maintenance locations within the Base and Camps if needed. On large, complex incidents, there is a need for additional portable radios, batteries, and possibly radio maintenance, depending on the duration of the incident. The Communications Unit must have sufficient radios and batteries to provide effective communications to various resources arriving at the incident.

The Communication Unit Leader ensures that an equipment accountability system is established. Due to the possible amount of communications equipment that will be under the control of the Communications Unit, some method of accounting for the equipment must be established.

It is the Communications Unit Leader's responsibility to provide technical information on:

- Adequacy of communications systems currently in operation
- Geographic limitations on communications systems
- Equipment capabilities and limitations
- Amount and types of equipment available
- Anticipated problems in the use of communications equipment

The Incident Commander and the General Staff must understand all the above in order to make effective decisions on communications needs and how that capability affects operations during the incident.

The Incident Dispatcher will obtain and review the Incident Action Plan and the Incident Communication Plan. Communications on large, complex incidents can present numerous problems to dispatchers. One cannot effectively communicate with various parts of the organization without knowledge of the Incident Action Plan and the Incident Communications Plan.

The Incident Dispatcher sets up an Incident Communications Center and checks out equipment. On large, complex incidents, it is often necessary to establish the center to maintain control of communications. There must be sufficient Incident Dispatchers to implement the Incident Communications Plan.

The Incident Dispatcher also sets up Message Center locations as required and receives and transmits messages within and external to the incident. In conjunction with the Communication Unit Leader he or she procures sufficient personnel to staff the Message Center and to retrieve and deliver messages using the ICS Form 213.

Medical Unit
Large, complex incidents require the establishment of medical resources that will be focused on providing medical care to injured response personnel. This means that there may be Medical Units placed at various locations around the incident.

Figure 7.5 Communications Unit

The Medical Unit Leader prepares the Medical Plan (ICS Form 206). To ensure uniform application and understanding by other incident personnel the Medical Plan must become part of the written documentation of the Incident Action Plan. As noted previously in this chapter, the Medical Plan is reviewed by the Safety Officer and possibly by the Logistics Section Chief. The Medical Unit Leader is responsible for briefing medical personnel on the current incident situation, anticipated medical needs, and required medical protocol including documentation. At large, complex incidents, understanding the current situation is critical to assessing medical needs for the present and the future. Effective and complete documentation is necessary for the good of the patients and the agency. One consideration is to be sure to have sufficient medical reporting forms on-hand.

On large, complex incidents, a Responder Rehabilitation Manager may be designated. This unit will establish a Responder Rehabilitation Area and announce that location on the radio using the as designation "rehab." It may also be necessary to create more than one rehab area. Food, drink, and other supplies may need to be delivered to these facilities as needed.

The Responder Rehabilitation Manager may request necessary medical personnel to evaluate the condition of personnel in rehab. On any working incident, a rehab area must be established for the safety of personnel. Medical checks are part and parcel of Rehabilitation.

Emergency Medical Services (EMS) should be provided and staffed by the most highly trained and qualified personnel on the scene (at a minimum Basic Life Support

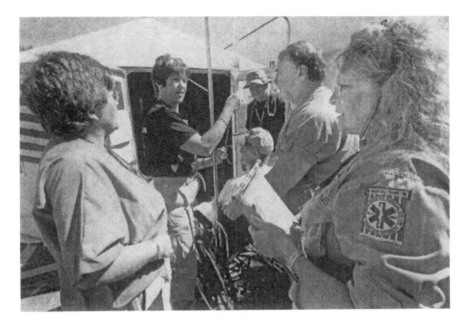

Figure 7.6 Medical Unit

Figure 7.7 Rehabilitation (Rehb) Area at an incident

(BLS). They evaluate vital signs, examine limbs, and make proper disposition (return to duty, continued Rehabilitation, or medical treatment and transport to medical facility).

Continued Rehabilitation consists of additional monitoring of vital signs and providing rest and fluids for rehydration. Medical treatment for people whose signs and/or symptoms indicate potential problems should be provided in accordance with local procedures. EMS personnel shall be assertive in an effort to find potential medical problems early.

To determine heat stress, the heart rate should be measured for 30 seconds as early as possible in the rest period. If the heart rate exceeds 110 beats per minute, internal body temperature should be taken. If the temperature exceeds 100.6 degrees F., the person should not be permitted to wear heat-restricting personal protective equipment. If it is below 100.6 degrees F. and the heart rate remains above 110 beats per minute, Rehabilitation time should be increased. If the heart rate is less than 110 beats per minute, the chance of heat stress is negligible.

All medical evaluations shall be recorded on standard forms along with the person's name and complaints and signed with date and time noted by the Responder Rehabilitation Manager or his or her designee.

The Responder Rehabilitation Manager must ensure that sufficient food and drink is available for a large number of personnel who will be rotated through Rehab. All

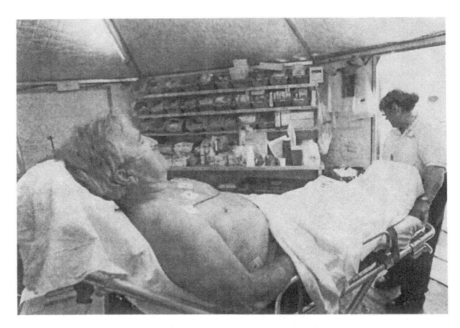

Figure 7.8 Emergency Medical Services transporting a patient from the Medical Unit

personnel released from Rehab will be sent to the Planning/Intelligence Section for reassignment based on the needs of the incident.

Food Unit
The last unit in the Service Branch is the Food Unit. The Food Unit Leader determines food and water requirements for the incident. On large, complex incidents there will be a great number of response personnel from numerous agencies. This unit is responsible for providing food and drink to all response personnel.

The Food Unit Leader will determine with the Service Branch Director the projected number of personnel who will need to be fed at each mealtime. This will allow him to order sufficient supplies to provide the Rehab Area with enough food and drink supplies for the Operational Period. He or she will determine the method of feeding to best fit each facility or situation. Feeding is often done at a Base or Camp or directly at the scene. The Food Unit Leader will coordinate this effort with the Facilities Unit and possibly the Supply Unit in the Support Branch.

The Support Branch
The Support Branch is composed of the Supply, Facilities, and Ground Support Units. The Support Branch Director ensures that sufficient personnel are requested to effectively support incident operations. On large, complex incidents a large number of support personnel may be required. This affords the Support Branch Director the ability to provide input into the capability of the branch, understand the needs of the incident, and request any additional personnel required.

Figure 7.9 Large Scale Incident Feeding operation

Supply Unit
The Supply Unit Leader determines the type and amount of supplies ordered and en route to the incident. Often, large amounts of supplies, materials, and resources have been requested prior to the establishment of a full Incident Management System organization. In order to reduce the possibility of duplication, the branch director must know the scope of previous orders.

The Supply Unit Leader reviews the Incident Action Plan for information on the operations of the supply unit. It is important that the Support Branch personnel understand the Incident Action Plan.

The Supply Unit Leader develops and implements safety and security requirements to set up operations. The Supply Unit Leader is responsible for ordering; receiving; and storing supplies; setting up facilities for response personnel; and providing effective maintenance and transportation of personnel. Security at all these sites and actions is a critical responsibility of the Supply Unit Leader.

The Ordering Manager when established and working under the direction of the Supply Unit Leader, establishes ordering procedures, names and telephone numbers of agency(s) personnel receiving orders, and names of personnel who have ordering authority. On large, complex incidents, it is necessary to control the ordering of supplies and other resources. Only through specific identification of persons designated to receive and order supplies can this be accomplished. Failure to control ordering will result in redundancy and excessive costs to the agency. The Ordering Manager identifies times and locations for delivery of supplies and equipment. It is critical to operational planning that the Logistics Section knows the predicted time of arrival of all supplies and equipment ordered for the incident. The ordering of resources, whether for an emergency such as a wildfire or for a planned project such as a Super Bowl, must follow the established ordering process.

Resources are divided into five categories:

- Overhead positions (single person)
- Tactical crews (such as firefighting) crews
- Supplies (such as weather kits and radios)
- Equipment (such as engines or bulldozers)
- Aircraft (such as tankers or helicopters)

All state and federal natural-resources agencies have designated ordering procedures for natural disasters, crisis incidents, and fire support and services. These established ordering channels provide for:

- Rapid movement of requests
- Efficient use of resources
- Safety of personnel during mobilization
- Cost effectiveness

The Ordering Manager will then coordinate with the Receiving and Distribution Manager, who will be responsible for operating a supply area. The supply operation may take numerous personnel to receive, store, and distribute supplies and equipment to the operational personnel.

The Receiving and Distribution Manager, under the direction of the Supply Unit Leader, will organize the physical layout of the supply area. If large amounts of supplies and equipment are needed, it is critical to have the receiving and storage areas effectively laid out and in an orderly condition.

It is important to develop security requirement for the supply area. The Receiving and Distribution Manager should obtain necessary police or other security personnel to protect supplies from misappropriation.

The Receiving and Distribution Manager will notify the ordering manager of supplies and equipment received. Close coordination is necessary to ensure that the correct supplies and equipment have been delivered. This coordination will enhance the effect of the operating personnel in accomplishing their tasks to control the incident.

Some incidents might require a Tool and Equipment Specialist, who would work for the Receiving and Distribution Manager and set up the tool storage and conditioning area. On any large incident, there must be a one definite place to drop off broken tools and equipment for repair and pick up spare and refurbished tools and equipment.

The Tool and Equipment Specialist establishes a tool inventory and accountability system to ensure that all equipment is returned to its rightful agency at the completion of the incident. Additionally the Tool and Equipment Specialist would assemble tools for issuance during each Operational Period per the Incident Action Plan. An Operational

Period may require specific tools and equipment be provided to personnel. The Tool and Equipment Specialist must have a copy of the Incident Action Plan for the next Operational Period in order to obtain and prepare the necessary equipment or tools.

The Facility Unit

The Facilities Unit receives a copy of the Incident Action Plan as it provides a blueprint of what facilities are needed and how active the unit will be. The Facilities Unit participates in Logistics Section, Support Branch planning activities.

The Facility Unit Leader determines the requirements for each facility. He or she is responsible for the Incident Command Post, Base, and Camp(s) for the incident. Each incident will require specific requirements for each facility. Each Operational Period may affect the number and types of facilities required.

To assist in the management of facilities, Base and Camp Managers may be needed. The Facilities Unit Leader will need a Manager to be accountable for any Base or Camp(s) that are established. The Manager will be responsible to set up and lay out the facility as directed by the Facilities Unit Leader.

The Facilities Unit Leader requests maintenance and support personnel and assigns duties. Maintaining and supporting an incident facility may take a number of personnel. These personnel do not have to be emergency response personnel but are needed to support the facility and may include trades persons such as plumbers and electricians and the like.

The Facilities Unit assembles and disassembles temporary facilities when no longer required and restores the area to pre-incident condition. It is critical to good relations with the public and/or other agencies that areas are left as they were found.

Since the Facility Unit Leader is responsible for incident facilities, he or she may enlist a Security Manager along with sufficient security officers. These people would establish contacts with local law enforcement agencies as required. Normally, law enforcement will be asked to provide the security needed at any facility.

If there are Base Managers, they ensure that all facilities and equipment are set up and properly functioning. Their job is to supervise the establishment of: sanitation and sleeping facilities. The Base is often a marshaling point for in-coming resources and a place to Rehab, get checkups, eat, rest, and sleep.

A high school often makes a good Base if one is available within a reasonable distance from the incident. This type of facility often provides a large parking area, classrooms that can be rearranged, showers, a cafeteria, an auditorium for sleeping, and phones, copiers, and fax machines.

When there is a need for remote logistical sites, the Facility Unit Leader may establish Camp Managers. All the functions provided at a Base can be provided at a Camp. The Camp is not usually within a structure such as a school but rather out in the field or other open area such as a parking lot near the incident.

The Ground Support Unit

The final unit in the Support Branch is the Ground Support Unit. The Ground Support Unit Leader will be able to provide input on this function and understand what must be provided during the Operational Period.

The Ground Support Unit Leader will develop and implement the Incident Traffic Plan. All large, complex incidents require an Incident Traffic Plan to accommodate efficient access and egress from the various facility and operational areas and to alleviate confusion.

The Ground Support Unit Leader must support out-of-service resources and provide necessary maintenance. This will include arranging for and activating fueling, maintenance, and repair of ground resources. The Ground Support Unit Leader must ensure that sufficient personnel with specialty job capabilities are requested at the scene.

The Ground Support Unit Leader maintains inventory of support and transportation vehicles (ICS Form 218). Accountability and control of resources assigned is mandatory at large, complex incidents. He or she will provide transportation services such as buses, rental cars, and trucks when necessary. Ground support may need a large number of utility and other transportation vehicles with drivers to move supplies and personnel to various locations within and outside of the incident scene.

Maintaining access or roads within the incident area is also a responsibility of the Ground Support Unit Leader. Various types of heavy equipment and operators may be required for this activity. In urban areas the use of tow trucks and heavy wreckers to clear parked or damaged cars may be more appropriate.

The Ground Support Unit may have an Equipment Manager, who will obtain a copy of the Incident Action Plan to determine locations for assigned resources, Staging Area locations, fueling and service requirements for all resources. This Equipment Manager must know where every piece of rolling stock is located, as well as other engine and motor powered equipment that may need service in the field. A schedule must be established for the fueling of vehicles and other equipment.

The Equipment Manager prepares schedules to maximize use of available transportation and provides transportation and support vehicles for incident use. Scheduling transportation is necessary at large, complex incidents in order to reduce the number of trips made and the number of transportation vehicles and drivers required.

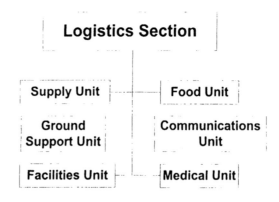

Figure 7.10 Logistics Section chart

CONCLUSION

It has been said that an army moves on its stomach; this may well be true, but the fact that troops were fed and rested was likely the result of a well oiled and fully functioning Logistics Section. This is as true in the case of a large and complex emergency incident as it is with a modern fighting unit of the U.S. military. The needs of an incident, regardless of its genesis, will likely become very complex in a rapid fashion, and a well trained and competent logistics section will literally mean the difference between a well managed incident and a "disaster within the disaster."

Chapter 8

CUSTOMIZING INCIDENT MANAGEMENT SYSTEMS FOR SPECIFIC APPLICATIONS

Many specific applications of Incident Management Systems have evolved since the early use of such systems. Some specific systems that are discussed in this chapter are the Hospital Emergency Incident Command System (also known as HEICS), law-enforcement-based systems, and business continuity systems. Each section was written by an expert who has firsthand experience and knowledge.

SECTION ONE—THE HOSPITAL EMERGENCY INCIDENT COMMAND SYSTEM (HEICS)

All communities are subject to both large- and small- scale disasters and environmental emergencies. The hospitals located in affected areas are magnets for the sick, injured, and worried, all of whom are looking for continued care and support. Just as first responders must do, hospitals and other healthcare agencies continue to serve and protect their communities, even if a catastrophic event occurs in or threatens the hospitals themselves. Along with the health department and Emergency Operations Centers, hospitals can effectively handle a potential large-scale emergency of nearly any type. They must provide an opportunity for people to collaborate and plan and allowing them to use their competence in the roles needed for disaster mitigation. This level of agency integration is most effectively demonstrated by command-structure-based disaster plans and lessons learned from prior incidents.

In the state of California the FIRESCOPE Program that was begun in the 1980s required that all state agencies use the state-established Incident Management System. A document called the California Earthquake Preparedness Guidelines for Hospitals

Figure 8.1 Large hospital

served as a cornerstone for the development of a Hospital Emergency Incident Command System, commonly referred to as HEICS (pronounced "hicks"). In 1992 after the Northridge Earthquake the Emergency Medical Services Agency in San Mateo, California developed the first draft of the current version, the Hospital Emergency Incident Command System.

Just as fire-service agencies saw the need for an organized method to guide personnel and communications, hospitals also needed a way to manage these same functions in any emergency that may overtax a community's resources, including the hospitals that serve the community. The functionality of hospitals is diverse and ever changing, a good learning environment for occasions with a mass influx of patients or because of community events.

Any incident that taxes available hospital and or medical community resources will require an integrated systems approach. Confusion and chaos are commonly experienced by hospitals at the onset of any medical disaster. However, these negative effects can be minimized if the hospitals management and staff respond quickly with structure and a focused direction of activities. The Hospital Emergency Incident Command System is one of the many integrated emergency Incident Management Systems that employ a logical management structure, defined responsibilities, clear reporting channels, and a common nomenclature and language to help unify hospitals and other emergency responders in reacting to an event that overtaxes local hospital and medical resources.

As stated earlier in this book, the use of Incident Management Systems has proven to be beneficial in many large-scale incidents. This same system has been used within the hospital with great success; airline disasters, tornados, and floods are a few examples of the types of incidents in which the effective use of an Incident Management System proved to be beneficial. Incident Management Systems are also useful in psychologically traumatic incidents that cause multiple casualties and generate a lot of patients exhibiting signs of post-traumatic stress disorder syndromes. Hospitals have seen this in situations such as the school shooting in Columbine, Colorado, and the Branch Davidian stand-off situation in Waco, Texas. These situations may inundate a medical system with patients who may or may not appear to be injured physically but in reality are more psychosomatically affected by the carnage of the incident that they just witnessed. The hospitals and the medical community still must respond to their needs, both physical and psychological in nature. The effective use of an Incident Management System will help to elevate the chaos associated with mass casualties regardless of the genesis of the casualty, physical or physiological.

A vulnerability assessment allows preparation efforts to be focused in those areas of higher danger probability. California hospitals prepare for earthquakes; in Kansas they prepare for tornados; and in Florida they prepare for hurricanes, all while still maintaining an "all hazards approach" to planning and overall preparedness. Medical disaster planners need to ask the following questions when beginning their planning process:

- Do you have a rail system in your community?
- Do you have chemical plants, manufacturing plants, or farms in your community?
- What potential disasters, natural and/or man-made, exist in your service area and in the areas adjacent to your facility?
- How your facility fit in with larger planning efforts?

The HEICS system is a byproduct of the California FIRESCOPE Incident Command System structure. The same systems, philosophies, policies, and structures are modified for use within the hospital setting regardless of its size or specialty in terms

Figure 8.2 Northridge earthquake

of patients cared for weather that hospital has very few inpatient beds as for those that are multifaceted, providing tertiary services.

NFPA Standards 1500 and 1561

NFPA 1500 recommends the following up-to-date standards:

- Written procedures for IMS
- All members must be trained for IMS.
- System must outline responsibility for safety at all supervisory levels.
- System must provide for accountability of personnel at all levels.
- System must provide sufficient supervisory persons to control all members operating at the scene.

NFPA Standard 1561 recommends and provides for a broad set of guidelines for what should be included in any Incident Management System. These systems are now commonly used nationally in hospitals, particularly those that are part of larger geographically diverse networks. When a Southern California hospital, part of a national network, had a significant shooting incident, members of the hospital's national leadership team were present in the emergency operations center.

The Hospital Emergency Incident Command System was implemented immediately for a complete evacuation of the facility. This effort necessitated collaborative efforts by all responding agencies. In talking to the individuals present there, an Incident Management System was used by all agencies including this facility, which imbedded the hospitals HEICS plan into its emergency-preparedness strategy.

The State of California Emergency Medical Services Authority (CAEMSA) describes its mission statement as follows:

> The CAEMSA mission is to ensure quality patient care by administering an effective, statewide system of coordinated emergency medical care, injury prevention, and disaster medical response.

> (http://www.emsa.ca.gov/newpg)

Standardized Emergency Management System (SEMS)

The State of California Multi-Hazard Mitigation Plan was developed and published in July 2004. It was produced in collaboration with multiple state agencies, although the Governor's Office of Emergency Services (OES) has lead responsibility for the development and maintenance of the document. The State Hazard Mitigation Team is truly the "owner" of the plan, while OES is its steward.

Initial response to an incident will be managed by the first responders, usually law enforcement, fire, and EMS. When these local resources under the standardized Emergency Management System (SEMS) system are exhausted, field operations would then rely on local government to expand the response area to mutual aid departments. Again, when resources at the local area are exhausted, the next level will be utilized. In the case of a major incident federal resources may be deployed more quickly, bypassing each step to more readily mitigate the incident. This was done in the case of the Oklahoma City bombing of the Alfred P. Murrah Federal Building for a variety of reasons at the time of the blast, but in hindsight the response was a more effective one due to the early involvement of federal resources (an anomaly in some ways due to the fact it was a federal building and the obvious fact that this would be a federal crime prosecution, regardless of motives for the act).

On February 28, 2003, a federal document outlining management of domestic incidents was produced and released for publication. Within this plan is the presidential directive Homeland Security Presidential Directive (HSPD) 5 (see http://www.whitehouse.gov/news/releases/2003/02/20030228-9.html for the text of the full document.).

This directive outlining agency responsibility and accountability is outlined below.

Homeland Security Presidential Directive (HSPD) 5

The White House
Office of the Press Secretary
February 28, 2003
For Immediate Release
Subject: Management of Domestic Incidents

Purpose: (1) This directive establishes policies to strengthen the preparedness of the United States to prevent and respond to threatened or actual domestic terrorist attacks, major disasters, and other emergencies by requiring a national domestic all-hazards preparedness goal, establishing mechanisms for improved delivery of Federal preparedness assistance to State and local governments, and outlining actions to strengthen preparedness capabilities of Federal, State, and local entities.

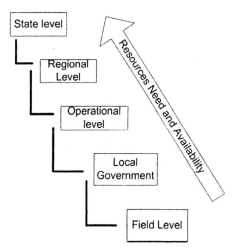

Figure 8.3 Standardized SEMS chart

Regulations and Standards

Hospitals are under strict guidelines to develop systems to ensure that a hospital continues to perform the service to its stated community to the best of its ability. Each hospital must create community relationships and inter-agency plans with the community in which the hospital provides service. It can no longer rely on the fire department or other service to come to its aid. The fire department and other first-responder agencies will be overwhelmed in the field and frankly need to stay there! The hospital cannot come to the scene of an incident, but responders can and do go to the scene, then move the injured and otherwise needy to the hospital. When the hospital is in the path of the incident, this complicates the situation even more.

The following are some California EMS standards governing hospitals:

Standard EC.4.10 addresses emergency management. An emergency in the hospital or its community could suddenly and significantly affect the need for the hospital's services or its ability to provide those services. Therefore, a hospital needs to have a management plan that comprehensively describes its approach to emergencies in the hospital or in its community.

Standard MS.4.110 grants emergency privileges when the Emergency Management Plan has been activated and the organization is unable to handle the immediate patient needs. During disaster(s) in which the Emergency Management Plan has been activated, the Chief Executive Officer or Medical Staff President or Designee(s) has the option to grant emergency privileges.

The individual(s) responsible for granting emergency privileges is identified, along with contingencies for alternates should the need arise. The responsible individual(s) is not required to grant privileges to any individual and is expected to make such decisions on a case-by-case basis at his or her discretion. The Medical Staff identifies in writing the individual(s) responsible for granting emergency privileges and the following:

- The Medical Staff describes in writing the responsibilities of the individual(s) granting emergency privileges.
- The Medical Staff describes in writing the mechanisms to manage the activities of and to identify individuals who receive emergency privileges.
- The Medical Staff addresses the verifications process as a high priority.
- The Medical Staff has mechanisms to assure that the verification process of credentialed individuals who receive emergency privileges begins as soon as the immediate situation is under control. This privileging process is identical to the process under medical staff bylaws for granting temporary privileges to fulfill important patient care needs)
- The Chief Executive Officer or the President of the medical staff or their designee(s) may grant emergency privileges upon presentation of any of the following:
  ◦ A current picture hospital ID card
  ◦ A current license to practice and a valid picture ID issued by a state, federal, or regulatory agency

- ○ Identification indicating that the individual is a member of a Disaster Medical Assistance Team (DMAT)
- ○ Identification indicating that the individual has been granted authority by a federal, state, or municipal entity to render patient care in emergency circumstances
- ○ Presentation by current hospital or medical staff member(s) with personal knowledge regarding practitioner's identity

The emergency management standards published in 2001 by the Joint Commission on Accreditation for Hospitals (JCAHO) have been modified in regard to the requirement for comprehensive emergency-preparedness efforts. These standards now include the need for an "all hazards" approach, which must include community involvement. JCAHO defines an emergency as a natural or man-made event that suddenly or significantly disrupts the environment of care, disrupts care and treatment, or changes or increases demands for the organization's services. Mitigation activities lessen the severity and impact of a potential emergency and include identifying potential emergencies that may effect the organization's operations or the demand for its services and implementing a strategy that supports the perceived areas of vulnerability within the organization.

Healthcare organizations are expected to address the four specific phases of disaster management:

- Mitigation
- Preparedness
- Response
- Recovery

Hospitals must also participate in at least two annual drills; it is recommended that one drill be conducted on a community-wide basis. Since JCAHO Accreditation is directly tied to federal health carefunding in many cases almost all community hospitals will be using these standards in their planning process as the base for their plans.

During an on-site visit, JCAHO surveyors will assess the ability of the facility plan to improve the institution's capability for various emergencies based on a vulnerability assessment and the ability to ensure competency by noting whether staff at all levels have been trained in their roles and responsibilities in the plan.

Underscoring the all-too-real potential for new and even large-scale terrorist attacks in the United States, the Joint Commission on Accreditation of Healthcare Organizations issued a report warning of a "brewing cataclysm" of under-funding, inexperience, and un-preparedness of emergency-response capabilities across America's communities. This vulnerability, the report stresses, must be urgently and effectively addressed by local, state, and federal authorities. "This report is about accountabilities in protecting the citizens of the United States—we need to start addressing the identified needs," says Dr. O'Leary, M.D. "The Joint Commission stands ready to work with others to meet our collective national preparedness goals" (JCAHO News, March 12, 2003).

Integration

It is helpful to gather multiple references in order to select best practice. Understanding the willingness of agencies to provide collaboration of resources will assist in creating a valued community plan. In the effort to create this network of resource opportunities, the building of community-wide relationships is an additional benefit. There is a tremendous amount of information regarding emergency management planning available at both state and federal levels, and increasingly even corporate America has such information available. Local public-health departments are now required to have community-based plans for mass casualty incidence and the potential to mitigate any small- or large-scale biological outbreak ranging from flu pandemics to outright attacks against the country as a whole.

It is important to have a guiding committee or task force to advise and facilitate the project and to keep the planning on task. In creating this task force for the establishment of a community "game plan" it is important to choose interested members, members with previous Hospital Emergency Incident Command System experience, and meeting times that meet the majority of the group's scheduling needs. The membership of this task force should also be multifaceted in various disciplines. People who would enhance this team would be from nursing, technical services, biomedical services, facilities management, social services, medicine, and all aspects of environmental services. All hospital employees must be familiar with the facility's Emergency Management Plan including the basic functionality of the Hospital Emergency Incident Command System and their potential roles if activated.

HEICS, like all Incident Management Systems is an operational emergency management system made up of positions on an organizational chart. Each position has a specific mission to address in an emergency situation. In the HEICS system each position has an individual checklist designed to direct the assigned individual in disaster recovery tasks. These positions are designed to be similar to that of the hospital's everyday function. The HEICS plan includes standardized forms to enhance this overall system and promote accountability for each position that is filled assuring that those critical functions of each position is fulfilled in each and every incident. This plan is designed to be flexible. Only those position functions needed should be activated for a given incident. The plan allows for the addition of needed positions, as well as the deactivation of positions at any time. This allows for the flexibility to respond to the needs of an ongoing, static, or evolving emergency event, including events that occur as part of the daily routine.

The HEICS Organizational Chart

The organization chart in Figure 8.4 may be fully activated for a large, extended disaster such as an earthquake with multiple casualties or a large-scale terrorist attack. However, full activation may take hours or even days to fully staff and accomplish even in the most robust of organizations. The majority of disasters or emergencies will require the activation of far fewer positions. It is important to remember to plan deactivation at the same time as certain roles are being activated. It is necessary to understand the criteria for a job having been completed when positions are being established.

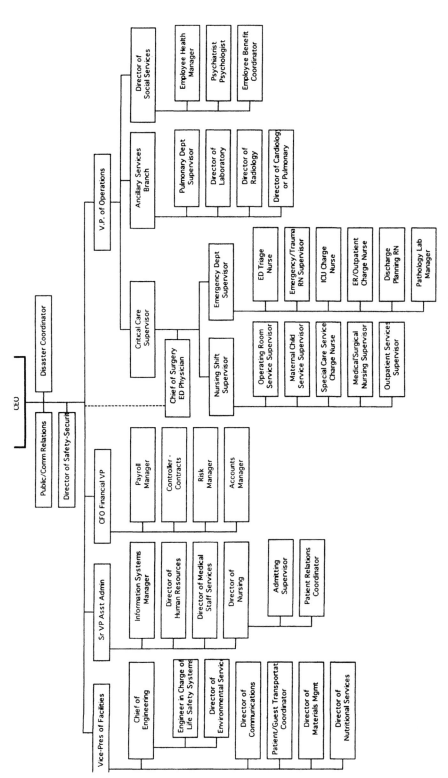

Figure 8.4 Complete HEICS organization chart

More than one position may be assigned to one individual; an individual may have to do multiple tasks until additional support is obtained. Individuals need clear guidance on the priorities of the incident from the outset as well as guidance as to how those priorities change from time to time, just as in normal day-to-day operations of the facility.

When some facilities incorporate the Hospital Emergency Incident Command System, they make the mistake of breaking down certain sections into subsections. That means more people and more chances for a gap in communication, in response to this Mr. Paul Russell, RN, the California HEICS Update, Project Coordinator, cautions. "The plan may work well in the facility, but if an outside agency has to respond to an emergency with the facility, it could mean a breakdown in communications."

Allow time for hospital exposure, practice, and more practice. The components that help to make this a successful model are:

- Flexibility
- Task-focused management
- Directed reporting for accountability
- Confidence
- Inspiration to the team
- Universality with community Incident Command System/Unified Command

Hospital, police, fire, and Emergency Medical Services relationships have had significant challenges in recent years. Significant networking will be needed to strengthen these relationships immediately. In California most hospital disaster plans are developed in accordance with the Governor's Office of Emergency Services. The OES seeks to create an organization that:

- Anticipates the impact of potential disasters and emergency situations
- Has the capacity to immediately mobilize a dynamic and variable response capability
- Uses a systematic approach to emergency management that maximizes OES resources for reducing hazards
- Preparing for emergencies
- Responding to and recovering from disasters

This system seems to work well in California, but not all states have such systems in place. With the advent of new funding streams for healthcare facilities, the advent of the National Incident Management System (NIMS), and the apparent tying of the funding to NIMS compliance, one of the major goals of the HEICS Update Project in California is to make HEICS more NIMS-compliant to serve as a tool for other hospitals around the United States and even the world to adopt.

"The concept of culture is particularly important when attempting to manage organization-wide change. Practitioners are coming to realize that, despite the best-laid plans, organizational change must include not only changing structures and processes, but also changing the corporate culture as well. Usually, this failure is credited to lack of understanding about the strong role of culture and the role it plays in organizations. That's one of the reasons that many strategic planners now place as much emphasis on identifying strategic values as they do mission and vision." McNamara (1999).

Utilization

The Hospital Emergency Incident Command System is "a tool for management to solve objectives during a crisis or unusual event," explains Russell, who drafted the original version of HEICS in 1991 with revisions from disaster drill coordinators in California. The update will hopefully have both the new features of NIMS compliance as well as the "feel" of the more tried and true Incident Management Systems models in place all over the nation today.

The system is devised with as many as forty-nine positions on an organizational chart, each with a specific mission to address in an emergency situation. Each position is also given a checklist—called a job action sheet—describing the actions assigned to the individual when disaster strikes. Duties assigned to each position are ranked "immediate," "intermediate," and "extended."

The hierarchy is the Incident Commander followed by a Section Chief, then a Director, Supervisor, Unit Leader, and finally Officer. The system is flexible so that more or fewer positions, depending on the needs of the emergency, may be implemented. "Any size facility could implement this," says Russell. If you don't have forty-nine staff members at your facility, you can always assign multiple roles to employees or better yet, prioritize actions and assign only those that are most important, Russell explains. When assigning multiple roles to one person, it is imperative that the Incident Commander make those priorities clear to that person and that the person know when to report conflicts back to the Incident Commander for reevaluation of the situation as the needs change over time.

The concepts and principles of the Hospital Emergency Incident Command System have been used as a planning tool for upcoming JCAHO surveys, staffing crises, and "flu season" with tremendous success. This makes perfect sense given the fact that for many years the Incident Command System itself and/or other forms of Incident Management Systems have been used as planning tools for any large-scale event with great success.

Case Study

In one case the system was used when two-hundred patients who had been evacuated from a nearby convalescent hospital arrived on a large hospital's ambulance dock. The hospital emergency department and various other responding departments were able to triage and place all those patients within an hour, quickly returning to normal operations.

Hospital operations continued and were not interrupted except for some business office tasks for a short duration. The emergency department operations went uninterrupted by placing triage physically outside of that area. Radios and runners were used to coordinate patient flow and placement. Quick thinking by the Incident Commander to transfer the majority of the patients to a long-term-care facility that the hospital owned, which happened to be immediately adjacent to the hospital, made the situation immediately more easily resolved. Empty in-patient units were opened for temporary operations. A telephone-call tree was initiated by the Incident Commander to staff the in-patient units for the time being. Only the Incident Commander, Operations Chief, Inpatient Area Supervisor, Outpatient Areas Supervisor, and Triage Unit Leader positions were initiated for this event.

Because of the organization that the Hospital Emergency Incident Command System provides and its organized approaches to continual events that tax our available resources, the system worked very well in the above case study. HEICS like all Incident Management Systems is nothing more then a complex resource management system in essence. Hospital management staff is very used to dealing with the management of sparse resources and higher than expected workloads that the HEICS allows. Normal day-to-day management can be rapidly expanded and escalated to meet crisis needs.

Documentation and Casualty Reporting Issues
It can be expected that many inquiries about casualties as well as general inquires as to whether a certain person was taken to a facility will occur, hence careful documentation of all casualties and their disposition is a must for all facilities in a given situation. A single facility or several facilities in a region may be involved, given the nature of the emergency. Much of this was seen firsthand in the events of September 11, 2001, where public transportations systems allowed patients to self-transport to facilities several states away from New York City and Washington, DC.

Cost Tracking
Tracking costs related to the mitigation and recovery will allow that information to be shared with the Health Care Financing Administration (HCFA) and other insurance entities so that your costs may be accurately reimbursed. HEICS has specific positions built into it to support these vital functions; they are all located in the Finance/Administration Section, where they will be able to mesh with their non-hospital-based counterparts from other responding agencies. This will allow higher authorities to take full advantage of any disaster monies allocated to the incident in the days after the response.

Triage at the Scene versus the Hospital
Triage plays a critical role in rapidly identifying patients with serious injuries, as well as determining their treatment priority. Historically, most patients arrive at the hospital in an uncoordinated manner within one to two hours of the disaster. Further, less than twenty percent of victims receive medical treatment and transport at the scene of an incident; the "walking wounded" patients can be expected to seek medical care at the closest hospital. However, the events of September 11, 2001 in both New York City and Washington, DC, showed that "escapees" can and will find transportation to medical-care facilities on their own. This emphasizes the importance of the use of regional hospital communications systems and networks as well as the use

of the media (judicially) to alert other medical facilities that they may be getting patients from an incident.

Re-triage at the hospital should occur because the victim's medical status may have changed; the priority for treatment may differ from that afforded at the scene and may well be dictated not by medical need but rather by available resources at the facility at that moment, which is likely to be changing as additional staff report to the facility and capabilities increase.

Why "Do" HEICS?

Outlined here is the basic incident command structures used by HEICS, which is similar to those used by fire, police, and military agencies. The terms used to describe these roles are the same throughout these agencies and in those acute-care facilities that use this or similar systems. The semantic differences are needed due to the uniqueness of hospital-based operations.

The benefits of the system were discussed earlier. Outlined here are those issues that help to make hospitals successful in mitigating disasters using HEICS. With the healthcare provider population so mobile in our current society, the fact that this plan can be universal allows those who provide needed resources "to be prepared" wherever they are employed. In essence, once they learn the base system they can pick up the slight deviation based on region or facility needs in a very short order.

Like most successful programs, it takes dedication and patience to implement a change. This is not a turnkey system. In most facilities with due diligence and effort on the part of management (buy in) it usually takes about a year to fully implement HEICS with ongoing training and practice for success. It may take much more time to implement a region-wide system that is effective, and the efforts needed will be greater as facility and jurisdictional lines are crossed; this is true in all cases of Incident Management Systems integration.

HEICS utilizes an organizational chart to enhance communication flow. In the older Incident Command and Incident Management Systems all information would come through the Incident Commander and be disseminated from there. This can become overwhelming for the Incident Commander in even a small-scale incident due to the sheer volume of information likely to be generated; communication efforts usually proved to be chaotic and unsuccessful.

The General Staff

The General Staff is responsible for the overall management of the incident. In a business model the General Staff might well be referred to as the Management Group. These are the folks that communicate directly to the Incident Commander. Their roles are that of Section Chiefs. They communicate needed information and data to incident command for further direction and leadership. The Incident Commander will designate times for these updates, usually depending on the type of incident. In longer-duration incidents the use of planning meetings and operational periods will ensure the timely flow of data and information to all that need such input for decision-making purposes. It is essential

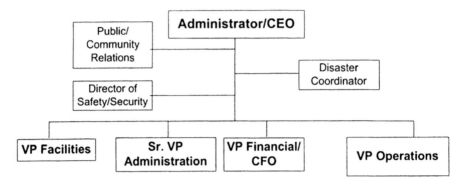

Figure 8.5 HEICS Command and General Staff chart

that all persons in these positions must have the authority to make decisions for the hospital.

The Incident Commander is responsible for organizing and directing the overall operations for the incident. He or she gives overall direction for hospital operations and if needed authorizes evacuation of the facility or portions thereof. In the existing HEICS system this function is designated by the colors black and white.

The Public Information Officer is responsible for providing information to the media. This person works closely with the Incident Commander for careful collaboration and may operate as part of a larger Joint Information Center if one is established in the community as a response to the incident.

The Liaison Officer functions as the contact person for representatives from other agencies. This position reports directly to the Incident Commander.

The Safety and Security Officer monitors and has authority over the safety of rescue operations and hazardous conditions, including those involving weapons of mass destruction. He or she organizes and enforces scene/facility protection and traffic security. The person in this position also reports directly to the Incident Commander.

The Logistics Section Chief is responsible for organizing and directing those operations associated with maintenance of the physical environment and adequate levels of food, shelter, and supplies to support the medical objectives. In the existing HEICS system this function is represented by the color yellow.

The Planning Section Chief organizes and directs all aspects of planning section operations. He or she ensures the distribution of critical information/data, completes scenario/resource projections from all section chiefs, effects long range planning, and documents and distributes the facility action plan. The nursing function in this section plans the resources needed to continue to care for patients and provide information to their family and friends. In the existing HEICS system this function is represented by the color blue.

The Finance/Administration Section Chief is responsible for monitoring the utilization of financial assets. He or she oversees the acquisition of supplies and services necessary to carry out the hospital's medical mission and supervises the documentation of expenditures relevant to the emergency incident. In disaster situations the Health Care Financing Administration often actually pays all these costs, or other means of cost recovery may be made available to a facility. In the existing HEICS system this function is represented by the color green.

The Operations Section Chief oversees the largest functional section found in any Incident Management System. This section is responsible for carrying out directives of the Incident Commander. Coordination and supervision of the medical, ancillary, and human services subsections are also within the section's responsibilities. The care of both existing patients and their families and the large number of patients arriving from the incident are where immediate activities are usually focused. In the existing HEICS system this function is represented by the color red.

The following four roles communicate directly with the Operations Section Chief. He or she will then in turn make sure that the Incident Commander is well informed. Information may also be distributed to other section chiefs for probable action.

The Medical Staff Director is responsible for the organization, prioritization, and assignment of positions to carry out medical care in designated areas. He or she advises the Incident Commander on issues related to the medical staff. This role can be carried out by the current Medical Staff Director. The Medical Staff Director must be a licensed physician with staff privileges; he or she is responsible for overall care that is delivered in areas of the hospital. This position reports directly to the Operations Section Chief.

The Inpatient Area(s) Supervisor(s) ensure that the patients currently within the facility are informed and moved if able to allow for further influx of the ill and injured. Depending on the size of the facility more then one Supervisor may be needed, with one designated as a Director, who then reports to the Operations Section Chief. Inpatient Area Supervisors must be Registered Nurse

The Ancillary Services Director is responsible for organizing the increased need for ancillary medical services. This role will assist in the optimal functioning of these services. In addition, this role must monitor the use and conservation of resources.

The Human Services Director will implement systems needed to support the social and psychological issues occurring for patients, staff, and their families.

Use of Colors in HEICS
The colors outlined in the above organizational chart are often used in community agency plans and other IMS systems. One cannot state that all systems and agencies will follow this color code, as IMS has evolved in different ways in different parts of the United States. There is currently not a "standard," but there is sincere movement in that direction. The use of a standardized color-identification system would greatly assist the incident participants in locating vital functions based on specific colors. Vests with job function and section colors are easily identifiable. The use of a color-code system also suggests that the support materials for that section are also identified by color for easy identification and usage during a widespread emergency. In addition, these color codes help to segregate and identify people occupying a Unified Command Incident Command Post.

Figure 8.6 Three-day exercise in biological terrorism—
Riverside County, California, May 1, 2004

Figure 8.7 HEICS train the trainer exercise, 2004, "Spreading the Word"—
Tucson, Arizona

The following information documents an example of community integration:

METROPOLITAN AREA HOSPITAL COMPACT

http://www.aha.org/aha/key_issues/disaster_readiness/resources/content/TwinCities
MetroCompact.doc

This Compact is made and entered into as of this 3rd day of April, 2002 by and
between twenty-two hospitals located in the twelve-county metropolitan area.
The undersigned hospitals will:

1.1 Communicate and coordinate efforts to respond to a disaster via their liaison
officers, public information officers, and incident commanders primarily.

1.2 Receive alert information via the EmSystem regarding any disaster or special
incident with radio notification by East and West Metro Medical Resource Control Centers (MRCC) as a back-up system.

1.3 Communicate with each other's Emergency Operations Centers (EOC) by phone,
fax, email, and will maintain radio capability to communicate with MRCC as
a minimum back-up.

1.4 Utilize a Joint Public Information Center (JPIC) during a disaster to allow their
public relations personnel to communicate with each other and release consistent community and media educational/advisory messages. Each undersigned

hospital should designate a Public Information Officer (PIO) who will be the hospital liaison with the JPIC. Depending on the event, this may be coordinated through the Minnesota Department of Health, Minnesota Division of Emergency Management, Minnesota Hospitals and Healthcare Partnership. If no umbrella organization assumes responsibility, Hennepin County Medical Center and Region's Hospital Public Relations departments will assume this responsibility.

ARTICLE II
ONGOING COMMUNICATION ABSENT A DISASTER

The undersigned hospitals will:

2.1 Meet twice yearly under the auspices of the Metropolitan Medical Response System to discuss continued emergency response issues and coordination of response efforts.

2.2 Identify primary point-of-contact and back-up individuals for ongoing communication purposes. These individuals will be responsible for determining the distribution of information within their healthcare organizations.

CONCLUSION

The Hospital Emergency Incident Command System is a part of a larger plan and system that assists hospitals in the necessary preparation to continue to provide essential services for its community, even in extreme conditions. Agency integration within the community both by business affiliation and geographical location is imperative to maintain the integrity of a community's homes, businesses, and loved ones. The use of and implementation of HEICS is beneficial mainly for its ability to integrate with community agencies essential for continued emergency care services.

Chapter 8

CUSTOMIZING INCIDENT MANAGEMENT SYSTEMS FOR SPECIFIC APPLICATIONS

SECTION TWO—LAW ENFORCEMENT-BASED SYSTEMS

There is no doubt that, during the response to a large-scale emergency incident; agencies will implement the Incident Command System or another type of Incident Management System. The fire service, as a whole, has been very successful in developing and implementing the concepts and practices of an Incident Command System, and it works very well for response to fire-related incidents of any magnitude. It will most likely be the fire department that will initiate an Incident Management System during the initial response to a mass casualty or high-impact incident. Is the law enforcement community able to integrate into the evolving and developing Incident Command/Management System and, perhaps more importantly, the Unified Command System and structures used in the region? An equally important question is, whether law enforcement agencies are prepared to utilize an Incident Command/Management System on smaller or more "routine" police calls for service.

With the proliferation of specialty and tactical units in law enforcement over the past few decades, there has been a demonstrated need and concerted effort for alignment of team structures with the concepts of Incident Command/Management Systems. Many law enforcement tactical teams are well versed in utilizing Incident Command/ Management Systems models for responding to incidents such as barricaded subjects, hostage situations, or dignitary protection assignments. With the implementation of the National Incident Management System (NIMS), law enforcement agencies are now mandated to integrate into an Incident Command/Management Systems/Unified Command structure for large incidents where state or federal assistance may be requested.

It has not been common, however, for an Incident Command/Management Systems to be routinely used at the street level by the rank and file of police organizations. Like other law enforcement skill sets, the effectiveness of Incident Command/Management System utilization will follow the old adage of "training like you fight and fighting like you train." It is the goal of this chapter to reinforce the importance of Incident Management Systems utilization for law enforcement agencies and responders at all levels in the agency command hierarchy, from a beat cop to the Police Chief, during incidents of all scopes and magnitudes.

THE MINDSET OF LAW ENFORCEMENT

The traditional mindset of the law enforcement officer is a critical element to consider in emergency incident response when attempting an evaluation of the effectiveness of any Incident Management System from a law enforcement perspective. During an emergency incident, the primary mission of a law enforcement agency is to identify the perpetrators of a crime and to build the prosecution's case for a successful conviction. Their priorities mainly consist of activities such as threat analysis and reduction, site security and responder protection, crime scene preservation, and witness interrogation. Needless to say, these priorities are not always consistent with those of the other involved responding agencies. When several agencies respond to a single incident, there will almost always be conflicting perspectives on incident response objectives and priorities. The National Incident Management System has been developed with the main goal of standardizing responders' organization structures to achieve one common overall mission goal, and law enforcement is no exception.

In order to successfully manage an incident from a law enforcement perspective during a large-scale event, agencies and officers at all levels must begin to alter their approach and overall management of more common, smaller incidents on a day-to-day basis. (Remember: "Train as you fight, fight as you train.") For an Incident Management System to succeed, all responders must have command and control of their own units with adequate internal communications and enhanced interagency/interdisciplinary interoperability. By utilizing an Incident Management System during routine day-to-day incidents, law enforcement functional groups will be better prepared to communicate more effectively during a large-scale event. History has demonstrated that internal and external communications failures are the most common complication and limitation during response to large or complex incidents. The goal of changing the operational mindset of the police officer to one that incorporates Incident Management Systems concepts in daily response activities will not only improve management of those day-to-day events but will also limit role and task confusion commonly seen in high-impact incidents, critical outcome likely incidents.

The scope and jurisdictions of law enforcement agencies vary widely from state to state. Even within a state, it is commonplace to find variations in responsibilities and jurisdictional boundaries and guidelines. More often than not, state police and highway patrols maintain jurisdiction on the nation's highways and interstates, while county and municipal police agencies usually have jurisdiction elsewhere. The jurisdiction of federal law enforcement officers and agents is usually determined by the nature of the crime, and generally they maintain law enforcement powers throughout the country. While this variation may affect how local operations are conducted, the goal of any Incident Management System as well as the overall goal of the National Incident Management System is to standardize the management as the incident escalates. No matter how the jurisdiction or inter-agency mutual-aid agreements are designed, an effective Incident Management System will overcome these classic barriers and obstacles through the concept of unity of command.

The classic model organizational structure for police agencies is a paramilitary command structure. Within these agencies, there has always been a core principle by design, which is a strict adherence to the concept of chain of command. At the scenes of the majority of calls for police service, law enforcement officers act autonomously, as they are trained and accustomed to making unilateral command decisions as a matter of conditioned response. In the past, law enforcement agencies often implemented their own command or scene management systems and organizational structures, which did not always compare to nor integrate well with model Incident Command Systems. One version known as, the Law Enforcement Incident Command System (LEICS), has been popular with many agencies. Though the officers and managers within law enforcement agencies have always possessed the necessary skill set to perform well within an Incident Management System, there has been a lack of movement towards using the system for law enforcement missions unless the incident escalates to the point of multi-agency response. In the post 9/11, NIMS era and based on historical data, it must be the goal of law enforcement agencies to operate within an Incident Management System before an incident escalates to that level.

As previous chapters have outlined, a critical concept for successful use of an Incident Management System is the effective use of its scalability and modular design. An Incident Management System must be just as effective for an apocalyptic terrorist event as it is for a "routine" assault investigation. If law enforcement agencies do not become completely trained and proficient in Incident Management Systems concepts and practices, then it is reasonable to expect that there will be a lack of unity of command at larger-scale emergencies. It is this lack of unity of command that breeds the inter-service conflicting priorities that we have seen between law enforcement, the fire service, and the other first responder/receiver disciplines in the past.

Past incidents have demonstrated that communities must be prepared for future events by coordinating the capability of first responder agencies. This coordination, in order to be successful, must entail the unified command ideology that comprehensively deploys all dispatched police, fire, and other first responder resources under one integrated command group with an emphasis on communications and interoperability.

THE SMALL-SCALE INCIDENT

As law enforcement assets arrive at a call for service, an Incident Management System should be implemented immediately, no matter how large or small the incident. Based on the initial clues of an emergency scene, the true scope and magnitude of the incident often cannot possibly be realized. The officer must be able to manage the incident in real time and respond to a rapid escalation of events. The size and complexity of the incident will drive the level of the incident management organization. Though it may only be implemented informally and conceptually, law enforcement officers need to shift their mindset and resolve that even if the event is small and simple in nature, an Incident Management System will still provide value and organization to the scene. If the incident escalates in size or complexity, the officer will be able to seamlessly adapt and expand the organizational structure.

As the first law enforcement officer arrives and establishes Command of the incident, he or she first evaluates whether there are victims requiring medical care; if so, care of the victims is the first priority. When a criminal act or an emergency incident occurs, the first arriving people at the scene will usually not be police, fire, or emergency medical personnel. Bystanders, security personnel, or members of other public-service disciplines who happen to be in the wrong place, at the wrong time, find themselves aiding the victims of the incident. They may be co-workers or complete strangers who attempt to render first aid or psychological care.

These civilians are an extremely valuable resource for the police during the initial phase of an incident, particularly before the arrival of firefighters or emergency medical personnel. In rural locations with small agencies or large geographic areas of coverage, they can be of vital assistance, as there may be a delay in the arrival of additional response assets. The police officer needs to manage these resources just as he or she would any other resource during an incident within the guidelines of the Incident Management System. The officer may wish to allow the continuation of care being provided by bystanders, especially if the bystanders have medical training, while the officer begins to set up a security perimeter. The officer may have security personnel as an initial resource that can be directed to assist in the effort to secure the scene until arrival of additional units.

Let us examine a minor vehicle collision as an example. A local police officer is called to a surface street location for a traffic collision involving three vehicles. Upon arrival, the officer finds that no other emergency personnel have arrived. The three-car collision has resulted in minor damage to the vehicles, gasoline leaking slowly onto the road, and bystanders assisting the three drivers who are all complaining of neck and back pain. The officer assumes the roles of Incident Commander and the Safety Officer and updates the dispatcher and other responding units, noting the hazards that he or she observes and requesting specific resources needed at that moment in time.

After determining that there is no immediate risk of fire or other significant hazards, the officer directs the bystanders (functioning as a member of the Operations Section) to ensure that the victims remain in the car and do not move in order to prevent further injury. The officer then sets up traffic-safety pylons to protect the scene and divert traffic away from the immediate site of the collision. He or she then advises dispatch to direct the other responding police officers to set up definitive traffic-control measures, which is an action point in the officer's mental Incident Action Plan. Remember that any Incident Management System is all about resource management, and initially the only resources available to the police officer may be bystanders. At this point in the incident, the Incident Management Systems organization chart may look like the example in Figure 8.8.

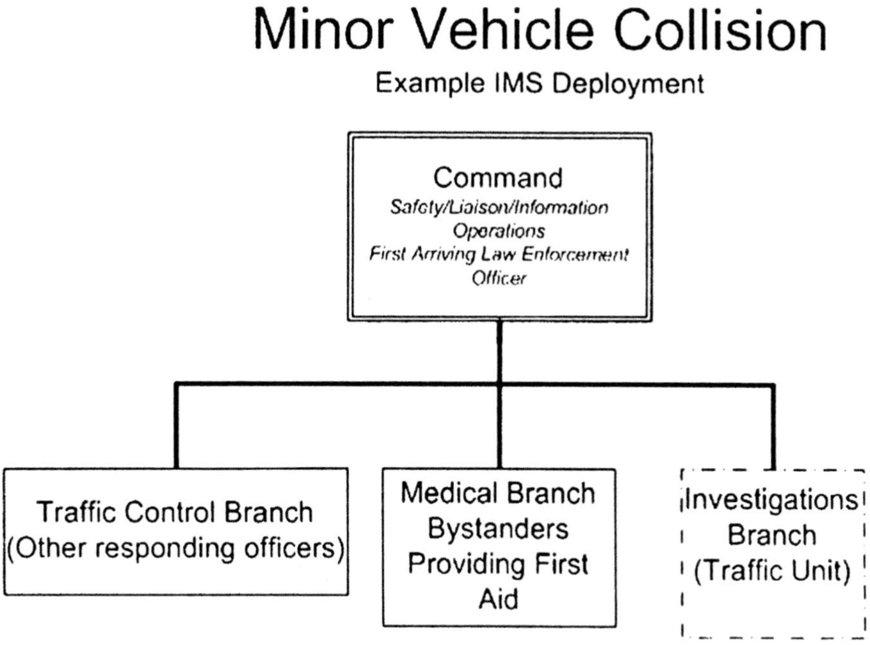

Minor Vehicle Collision
Example IMS Deployment

Figure 8.8 Minor motor-vehicle-collision IMS chart

As additional resources arrive on the scene, this structure will likely change quickly. The Incident Command, Safety Officer and Operations functions will likely be transferred to the first arriving firefighting or EMS unit, and the first arriving police officer may then assume the sole duties of the Investigation Branch while assisting the other responders in achieving the overall goal of the incident (safe extrication, treatment, and transportation of the victims; collision investigation; vehicle removal; and restoration of normal vehicle traffic flow).

Even on this small incident, the first responders ascertained the nature of the event and realized the importance of unity of command. Though the responders on this incident may not even realize that they are utilizing an Incident Management System, law enforcement officers are strongly encouraged to consider even small events in these terms so that if the incident escalates, the framework will already be in place and the Incident Management System structure can be expanded as necessary without additional confusion. A larger, more complex vehicle collision requiring specialized equipment may require an Incident Management System structure such as the example in Figure 8.9.

After life-safety considerations have been completely managed and the vehicle collision incident no longer requires the presence of the firefighters and emergency medical personnel, the Incident Management System structure should seamlessly scale back to a less complex organization, which, at that phase in the incident, may include other resources such as tow trucks, utility service vehicles, or transportation officials.

With the implementation of the National Incident Management System and other Traffic Incident Management Systems (TIMS), the role of the police officer at the

Complex Vehicle Collision
Example IMS

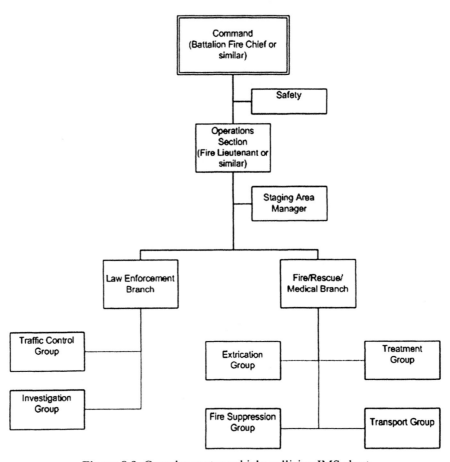

Figure 8.9 Complex motor-vehicle-collision IMS chart

scene of motor-vehicle collisions will likely change significantly in the very near future. Law enforcement agencies will no doubt institute new procedures and training for relieving traffic congestion and quickly clearing highway incidents. Traffic incidents are very dangerous for law enforcement officers and other emergency responders, and the safety incentive in implementing TIMS to manage vehicle collision scenes has been a driving force behind this movement for change in the procedures at crash sites. TIMS is a component of NIMS and integrates well during the management of traffic incidents. It should be implemented to control potentially adverse highway incidents such as structural damage, collisions, inclement weather, evacuations, or planned events such as mass gatherings and construction.

When the police are in the supporting role, such as on the scene of a hazardous materials incident, they are typically responsible for site security at the scene and at all the Incident Management System facilities. Specific responsibilities may include communications, traffic control, crowd control, and investigation. If traffic will be delayed for a significant length of time, the local, state, or federal Department of Transportation is often called in to integrate into the command structure by providing services such as barricades, detour signs, motorist assistance, and emergency infrastructure repair.

Once an incident occurs, the responsibilities of the police officer for the duration of the assignment as Incident Commander will include immediately taking action to stabilize the incident, provide for life safety, and establish traffic control. As soon as possible, a scene perimeter must be identified and established to control and deny entry and departure of potential witnesses or contaminated victims as needed based on the incident.

The most critical and time-sensitive action to take for the law enforcement Incident Commander is to evaluate the situation and call for appropriate additional assistance.

Control, triage, and treatment of any injured persons should be the next priority, utilizing any possible resources already available at the scene. Depending on the nature of the incident, it may be necessary to extend the operational area in order to ensure safe traffic flow. It will remain a command responsibility throughout the incident to provide for the safety, welfare, and accountability of all personnel at all times. Law enforcement will then be called upon to assist in restoring the roadway to normal operations once the incident has been cleared.

An important concept of TIMS, derived from Incident Management System, is that when feasible, all equipment and personnel from all the different disciplines and agencies arrive at the staging area should be quickly redirected so that the necessary personnel and equipment are dispatched to a specific location for assignment as needed. This will aid in avoiding large numbers of response and recovery equipment congesting the scene and will result in a more efficient operation when clearing the scene of equipment. Ingress and egress to and from the scene has historically been problematic at incidents where poor staging practices have been utilized.

A key point for law enforcement officers to remember is that obtaining the right resources for the tasks at hand is critical to the success of managing incidents regardless of scope and/or magnitude. To assist in the removal of vehicles, the correct towing and recovery vehicles must be at the scene. As there are a variety of needs, there are published reference tools with schematic descriptions of many of the different vehicles, towing configurations, and the best suitable equipment for the task. Law enforcement officers, particularly those who work frequently on highways, should have such a guide available in their police vehicle so that police at the scene can call for the appropriate resource the first time. Police officers should also be familiar with different types of fire suppression, EMS, and public works resources commonly needed for emergency incidents.

THE LARGE-SCALE INCIDENT

The key to the effective use of an Incident Management System is that it is designed to handle all types of incidents, including law enforcement calls for service. Incident Management Systems can be an especially effective tool for incidents such as large-scale homicide investigations and missing persons that have multiple crime scenes or are spread out over a large geographic area or multiple jurisdictional areas. Missing person's events benefit greatly from IMS, especially with regard to the use of area commands and/or geographic divisions to divide large areas into manageable search areas.

For routine and less complex incidents, the first arriving police officer may assume and maintain command throughout the incident. However, at more complex incidents, the command structure will likely expand and contract as needed. Establishment and maintenance of interoperable communications (across agency/jurisdictional lines) throughout the incident is absolutely critical for the larger-scale incident, and this needs to be thoroughly planned well in advance with contingency plans that take into account the unthinkable events that may occur without notice. Under a Unified Command scenario that is most likely to occur during the larger-scale incident, the ranking officials from all the representative agencies should coordinate efforts as a team and function as Incident Commander and Deputy Incident Commanders. Throughout the incident response phase, the roles of the Incident Commander and Deputy Incident Commanders may change and will eventually evolve as the goals of the incident change from preserving life to investigation, scene control, and crime scene processing. The Incident Commanders must mutually determine the objectives, strategy, and priorities for handling the incident.

For law enforcement officers, there is no greater responsibility at a large-scale incident than ensuring the safety and well-being of responders, bystanders, passing motorists, and other citizens. Operations must ensure that personnel, equipment, and vehicles are safely moved, depending on the location and duration of the incident. Because the vulnerability to emergency responders is significantly higher as the separation between moving traffic and responders decreases, it is important that adequate resources are dedicated to traffic assignments that will keep bystanders and motorists a safe distance from the incident operation. During these efforts, it is vitally important to prevent secondary incidents by warning motorists who are approaching vehicles that have slowed or stopped to look at what is going on.

To illustrate how an effective Incident Management Systems can be employed for a larger-scale criminal incident, we'll examine a hostage situation with five suspects holding fifteen hostages with confirmed automatic weapons and possible explosives. As the first arriving officer assesses and surveys the scene, she notes that the involved building is a one-story, 10,000-square-foot commercial facility with about twenty people outside, who appear to be very upset but ambulatory in the south parking lot. The officer advises her dispatcher that she is setting up an Incident Command Post at the southeast corner of the parking lot, and the other police units continue to respond. The officer calls for the agency's tactical team, emergency ordnance team, and K-9 units and then requests the fire department and EMS. To avoid congestion, the officer advises all the responding units to report to a specific staging area, which she has designated to be a vacant parking lot across the street on the southern end of the facility. Though no other units have arrived at the scene or staging area yet, the first arriving officer has already implemented an Incident Management System that will provide a framework that can be expanded as more responders arrive and the magnitude of the incident expands.

The officer also sketches a quick diagram of the incident to pass along to the future Incident Commander and tactical team. See Figure 8.10.

The lone private security officer from the facility advises the police officer that the initial dispatch information is accurate and that there have been shots fired inside the facility. Based on the only resources available at the time, the initial Incident Commander (the first arriving and only officer at the scene) takes immediate action to achieve the urgent and highest-priority goals of evacuating the outside people to a safer location and setting up a security perimeter around the facility to deny entry and egress. Though the command structure at this point in the incident may not be written down, conceptually and operationally it should look similar to Figure 8.11.

The complexity of this incident will escalate extremely quickly as other responders (law enforcement and other first responder disciplines) converge at the scene. It is critical that the Incident Commander reinforce the direction of responders to the staging area rather than the scene. Law enforcement officers will have an overwhelming sense of obligation and duty to respond directly to the scene to help "save the day," but this must be resisted. Freelancing has historically been a key obstacle and lesson learned from large incident responses in the past. As other officers arrive, they will begin to fill the vacant roles as needed in the organizational structure as directed by the Incident Commander, such as scene security/perimeter, traffic control, and evacuation.

The Incident Commander, by this point, based on the presence of automatic weapons and possibly explosives, should recognize that the victims and responders are too close to the immediate scene and are potentially in harm's way. The order should be given to evacuate everyone to a safer location, which in this case is at least 2,750 feet (based on the presence of a suspicious van-sized vehicle illegally parked at the front entrance of the building and the mention of explosives as a possibility). The officer quickly directs the evacuation, relocates the Incident Command Post, and communicates this information to dispatch and incoming units. It is highly recommended, if possible, that all responding units be directed to use an alternate communications channel designated for this incident, separate from the routine police communications that will still need to continue for the officers not assigned to the hostage situation to serve the remainder of the community.

As higher-ranking officials or more appropriately trained and experienced managers arrive at the scene, a change in Command may occur several times until the most appropriate manager arrives to assume command for the remainder of the Operational Period. These changes must be seamless and fluid with clear reporting to the incoming commander and appropriate reassignment to another function for the outgoing commander. As other law enforcement resources arrive, the Incident Management System structure will rapidly expand, similar to Figure 8.11. The diagram of the

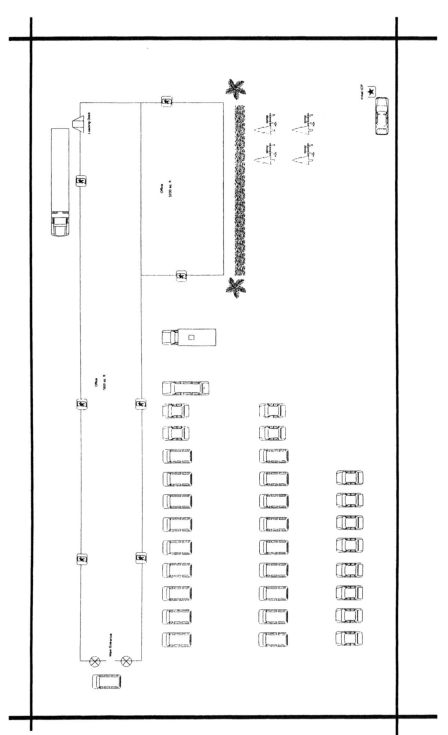

Figure 8.10 Scenario map

Hostage Situation
Example IMS

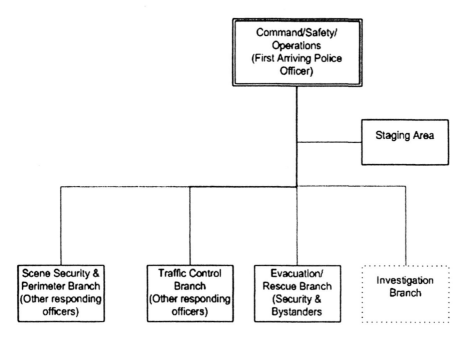

Figure 8.11 Hostage situation IMS

incident scene, reflecting the Incident Management Systems facilities and safer standoff distances, is shown in Figure 8.12.

When an incident becomes this complex with multiple agencies involved, unity of command becomes the paramount concept to maintain through out its duration. The expertise that each of these agencies brings to the incident is rarely challenged; however, because of different priorities, the conflict between emergency agencies has been a substantial problem during past larger-scale events. Planning and exercising before an incident occurs are critical steps in avoiding this conflict.

Within the Command and General Staffs, during the planning and execution phases of the incident, it is critical that law enforcement managers know the Incident Management System very well so that they are able to follow Unified Command procedures for the integration and implementation of each Incident Management System from all the disciplines. It is through frequent exercise and training that they become proficient with how the systems integrate and support the incident. Depending on the incident, a law enforcement commander must be able to recognize the need for and assist in implementation of a Unified Command structure. Because of the potential for mass

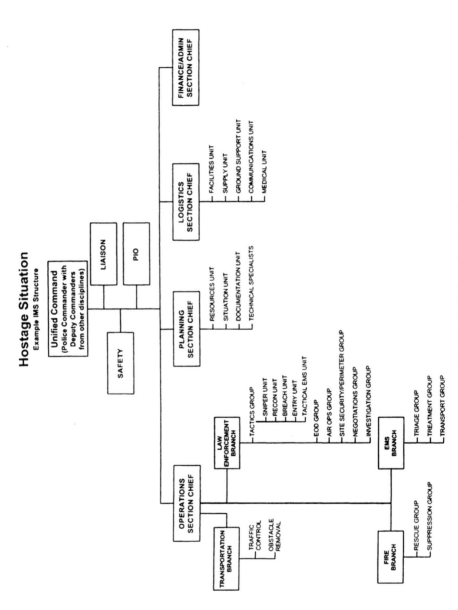

Figure 8.12 Hostage situation IMS Organization Chart, phase 2

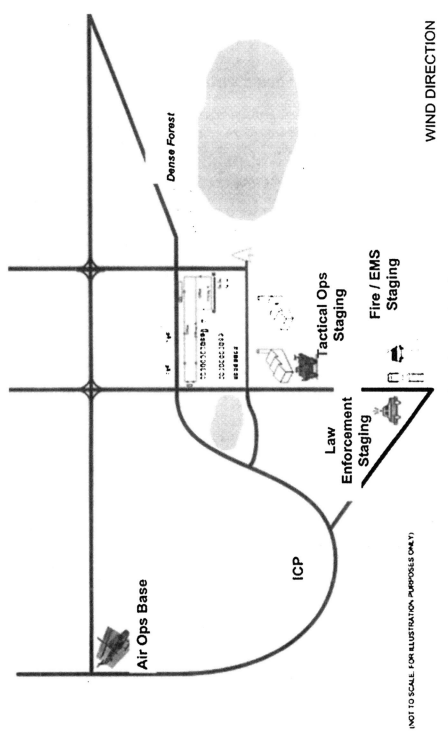

Figure 8.13   Scenario map, phase 2

victims, explosion, and fire in the hostage-situation example, the use of a Unified Command structure is critical in the planning and execution of the Incident Action Plan.

THE "BIG ONE": CATASTROPHIC AND APOCALYPTIC EVENTS

A responsible law-enforcement commander is always preparing for the "Big One." He or she will address the training requirements for law-enforcement officials who are expected to be part of the leadership and management team that will likely respond to an event involving Weapons of Mass Destruction (WMD). Law enforcement managers will be responsible for on-site planning for and managing scene security services that will face the challenges unique to terrorist events. They will coordinate the crime scene investigation and evidence-gathering activities. These managers are expected to efficiently manage on-site law enforcement resources and assist the Incident Commander/Unified Commander in bringing the event to a successful conclusion. Almost without exception, all the activities of these law enforcement managers should be conducted from outside the Hot Zone (any area deemed unsafe due to dangers such as hazardous materials or an active shooter). There is little benefit from managers accessing the hot or warm zones, and it will only create an additional risk to personnel and the operation. However, as the incident is stabilized, access will be provided to personnel such as investigators and crime-scene technicians to conduct their investigations and collect evidence.

Law enforcement managers are obligated to integrate into the Unified Command structure in order to supervise resources and assets being utilized to recover from the WMD event. Ideally, law enforcement managers will have successfully completed WMD training in awareness and operations, as well as performance, and management levels for events involving hazardous materials and WMD. This will be very beneficial to supplement their knowledge of Incident Management System and Unified Command procedures and the steps required for their implementation. This will make more efficient use of time, as the prepared law enforcement manager will know what information the Incident Commander needs to make responsible incident decisions. Because of the possibility of loss of agency command staff during a catastrophic event, law enforcement officials at all levels of command must be familiar with all Incident Management System and incident Commander Level functions, and be able to fulfill any functions related to law-enforcement operations as needed. A solid and verifiable chain of command is needed in order to ensure that operations will continue as smoothly as possible, regardless of the situation. A commitment to the Incident Management System and the chain of command will allow for continuity of emergency operations, even in the event of overwhelming casualties or responder losses.

During a catastrophic event, specific protocols for securing and maintaining control of the scene must be employed and strictly adhered to. A key challenge to managing these events is allowing only authorized persons actively involved in the emergency incident operations to gain access to the scene of the event. Early during the response phase of a high-impact event, agencies need to implement a credentialing and accountability system that will afford the law enforcement managers the ability to know and document all personnel entering and exiting the scene of the event. Considerable preparation and technology should be employed during the development of site perimeter and security plans.

Additionally, it is the responsibility of law enforcement managers to know and follow self-protection measures and protective measures for operational personnel. Close coordination with technical experts from the other disciplines, such as hazardous materials or EMS, will allow the law enforcement manager to make sound decisions with regard to personnel protection. Conversely, law enforcement managers should proactively reinforce the principles of crime scene preservation to the other disciplines whose priorities may not include crime scene management. As a catastrophic event unfolds, there are many other plans and assets available for the crime scene investigation and the control of events for securing and retaining evidence removed from the scene. On such a large-scale incident, it is certain that there will be federal assets in the lead role of evidence collection; however, it is critical that local law enforcement agencies are prepared to support such activities by being familiar with their role and its integration into the Unified Command.

CONCLUSION

Historically, law-enforcement and other responder disciplines have aggressively competed for priorities during the management of emergency incidents. Unfortunately, this practice, though maybe only subconscious, has demonstrated weakness in our abilities to efficiently manage incidents. The NIMS all-hazard resource-utilization-systems approach is designed to minimize this competition and afford managers from all disciplines the opportunity to coordinate their efforts for attaining a common overall goal.

Law enforcement, because of the changing threat to our societies, has begun to recognize the need for Incident Management Systems as the core to preparing to manage large-scale criminal events. Equally as important, law enforcement agencies have also begun to meet the challenges of integrating law enforcement operations in support of a unified command. By continuing to train and exercise, especially during smaller live events, agencies and personnel will be best prepared to meet the challenges of future threats and emergency events.

Chapter 8

CUSTOMIZING INCIDENT MANAGEMENT SYSTEMS FOR SPECIFIC APPLICATIONS

SECTION THREE—CORPORATE INCIDENT MANAGEMENT SYSTEMS (BUSINESS CONTINUITY PLANNING)

The need to address business continuity and incident response is growing in the public awareness. Fines for environmental incidents are increasing, and negative news coverage of the consequences of such incidents, especially with newer forms of media such as the Internet, rapidly deliver information into potential customers' homes and offices. Ineffective or tardy emergency incident management can quickly damage corporate reputations. The effect on businesses of all sizes is potentially disastrous, and there are several instances in which multi-national mega-corporations have never really been able to recover from the negative consequences.

The allowable time frame to recover from incidents grows shorter seemingly with each incident in the news. Often business partners, suppliers, and vendors are requiring incident management and continuity assurances in contracts, especially for critical services. There is a great deal of current and pending national and international legislation that calls for organizations to exercise due diligence and demonstrate an acceptable standard of care in how they manage their computing infrastructures and the information that such networks and systems create, transmit, and store, particularly when connected to the Internet. The tragic events of September 11, 2001 demonstrated the need for human elements of the business to be incorporated with the business continuity aspects of such plans.

There are an ever-growing number of standards, guidelines, checklists, and assessment instruments with which organizations are expected to demonstrate some level of compliance. An organization's ability to achieve and even institutionalize an incident management and business continuity system starts with executive sponsorship, enacted and sustained by governance. Those who lead, manage, set strategy, and are held accountable for an organization's success set the direction for how enterprise business continuity is perceived, prioritized, managed, and implemented. When business continuity is relegated to a secondary role in the organization, the organization is at greater risk when incidents occur. A practical extension of business continuity planning commonly referred to as BCP is to integrate the organization's plan with local emergency authorities and to utilize the same concepts of the Incident Command System found commonly in public safety response for business continuity incidents. Further, with the development of the National Incident Management System there is a need to institutionalize these Incident Management Concepts across public and private barriers in both governmental and non-governmental entities.

Benefits of this approach derive from the scalability of Incident Command/Incident Management Systems for incidents both large and small, along with the added communications and preparation that arise from integrated organization and local authority continuity planning. Integrated planning, in its full extension and implementation, leads to the National Incident Management System. By integrating the organization's business continuity planning with use of the Incident Command/Incident Management Systems and with coordination with local emergency authorities, the organization benefits from:

- Better teamwork within the incident response staff, resulting in a higher likelihood that essential response functions will be executed efficiently
- Reduced disruptions to operations due to higher response efficiency
- Reduced time to recovery and resumption of normal operations

THE HISTORY OF THE INCIDENT COMMAND SYSTEM

As stated elsewhere in this book, the concepts of the Incident Command System were developed in the 1970s as a result of problems in responding to a wildfire in California. At one point the overall cost associated with these fires totaled $18 million per day. Although all the responding agencies cooperated to the best of their ability, numerous problems with communication and coordination hampered their effectiveness. Later in the 1970s Congress mandated the creation of a coordinated fire response from the United States Forest Service. In California the resulting system became known as FIRESCOPE (FIrefighting RESources of California Organized for Potential Emergencies) and was designed so that it could respond to the hazards common in that state. By the mid-1970s, the basic structure of FIRESCOPE had been determined.

Ten years after the wildfires that initiated the development of FIRESCOPE; its command structure was in use for both wildfire and urban fire incidents. FIRESCOPE was researched, refined, and extended to non-fire incidents. The agencies adopting its use were initially in southern California, but by the mid-1980s its use spread widely across the United States. FIRESCOPE spread because it provided flexible, scalable, and coordinated incident management for a wide range of situations, including floods, hazardous-materials accidents, earthquakes, and aircraft crashes. It was flexible enough to manage catastrophic incidents involving thousands of emergency response and management personnel. In 1982 FIRESCOPE documentation was revised and adopted as the National Interagency Incident Management System (often referred to today as "two I'd NIMS). This document would later serve as the basis for the latest incarnation of the National Incident Management System (referred to as one I'd NIMS).

In March 2004 the United States Department of Homeland Security issued the National Incident Management System. It incorporates many existing best practices into a comprehensive national approach to domestic incident management, applicable at all jurisdictional levels and across all functional disciplines. The National Incident Management Systems was developed through a collaboration of local, state, and national authorities from a variety of disciplines affected by disasters and other events of consequence such as terrorist attacks against the nation. It represents a core set of doctrine, principles, terminology, and organizational processes to enable effective, efficient, and collaborative incident management at all levels and provides a framework for cooperation between emergency units without stifling best practices or new ideas.

The recommendations of the National Commission on Terrorist Attacks upon the United States further highlight the importance of the use of an Incident Command System. The Commission's report recommends national adoption of the basic concepts of the Incident Command System to enhance command, control, and communications capabilities.

The National Incident Management system provides a consistent, flexible, and adjustable national framework within which government and private entities at all levels can work together to manage domestic incidents, regardless of their cause, size, location, or complexity. This flexibility applies across all phases of incident management: prevention, preparedness, response, recovery, and mitigation. It further provides a set of standardized organizational structures, including the tenets of the Incident Command System, the Multi-Agency Coordination System, and a public information system as well as requirements for processes, procedures, and systems to improve interoperability among jurisdictions and disciplines in various areas.

NATIONAL INCIDENT MANAGEMENT SYSTEM

The Department of Homeland Security originates and maintains the structure of the NIMS. This structure flows through state and local incident command, providing interoperability during incidents. Some of the foundation concepts in NIMS are:

- Common terminology
- Modular organization
- Management by objectives
- Reliance on an Incident Action Plan (IAP)

- Manageable span of control
- Predesignated incident mobilization center locations and facilities
- Comprehensive resource management
- Integrated communications
- Establishment and transfer of command
- Chain of command and unity of command
- Unified Command
- Accountability of resources and personnel
- Deployment
- Information and intelligence management

These are common to previous versions of the Incident Command System and mirror such implementations of Incident Management Systems across the United States that were in use at the time of the development and evolution of the National Incident Management System.

Several areas defined by the Incident Command System model intersect with the business continuity planning model and must be considered by continuity planners as they prepare for incidents. One of the first steps for becoming compliant with National Incident Management System is to integrate the current version of the Incident Command System into the organization's Incident Management System. This process has become known as the "institutionalizing of the NIMS concept"

Two areas of interest to continuity planners are Unified Command and Incident Action Plan. Unified Command provides guidelines to enable agencies with different legal, geographic, and functional responsibilities to coordinate, plan, and interact effectively with each other and with public and private entities.

Unified Command overcomes much of the inefficiency and duplication that result when multiple agencies and business entities respond to an incident. In a Unified Command structure, the individuals designated by their jurisdictional authorities jointly determine objectives, plans, and priorities and work together to execute them. In other command structures such as a single Incident Commander structure, command of the incident may not be able to expand effectively as the incident evolves. The multiple agencies present may not work together efficiently, especially if joint exercises are not routinely held.

The Incident Action Plan includes the overall incident objectives and strategies established by Incident Commander working together in the Unified Command structure as a collaborative team. The Incident Action Team addresses the overall incident objectives, mission, and assignments of the business or businesses involved and each responding agency. The Incident Action Plan is also a tactical operations plan, generally addressing a 12- to 24-hour "Operational Period" in addition to longer-term items (tasks that will likely take multiple operational periods to accomplish, such as debris removal). A well-written Incident Action Plan prescribes a process for "lessons learned" activities both during and after the incident. In the case of a formal Incident action Plan these lessons will be captured as the Incident Action Plan is updated. In the case of an informal Incident Action Plan, the lessons learned will need to be captured by the Incident Commander in the after-action reporting process.

ADVICE FROM LOCAL GOVERNMENTS TO BUSINESS CONTINUITY PLANNERS

The City and County of Denver offers advice in Appendix I of a document entitled "Continuity Advice to Business Organizations their Emergency Operations Plan" distributed to and designed for business organizations to prepare for business continuity disrupting incidents:

- Business Continuity Planning according to BS 7799 and other standards
- Business Continuity Planning compliance with British Standards 7799 and 15000 (more commonly known as the Information Technology Infrastructure Library or ITIL).
- Other national standards such as NFPA 1600 (United States) and PAS 56 (United Kingdom)

International standards related to business activities emphasize a plan-do-check-act cycle; and this cycle is apparent in the BS 7799 Business Continuity Planning framework. Five clauses of BS 7799[a] describe a minimum BCP structure. These five clauses call for:

- A managed process for developing and maintaining Business Continuity Planning throughout the organization
- A strategic plan, based on risk assessment and business impact studies, that defines the high-level approach to Business Continuity Planning
- A tactical plan that provides details regarding time to restoration, order of restoration, contact lists, resources required, and similar aspects of recovery after a business disruption has occurred
- A framework to enable the organization to mature in its implementation of Business
- Continuity Planning and to identify high risk/high priority areas for improvement
- Testing of Business Continuity plans on a regular basis and management-level review of the results, with corrective action taken when test results do not meet expectations

BS 7799 defines continuity in terms of both disasters (natural disasters, accidents, equipment failure, sabotage) and security breaches. Compliance with the standard requires the continuity planner to assess risk from both traditionally defined disasters such as landslides and security issues such as theft. Continuity and security both begin with a risk assessment, which may also require an analysis of the assets in the organization. These assets may be tangible (buildings, people, and office equipment) or intangible (information). Once assets are defined, the business-continuity-planning team lists the risks to those assets.

BS 7799 calls for the "owners of business resources and processes" to be part of the Business Continuity Planning team. The asset inventory and risk assessment lead the team to publish a strategic plan for business continuity. This plan will provide an overarching framework from which lower-level plans may be derived.

NFPA 1600 integrates business continuity with both emergency and disaster management. It describes a program that covers the entire continuity lifecycle from preparation of plans to incident recovery.

The NFPA standard requires planning for continuity incidents, starting with a vision for business continuity from top management. The standard requires that a program coordinator be assigned to the continuity planning and execution process, with an advisory committee to assist the coordinator by providing ideas, resources, and other organizational support. Like BS7799 and other standards, NFPA requires evaluation of the continuity program with a view to continual performance improvement and evolution to meet the organization's needs.

NFPA 1600 also includes program elements that integrate an organization's incident response with its business continuity activities. The prescribed elements are:

- Laws and authorities—compliance with existing law and structure to evolve and comply with changing law
- Hazard identification, risk assessment, and impact analysis—both natural and human-caused, including impact to the business continuity
- Hazard mitigation
- Resource management—create objectives for the human, tangible, and intangible assets involved in continuity and incident management
- Mutual aid—determine needs
- Planning—strategic, tactical, mitigation, recovery, and continuity
- Direction, control, and coordination—develop the capability to respond through development of an incident management system, with roles and responsibilities communicated to appropriate resources as identified in resource management
- Communications and warning—implement and test warning systems with those under threat of an incident
- Operations and procedures—develop tactical response and recovery procedures
- Logistics and facilities—establish logistics capabilities; including primary and alternate sites
- Training—establish a personnel training program by defining the content and frequency of necessary training
- Exercises, evaluations, and corrective actions—perform periodic exercises of response and continuity elements up to and including the entire system for response and continuity; evaluate the results; and take corrective action against areas that do not meet objectives

- Crisis communication and public information—develop a capability to publish incident-related information and to respond to inquiries from the public and the media
- Finance and administration—develop a capability to make expeditious financial decisions to support response and continuity efforts before, during, and after an incident

Business-continuity planners will recognize that the steps in NFPA 1600 mirror the usual process through use of business impact analyses (section 5.3 in the standard), risk assessment and asset management (section 5.5), and other sections that include planning, training, and testing the continuity and incident-response programs. The NFPA standard also provides details regarding the content of an organization's business continuity plan (see section A.5.7.2.5). Section A.5.8.2 prescribes the use of an Incident Management System "to systematically identify management functions assigned to various personnel."

Integrated response, in which both the organization's staff and local emergency authorities are involved, needs a functional command center. It may be fixed or mobile but more often than not is a mobile Command Post as simple as the back of a police car or as complex as a fully functional mobile command center designed for such operations.

Figure 8.14 Small Incident Command Post

Figure 8.15 Large mobile ICP

While other references describe the geographic sitting and initiation of the command centers, there are nuances associated with integrated response. As with other aspects of incident response, it is imperative that the organization regularly exercise its staff both singly and jointly with local emergency responders. This will identify weaknesses on both sides and foster cooperative solutions that can benefit both parties.

In setting up the Unified Command center, a main room with a central table for the senior leadership, supported by their direct reports in an adjacent area, works well. The room should allow incoming information to be received without causing interruption and the input of all members of the unified command to be recorded. Telephone communications should be physically separated from radio communications to avoid distraction. When the facility includes tactical operations response staff, the following physical setup works well:

- Incoming communications received by assigned message taker
- Immediate response given if appropriate message created
- Messages are logged per procedure.
- Messages are assigned to resources for resolution.

Rooms adjacent to the main room and made available to the various Incident Command System departments (Operations Section, Public Information Officer, etc.) are an additional bonus for large-scale deployments.

Figure 8.16 EOCD-type operations

INCIDENT MANAGEMENT SYSTEMS/ BUSINESS CONTINUITY PLANNING GOVERNANCE

While responsible staff may initiate tactical business-continuity programs for integrated continuity and incident response, these programs will be short-lived and ineffective without senior management support. Organization governance, as it relates to business continuity and incident response, involves directing and controlling an organization to establish and sustain a culture (beliefs, behaviors, capabilities, and actions) of business continuity with a structured Incident Management System The Organization for Economic Development (OECD) defines corporate governance as follows:

"Corporate governance involves a set of relationships between a company's management, its board, its shareholders, and other stakeholders. Corporate governance also provides the structure through which the objectives of the company are set, and the means of attaining those objectives and monitoring performance are determined. Good corporate governance should provide proper incentives for the board and management to pursue objectives that are in the interests of the company and its shareholders and should facilitate effective monitoring."

The leader of the Marriott Corporation Business Continuity Organization states: "As a company, we're very proud of the fact that we go above and beyond what many companies do, and we're doing it not because it's required by regulations or legislation. We do it because, for us, it's a part of our culture."

Governing for business continuity builds upon and expands commonly described forms of governance, such as financial accounting and reporting. Corporate governance and business continuity/incident response overlap when the definition is expanded to include the "structure through which the objectives of the enterprise are set, and the means of attaining those objectives and monitoring performance are determined," as noted above. Structures and means may include, for example, policies (and their corresponding standards, procedures, and guidelines), strategic and operational plans, awareness and training, risk assessments, exercises, and evaluation of exercise results. While these definitions speak most often to large commercial corporations, they can also be interpreted and appropriately tailored for government, education, and non-profit institutions as well as organizations of any size.

Organizations such as the Marriott Corporation have found good continuity/incident-response governance through a structure that combines executive-level support with grassroots participation. A good continuity governance model contains these elements:

- Awareness and understanding—governing boards and senior executives are aware of and understand the criticality of governing for incident response and business continuity
- Protection of shareholder (or equivalent) value (reputation, brand, customer privacy)
- Customer satisfaction (sustaining marketplace confidence in comparison to competitors in order to retain existing and attract new customers)
- Strategies and plans for incident response and business continuity demonstrate how they support business objectives:
- Investments aligned with and allocated so as to meet strategies and plans, taking risks and costs into account
- Reporting measurement of performance and corrective actions taken
- Policies, standards, guidelines, procedures, and measures for incident response and business continuity are regularly reviewed and enforced
- Responsibility and corresponding accountability and authority for incident response and business continuity are clearly defined; communication and integration with local authorities and responders are institutionalized, exercised, and encouraged
- Internal controls are defined to effectively protect life, the environment, and organizational assets—people, information, hardware, software, processes, services, physical facilities, knowledge, and reputation.
- Risk-management processes guide the allocation of resources to and within business continuity programs.
- Oversight—the enterprise is regularly evaluated and audited to ensure an acceptable level of compliance with requirements, both internal and external, such as regulations, standards, audit criteria, market sector requirements, and continuity/response objectives.
- The enterprise's continuity and incident response plans are open to public disclosure to aid integration with local authorities and responders.

A short checklist for senior managers when considering the state of business continuity and incident response in their organization would include:

- Does the organization have an Incident Management System that can manage, in a uniform and process-oriented manner, all incidents that the organization may face?
- Are business continuity and incident response distinct functions in the organization, with adequate resources and a focus on risk reduction through teaming with organizational departments and outside agencies?
- Is business continuity spread across the organization with no established focal point for uniformity and performance monitoring?
- Is the planned response (if there is one) to incidents and continuity-impairing events process- and system-driven, or is it dependent on particular staff members for success?

Affinity Groups and Employee Awareness Programs

Affinity groups were initially developed by the Department of Energy to improve the performance of cross-functional and cross-agency teams. They can also be used to improve the effectiveness of integrated teams for continuity and incident response.

Affinity groups consist of persons from the organizations that wish to team with those who share similar job roles and responsibilities. Affinity groups allow their members to share information, solve problems, and create a vision. Affinity groups work best when the members develop informal communications channels, preventing organizational barriers, politics, and other ineffective behaviors from undercutting the cross-organization continuity-team effectiveness. Some ground rules for a continuity/incident response affinity group include:

- Formal roles with informal communication
- Regular meetings
- No supervisors (relative to the affinity-group members) in the meetings
- Self-managing and responsible for its mission, with oversight
- Group mission and vision as part of the larger cross-organization mission and vision

The affinity-group structure builds cohesiveness in environments where persons are stressed due to the gravity of the situations they face (such as disasters) and conflicting or multiple priorities (such as a municipal fire department with many businesses to protect). Affinity-group members tend to be respected, not because of their organizational affiliation, but because they are responsible group members. The members grow to a point where it is beyond the norms of the group to fail to perform a task. Groups that evolve to this level are efficient and have a higher degree of success in stressful situations.

Staff awareness is an important part of business continuity. Keeping internal Incident Management staff as well as external agency staff in constant communication promotes camaraderie, a spirit of "belonging", and new insights into improving the continuity/ response systems. This immediate training is often high-level and is later expanded into the tactical detail.

The detail training may be different for different staff members, depending on their responsibilities in the various integrated organizations. All participants in incident-response exercises should be asked to participate in post-exercise evaluations to empha-size the importance of continuity and response plans as well as to build a feeling of worth and teamwork. Senior management should ideally participate in post-exercise debriefings to reinforce the importance of exercises and of business continuity.

City and County of Denver Continuity Advice to Business Organizations

A goal of a staff awareness program is to immediately make new staff members at the organization and its incident-command partners (external agencies) aware of the Busi-ness Continuity Plan and Incident Management Systems based approach.

Step 1: Establish an Internal Planning Team
The size of this team will depend on the size of your operations, requirements, and resources. Having a team will encourage participation and interest in this process as well as providing a broad perspective on those issues relevant to your department.

Establish Authority
Upper-management involvement promotes an atmosphere of cooperation by author-izing the planning group to take appropriate steps to develop the plan. Within the group establish clear lines of authority, without being too rigid as to limit a free flow of ideas!

Develop a Mission Statement
The mission statement should define the purpose of the emergency plan and indicate that it will involve the entire organization. Define the structure and authority of the planning group.

Establish a Schedule and Budget
Establish a work schedule and planning deadlines. Timelines can be modified as pri-orities change. Develop an initial budget for research, printing and other costs associ-ated with developing your emergency operation plan.

Step 2: Analyze Capabilities and Hazards
Review internal plans and policies. Documents you should review include any building evacuation and fire plans, as well as risk-management plans. Review any documents relating to health, safety, and security procedures.

Identify Codes and Regulations
Identify applicable federal, state, and local regulations such as Occupational Safety and Health Administration (OSHA) regulations, fire codes, and environmental regulations.

Identify Critical Products, Services, and Operations
Determine the potential impact of an emergency on your organization as well as possible backup systems. Identify lifeline services (e.g., electricity, water, telecommunications, computer networking) and critical personnel and equipment necessary to continue your operations.

Identify Internal Resources and Capabilities
Examples of internal resources and capabilities include identifying first-aid supplies, emergency power equipment, and first aid stations as well as personnel resources available to assist in evacuation or provide public information.

Identify External Resources
Examples of external resources include 9-1-1, utilities, and contractors.

List Potential Emergencies
Consider emergencies that have occurred in your organization or the community that could affect you. Start by reviewing the history of emergencies that have taken place. Look at technological failures (e.g., HVAC, power, computer systems, telecommunications, and internal notification systems) that could impact your group. In addition, evaluate emergencies that may occur due to human error or the physical construction of a facility (e.g., lighting, layout of equipment, evacuation routes and exits). Then identify the probability or likelihood of the emergency occurring and the impact to people, property, and the continuation of operations. Finally, assess the internal and external resources you listed. Look at possible multiple impacts of disasters. A slow-rising flood can cause additional problems such as a loss of power, contamination of the water supply, or interruption of transportation.

Step 3: Develop the Plan
Your plan should include an executive summary describing the purpose of the plan, the types of emergencies that could occur, and the authority and responsibility of key personnel within your organization.

Emergency Management Elements
Describe your organization's approach to key elements of your plan: direction and control, communications, life safety, property protection, recovery and restoration, and administration and logistics.

Emergency Response Procedures
The procedures spell out how your department or agency will respond to emergencies. One approach is to develop a series of checklists that can be quickly utilized by

personnel at your location. You need to determine what actions should be taken to assess the situation; protect employees, customers, visitors, vital records, and other assets; and keep operations going. Specific procedures may be needed for a variety of situations including:

- Tornadoes
- Bomb threats
- Conducting an evacuation
- Fires
- Protecting vital records
- Warning customers and employees
- Flooding
- Restoring operations
- Communicating with responders

Supporting Documents
Documents that should be cataloged include:

- Emergency call lists—personnel in your organization who should be involved in responding to emergencies
- Building and site maps—indicate utility shutoffs, electrical cutoffs, gas and water main lines, floor plans, alarms and enunciators, fire-suppression systems, exits, stairways, designated escape routes, and restricted areas

Step 4: Exercise the Plan
Prior to implementing the plan, we recommend that your agency look at holding a tabletop exercise to test the plan. The following is an approach you may want to use:

- Select several objectives that you want to evaluate; you can inform the players at the tabletop exercise of those objectives.
- Select a relevant scenario, such as reduced or complete failure of electrical power, file-server or personal-computers failure, telecommunication-lines/disruption, and corruption of data
- Agenda—review the objectives, introduce the participants (and the roles they're playing), present the scenario, evaluate the plan and strategies as used by the players during the scenario, and review key issues brought up during the exercise including corrective actions and responsible parties
- Under tabletop rules everyone is free to contribute; the scenario can be changed as needed, and the facilitator can table any issue for later resolution.
- A facilitator's primary responsibility is to ensure the tabletop exercise proceeds on schedule and covers the objectives outlined.

If planned properly, tabletop exercises are usually a cost-effective method of testing plans and procedures. The most common feedback is that the exercise either demonstrated the viability of the plan or captured issues that will improve the plan. In addition, attendees appreciate the clarification of roles and responsibilities during exercise play. The tabletop exercise will often show a need to update or modify the plan. You may determine a need to modify duty assignments or procedures and update call-down and resource lists. It may even be necessary to modify a person's duties and responsibilities or add backup personnel.

Step 5: Implement the Plan

Implementation means more than simply exercising the plan. Act on recommendations made during the vulnerability analysis, train employees, and evaluate the plan.

CASE STUDY—THE EXXON VALDEZ OIL SPILL

The Exxon Valdez Oil Spill incident caused the entire oil industry to review its incident response procedures with fundamental and strategic new ideas. Oil-industry participants derived a series of plans that enabled them to respond effectively to incidents of any size and type. These response plans called for an integrated response to any incident. One pillar of these response plans was industry-wide adoption of an Incident Command System based Incident Management System as an industry standard. In 1990, oil-industry participants began to use the principles of Incident Command Systems when responding to oil spills. The experience from the Public Safety arena showed it to be applicable to any type of incident and that even when an incident crossed multiple jurisdictional boundaries it could be used effectively.

The use of an Incident Management System was valuable when responding to any incident, regardless of size, location, or complexity of the response. Initially, the oil industry found that it did an inadequate job addressing disruptions to business activities in the incident area. This was due to the fact that the Incident Command System as it was first developed as a tool for response agencies was never designed nor intended to deal with these issues. Oil-industry users of "their" Incident Management System simply added a Business Continuity Planning section to the standard Incident Management Systems organization.

The oil industry went one step further to improve its incident response and continuity activities by adopting a common terminology related to incident management for use within the industry itself. As discussed by other business-continuity practitioners, ambiguous jargon and use of nonstandard terms can hamper incident response, leading to larger loss of life and property as well as longer recovery time. The oil industry extended common terminology frameworks beyond the traditional definition of incidents (fire, spills, and rail accidents) to information technology, which requires a business-oriented response.

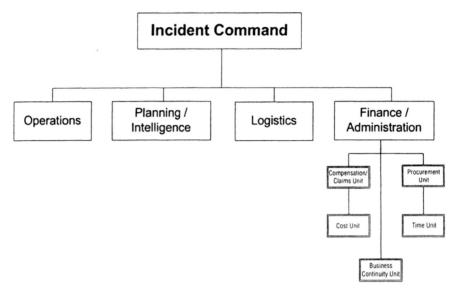

Figure 8.17 IMS Organization chart with continuity section added to Finance/Administration Section

The new framework defines the following parameters:

- An incident is the entire set of unplanned events that can affect the company.
- Emergencies are incidents that require an immediate response to protect life, health, property, or the environment.
- Crises are a subset of incidents and emergencies that can significantly affect company operations or reputation.

Oil-industry participants in this new integrated Incident Management/Business Continuity system set up "centers of excellence" where best practices were to be developed and models created to help others in the industry to institutionalize the use of the new system. This use of corporate governance to develop the integrated system and continually improve it is an exemplary model. In the case of oil companies, different departments had traditional responsibility for various aspects of incident response:

- Information technology for computer-related incidents
- Safety for life-threatening incidents
- Environment for incidents that threatened an ecosystem

In many cases a dedicated Business Continuity Planning staff was developed to integrate these diverse functions for both preparation and the actual incident response when needed. Staff members in these new roles took a wide view of incident management.

The integrated department generally reported to an executive who reported to the Chief Executive Officer of the organization. The new integrated response organization implemented a common response framework. This framework provided a common method that is used for all business units, regardless of size or mission. The framework also includes mutual aid enablers, headquarters and field "Strike Teams," and an organization-wide response plan. Individual departments were made responsible for incident management within the framework and provided resources to aid the planning and test/exercise phases of implementation.

CASE STUDY—THE UNIVERSITY OF COLORADO, BOULDER, AND COORS BREWING COMPANY

A captain in its internal police department introduced the University of Colorado (CU) in Boulder to the use of the Incident Command Systems model in ·1991. A series of seminars at the university demonstrated the need to implement a flexible incident command to avoid well-meaning responders from "clogging the arteries" preventing effective response to incidents.

Planning has proven to be an important part of the incident-management process. Responsibilities and organizational structure were defined early in the implementation of IMS at the university. An emergency management group, consisting of area managers on the campus, met (and still meets) monthly to perform scenario planning.

The university is like a city with many small businesses (departments) and residents (faculty, administration, and students). Coordinating incident response with little authority over these "small businesses" can be challenging. University departments include such diverse units as the bookstore, student housing, classrooms, foodservice, and teaching offices. The university has found that frequent testing exercises help to involve the many managers and staff and to secure their interest in the business continuity incident response program. A full-time position of Business Continuity Planning Coordinator was added to the staff to manage the many departments on campus.

The Business Continuity Planning Coordinator assisted the departments in moving business continuity from being an insurance issue to a mindset. The Business Continuity Planning Coordinator and an Emergency Planner prioritize the top twenty most vulnerable and critical functions on the campus (such as telecommunications, enrollment, and larger teaching departments) and work with the area managers to formulate and test business continuity plans. A policy group, consisting of the provost and other senior staff on campus, provides guidance and resources to the emergency planning and continuity effort. The emergency-management staff members provide information and recommendations to the policy group, which then provides for emergency management needs.

The university tests its plans frequently with local authorities from the city, county, hazardous materials response agencies, and agencies such as the American Red Cross. The university and other local authorities develop an understanding of each other's capabilities through these exercises and enhance camaraderie and teamwork. The capabilities of the various authorities are often complementary; for example, the university has radiation safety expertise not commonly found in police, fire, and hazardous materials departments. The Incident Management Systems is used by the university to avoid situations in which numerous, possibly unnecessary responders move into an incident site and prevent needed resources from effectively performing their duties. The use of an Incident Mismanagement System also provides a flexible response framework that can grow or contract to meet the needs of a particular incident. It is often used in preplanned events such as graduations to control what would be expected to be a chaotic situation due to the influx of outside persons into the community.

The Coors Brewing Company has found similar benefits from use of the incident Command System. Like the University of Colorado, Coors has a large physical plant (12,500,000 square feet under one roof) and a diverse group of departments to protect. Coors uses a Unified Command structure, in which Coors staff remains under the command of a designated Coors employee and outside agency responders are under the command of an individual from one of the outside agencies. The Coors staffer and the Incident Commander from the responding agencies are always working together to effectively mitigate the incident for mutual satisfaction.

Several related departments, such as public information and finance, are managed closely, but tactical response operations are managed separately to assure that no conflicts are created in the response to the incident. At the incident site, both Coors and the responding agencies use their own communications apparatus, again with a coordinated response so that all parties can "talk" to each other effectively.

Coors has found that the only downside of Incident Command System with Unified Command is a difference in jargon in use by Coors and the many local responders.

Differences in jargon and terminology create inefficiencies in the incident response; considerable training and exercise time are expended to overcome this language barrier. The intense training has positive effects. By the time that local fire departments respond to exercise incidents, Coors staff often has many of the traditional fire-service duties underway or completed, such as area evacuation and search and rescue operations (via its in-plant response brigades, which are commonly found in any large industrial occupancy).

Experience from incidents and exercises result in feedback to an NFPA 1600 based business-continuity system. Coors management is aware of the importance of incident response and business continuity. Management staff reviews the evaluations from incidents and exercises, including videotapes of actual incidents as well as of post-exercise analyses. This feedback results in changes to the plan and training where needed; these are then further exercised in a continuous loop.

CONCLUSION

After the events of September 11, 2001 the need to assure that business assets of all nations continue as well as to assure that continuity of operations and continuity of government has been magnified globally for both businesses and governments. The world that we live in relies on a truly worldwide economy, and if we fail to protect that worldwide economy the results can be disastrous for all involved. Development of Business Continuity Plans that utilize the concepts of sound Incident Management Systems that are designed to interact with the existing Incident Management Systems of the community as a whole are a major step in protecting that worldwide economy.

Chapter 9

ADVANCED INCIDENT MANAGEMENT SYSTEMS CONCEPTS

This chapter will discuss the still evolving concepts of the use of Incident Management Teams at major large scale incidents such as hurricanes and terrorist incidents which may have potentially global impact.

THE INCIDENT MANAGEMENT TEAM CONCEPT

The concepts of the use of an Incident Management Team are not new to emergency management in the United States. There have been teams in existence for decades, organized to handle incidents specific to their areas, particularly California earthquakes and wildfires or urban search and rescue operations. Each member of these teams serves a particular function and may be highly trained and qualified in areas such as communications, safety, technology, tactics, operations, or command.

Since the World Trade Center event it has become apparent that more Incident Management Teams are needed in local and regional areas to effect as immediate a response as possible to incidents, events, or disasters. This recommendation has been made by the Department of Homeland Security and state and local governments as well. The depth and extent of damage done by certain types of disasters may be minimized through fast, safe response of emergency personnel. Incident Management Teams are welcomed by those who mitigate emergencies, insurance companies, and others who stand to lose the most financially in an incident.

Background

The first Incident Management Teams were known as Large Fire Organizations and were developed in the 1950s to combat large wildfires. These groups evolved into what became known as Overhead Teams with the development of FIRESCOPE ICS in the 1970s and lastly Incident Management Teams in the 1990s.

The original teams were organized to enable a group of qualified specialists to be readily available for rapid assembly and deployment to a disaster area. The mobilization and use of an Incident Management Team provides a significant capability for disaster response and mitigation. The multi-disciplinary team brings with it organization and technical capabilities, as well as a wide variety of services for catastrophic events. Members of the Incident Management Team must be available on short notice to mobilize within four hours of the request. Incident Management Teams members must be self-sufficient for at least twenty-four hours and prepared for a response assignment of up to fourteen days.

An Incident Management Team equipment cache is pre-organized into functional kits and is available for dispatch with the team. In the case of a wildfire control area the National Interagency Fire Center at Boise, Idaho maintains kits to support the teams with communications equipment (including telephones and radios), computers, printers, and other administrative supplies. The Logistics Section Chief is responsible for facilitating orders and accounting for equipment and supplies. Team members request logistics support through the Logistics Section Chief upon arrival at the Base of Operations (BOO).

Definition

An Incident Management Team is comprised of appropriate Command and General staff personnel assigned to an incident or major event. The team is structured to provide assistance to complement and support the existing Incident Management/Incident Com-

Figure 9.1 Incident Management Team deployed at WTC, 2001

mand organization for events that exceed local capabilities or for other reasons. A local, state, or federal agency may request a team to provide incident support or completely manage the overall emergency if necessary. A local, state, or federal agency may request through appropriate channels all or part of a team.

The purpose of the Incident Management Team is to provide a pool of organized, highly skilled and qualified personnel to respond to complex emergency incidents. Emergency response and public safety must always be the first priority on any incident.

The mission of the Incident Management Team is to provide federal, state, and local officials with organizational and technical assistance and the acquisition and utilization of resources. The Incident Management Team also offers advice, assistance, management, and coordination.

Many emergency-response organizations recognize that large-scale incidents, disasters, and preplanned events will overwhelm their incident-management and staffing abilities. Development of an Incident Management Team will aid in the emergency management of the following situations:

- Natural disasters such as tornadoes, hurricanes, floods, earthquakes, or tsunamis
- Train derailments, airplane crashes, and other large-scale accidents
- Public or civil unrest (such as college disturbances and celebrations)
- Terrorist incidents

It is critical that team members train and work together to maximize limited resources and capabilities of local, state, or federal law enforcement, EMS, fire service, and related community assets.

Typing of Incidents

The types of Incident Management Teams are based on the types of incidents they will respond to. Incidents are categorized by five types based on complexity. Type V incidents are relatively simple emergencies, while Type I incidents are the most complex. Teams are designated to correspond with incident complexity.

Type V Incidents consist of an initial response with a single or multiple law enforcement, EMS, or fire unit(s). Generally this type of response will require only one to two resources and a few hours to resolve. An example might be a traffic citation, motor-vehicle accident, or medical emergency call. These are common, everyday occurrences.

Type IV incidents include an initial response that requires multiple resources and is usually resolved in one operational period. An example might be a large single-family-dwelling fire that requires multiple alarms with EMS, law- enforcement, and perhaps public-works support.

Type III incidents are characterized by an extended response. Initial actions may have failed, requiring multiple resources and perhaps multiple Operational Periods to be resolved. Some or all of the Command and General Staff may be activated. An example might be a train derailment.

Type II incidents are designated as an extended response requiring most or all of the Command and General Staff along with their functional staffs. The incident extends

Figure 9.2 Day-to day-response

Figure 9.3 USAR Team in a hurricane deployment

into multiple Operational Periods and requires written Incident Action Plans, Planning Meetings, and Briefings. Operational personnel should not exceed two hundred in an Operational Period, and total personnel committed to the incident should not exceed five hundred. An example of a Type II incident might be a large winter storm that halts electricity over a large area for days or weeks.

A Type I Incident is the most complex, requiring all the Command and General Staff positions to be activated along with their functional staffs. The management of the incident is often subject to great public and political scrutiny. Operational personnel often exceed five hundred per Operational Period, and total personnel may exceed one thousand. An example might be a large flood or hurricane incident that affects a large geographic area covering multiple states.

TYPING OF INCIDENT MANAGEMENT TEAMS

In July 2003, the U. S. Fire Administration convened a focus group of stakeholders and experts from across the country to develop all-hazards Incident Management Teams across the country. In the wildfire community, the United States Forest Service and the National Wildfire Coordinating Group currently recognize five types, or levels, of incident management teams (corresponding with the five types of incidents described above).

Figure 9.4 WTC site, 2001

The focus group agreed to stay with this model for the all-hazards emergency-response community. The team types, including certifying level and basic makeup are:

- Type V: city and township level, locally certified, jurisdiction-specific or by mutual agreement
- Type IV: county or fire district level, county or regionally certified, multi-agency jurisdiction
- Type III: state or large metropolitan area level, state certified, state, region, or area with multiple jurisdictions or mutual aid agreements
- Type II: national and state level, federally or State certified, less staffing and experience than Type I, smaller-scale national or state incident
- Type I: national and state level, federally or state certified, most experienced, most equipped

This recommendation was made in part as a result of a Memorandum of Understanding between USFA, the International Association of Fire Chiefs (IAFC), and the National Fire Protection Association (NFPA) Metropolitan Chiefs. The memo, signed in 2002, is designed to:

- Establish metropolitan-area Incident Management Teams, regional overhead teams based on the U.S. Forest Service (USFS) models
- Develop Incident Management Team capability
- Develop and train Incident Management Teams to support local level commanders
- Provide mutual aid staff and Unified Command Staff training and development

Type I incident management teams are currently part of the federal and state land-management agencies and are composed of federal, state, and local responders. These teams are considered national assets; there are seventeen teams in nine regions across the U.S. The teams respond with approximately fifty to sixty personnel. All Command and General Staff must successfully complete advanced Incident Management training.

Type II teams are also a part of the federal and state land- management agencies composed of federal, state, and local responders. Teams are national, state, and regional assets, with over fifty across the U.S. These are state and regional national interagency teams. These teams generally respond with ten to sixty personnel based on the size, type, and complexity of the incident.

Type III teams and lower are local and state assets. The composition of teams is currently determined locally. There is currently no national standard for staffing, but a standard is being developed at this time.

INCIDENT MANAGEMENT TEAM TRAINING

Out of the 2003 focus group, an all-hazards Incident Management Team training course was developed by the United States Fire Administration via its National Fire

Academy. The goal of this training was to provide the necessary tools for individuals to perform as members of a Type III Incident Management Teams immediately upon completion. The focus group also recommended a framework upon which departments or groups of departments can build Type IV and Type V Incident Management Teams. Members should come from all disciplines; they are trained to function in appropriate Command and General Staff positions during local incidents and to transition to a higher-level, more robust team if necessary after the first Operational Period to assist in managing major incidents. Training recommendations for Type IV and Type V Incident Management Teams include existing National Fire Academy courses that will primarily be provided through state fire-training networks.

The USFA developed an "Incident Management Team Training Roadmap," which identifies the training needed to develop Types III, IV, and V Incident Management Teams. The roadmap recommends that all emergency service personnel should take training equivalent to the National Wildland Coordinating Group courses I-ICS 100, I-200, and I-300, either web-based or classroom.

Training for Types I and Type II Incident Management Teams is already in place through the National Wildland Coordinating Group and currently is considered likely to remain as is for the foreseeable future.

The Incident Management Team training roadmap specifies the following:

1. Recommended for ALL emergency service personnel:
   - National Wildland Coordinating Group courses I-ICS 100, I-200, and I-300 (web-based or classroom) or equivalent
2. Recommended for emergency-service responders who may serve in Command and General Staff positions during the first Operational Period of a major incident, including those who may serve on a Type IV or Type V Incident Management Teams:
   - Introduction to Command and General Staff (NFA self-study course)
   - Command and General Staff Functions in the Incident Command System (NFA six-day course)
   - Introduction to Unified Command for Multi-Agency and Catastrophic Incidents (NFA two-day course)
3. Recommended for assigned members of Type III Incident Management Teams:
   - Appointed by metropolitan, regional, or state authority having Jurisdiction
   - Meet the requirements of Type IV/Type V Incident Management Team member (under development)
   - All-Hazards Incident Management Team Course (NFA seven-day course)
   - Position-specific training (as identified by the US Forestry Service)
   - Shadowing/mentoring process (as described by the US Forestry Service)

This Incident Management Team training roadmap, developed in partnership with US Forestry Service, also supports Homeland Security Presidential Directive 5 (HSPD-5), which states: "To prevent, prepare for, respond to, and recover from terrorist attacks, major disasters, and other emergencies, the United States Government shall establish a single, comprehensive approach to domestic incident management. The objective of the United States Government is to ensure that all levels of government across the Nation have the capability to work efficiently and effectively together, using a national approach to domestic incident management.

Several states and metropolitan areas already utilize the equivalent of Type III Incident Management Teams. The Fire Department of the City of New York (FDNY) recently completed the development of Incident Management Teams through training and shadowing in partnership with US Forestry Service.

Departments in the National Capital Region, through the Washington Metropolitan Council of Governments, Fire Chiefs and Senior Operations Committee, are preparing to take the US Fire Administration Type III Incident Management Team training regimen.

The Commonwealth of Pennsylvania, through the Pennsylvania Emergency Management Agency, has developed a Type III Incident Management Team to support major local operations. Several other states and metropolitan areas have or are making plans to develop Type III Incident Management Teams following the training roadmap.

Incident Management Teams have been designed to assist local emergency services and support unusually large, complex, or long-term emergency incidents when requested. An all-hazards team consists of emergency service officers from appropriate disciplines (fire, rescue, emergency medical, hazardous materials, law enforcement) trained to perform the functions of the Command and General Staff of the incident management system. These functions include Command, Operations, Planning, Logistics, and Finance/Administration, as well as Safety, Information, and Liaison. (These functions are described in detail in earlier chapters of this text) Members of the initial responding departments often fill these functions; however, the size, scope, or duration of an emergency incident may indicate the need for an Incident Management Team to support them. The local Incident Commander can request, through standard mutual-aid procedures, an Incident Management Team to help support management of the incident.

The operations of Incident Management Team are highly dependent on local community needs, available resources, and the level of training and experience. Local jurisdictions may establish, train, and control Incident Management Teams at their respective levels.

The US Fire Administration and US Forestry Service will work together in delivering training to develop Incident Management Teams.

The US Fire Administration and US Forestry Service are also working together to deliver specialized training for Type III Incident Management Teams for regional- or state-level incidents. Type III Incident Management Teams are recommended for states and large metropolitan areas with multiple jurisdictions and mutual aid agreements (such as the Department of Homeland Security, Urban Area Security Initiative locales). Members of Type III Incident Management Teams are appointed by a state or metropolitan authority having jurisdiction and respond as a team to support or assist a local team at major incidents that may have national implications.

Incident Management Teams are formed from pools of highly qualified personnel; these individuals meet a very high standard of emergency response training, qualifications, and experience requirements. Personnel selected for a team position agree to a commitment of three years. When an emergency exceeds the capabilities of an agency, a team may be requested to manage the event. Teams manage an emergency event through a delegation of authority written by the agency that has jurisdictional control.

UNIFIED COMMAND

Although a single Incident Commander normally handles the Command function at most incidents, an Incident Management System may be expanded into a Unified Command structure for certain types of complex incidents. Unified Command is a structure that brings together the Incident Commanders of all major organizations involved in the incident in order to coordinate an effective response while at the same time carrying out their own jurisdictional responsibilities. The Unified Command structure links the organizations responding to the incident and provides a forum for these entities to make consensus decisions and control limited or unique resources over wide geographic areas if necessary. Under the Unified Command structure, the various jurisdictions and/or agencies, both governmental and non-governmental, may blend together throughout the operation to create an integrated response team.

Figure 9.5 Incident Management Team support vehicle

Unified Command is responsible for overall management of the incident just as a single-entity Incident Commander would be. The Unified Command directs incident activities, including development and implementation of overall objectives and strategies, and approves ordering and releasing of resources. Members of the Unified Command work together to develop a common set of incident objectives and strategies, share information, maximize the use of available resources, and enhance the efficiency of the individual response organizations.

The four fundamental questions that need to be answered in choosing to move an incident to a Unified Command structure are:

- When should a Unified Command be used?
- Who is in a Unified Command?
- How does the Unified Command make decisions?
- What if your agency is not a part of the Unified Command?

When a Unified Command Should Be Utilized

Unified Command may be used whenever multiple jurisdictions or agencies are involved in a complex and/or large response effort. These jurisdictions or agencies could be represented by:

- Large or otherwise complex geographic boundaries are involved
- Multiple governmental levels are impacted
- Multiple functional areas are needed
- Statutory responsibilities overlap significantly
- Some combination of the above situations

Unified Command Membership

Actual Unified Command makeup for a specific incident will be determined on a case-by-case basis, taking into account:

1. The specifics of the incident
2. Determinations outlined in existing response plans
3. Decisions reached during the initial meeting of the Unified Command

The makeup of the Unified Command may change as an incident progresses in order to account for changes in the situation. Unified Command is a team effort, but to be effective, the number of personnel should be kept as small as possible.

Frequently, the First Responders to arrive at the scene of an incident are emergency-response personnel from local fire and police departments. As local, state, federal, and private responders arrive on-scene for multi-jurisdictional incidents, responders should integrate into the Incident Management System organization in place and begin to establish a Unified Command to direct the expanded response efforts. Although the role of local and state responders can vary depending on state laws and practices, local responders will usually be part of the Unified Command.

Members in the Unified Command have decision-making authority for the response. To be considered for inclusion as a Unified Command representative, the representative's organization must:

- Have jurisdictional authority or functional responsibility under a law or ordinance for the incident
- Have an area of responsibility that is affected by the incident or response operations
- Be specifically charged with commanding, coordinating, or managing a major aspect of the response
- Have the resources to support participation in the response organization

In addition, Unified Command representatives must also be able to:

- Agree on common Incident Objectives and priorities
- Have the capability to sustain a 24-hour-a-day, 7-day-a-week commitment to the incident
- Have the authority to commit agency or company resources to the incident
- Have the authority to spend agency or company funds
- Agree on an incident response organization matrix
- Agree on the appropriate Command and General Staff position assignments to ensure clear direction for on-scene tactical resources and unity of Command
- Commit to speak with "one voice" through the Information Officer or Joint Information Center, if and when established by the Unified Command
- Agree on logistical support procedures
- Agree on cost-sharing procedures, as appropriate

Unified Command members bring their authorities to the Unified Command, as well as the resources to carry out their responsibilities. Unified Command members may change as the response transitions out of emergency response and into long-term cleanup. Members in a Unified Command have a responsibility to the Unified Command and also to their agency or organization. These individuals in the response-management system do not relinquish agency authority, responsibility, or accountability. The addition of a Unified Command to the Incident Management System enables responders to carry out their own responsibilities while working cooperatively within one response-management system.

How the Unified Command Makes Decisions

The decision-making process that takes place in the Unified Command structure is not "decision by committee." The principals in the decision-making process are there to command the response to an incident just as they are in the case of a single Incident Commander. Time is generally of the essence. The Unified Command should develop synergy based on the significant capabilities that are brought by the various representatives. There should be personal acknowledgement of each representative's unique capabilities, a shared understanding of the situation, and agreement on the common objectives. With the different perspectives on the Unified Command comes the risk of disagreements, most of which can be resolved through an understanding of the underlying issues. Contentious issues may arise, but the Unified Command framework provides a forum and a process to resolve problems and find solutions. If situations arise where members of the Unified Command cannot reach consensus, the Unified Command member representing the agency with primary jurisdiction over the issue would normally be deferred to for the final decision.

The bottom line is that the Unified Command has certain responsibilities as noted above. Failure to provide clear objectives for the next Operational Period means that the Command function has failed. While the Unified Command structure is an excellent vehicle (and the only nationally recognized vehicle) for coordination, cooperation, and communication, the duly authorized representatives of each jurisdiction and agency represented must make the system work successfully. A strong Command function, be it in the single Incident Commander or a Unified Command structure, is essential to an effective response and control of emergencies.

Representatives Outside of the Unified Command Structure

To ensure that your organization's concerns or issues are addressed if your agency is not represented within the Unified Command structure, your organization should assign representatives to:

- Serve as an agency or company (private-sector) representative
- Provide input to your agency or company representative, who has direct contact with the Liaison Officer
- Provide stakeholder input to the Liaison officer
- Serve as a Technical Specialist in the appropriate section(s)
- Provide input to any Unified Command member who needs such input from your agency or jurisdiction

Advantages of a Unified Command

Unified Command has been used to manage local, state, and federal responses to complex multi-agency, multi-jurisdictional incidents. The best example of this system at work over time has been in the wildfire arena as well as in the area of international oil-spill response. The most public and graphic example of this system was used during the events of September 11, 2001.

The following is a list of the advantages of an Incident Management System with a Unified Command structure:

- Uses a common language and response culture
- Optimizes combined efforts
- Eliminates duplicative efforts
- Establishes a single Command Post
- Allows for collective approval of Operations, Logistics, Planning/Intelligence, and Finance/Administration activities
- Encourages a cooperative response environment
- Allows for shared incident facilities, reducing response costs, maximizing efficiency, and minimizing communication breakdowns
- Permits responders to develop and implement one consolidated Incident Action Plan

The Unified Command structure outlines responsibilities and functions, thereby reducing potential conflicts and improving information flow among all participating organizations. The Incident Management System maintains its modular organizational structure so that none of its advantages are lost by the introduction of a Unified Command.

Unified Command Meeting

Once the decision has been made to move to a Unified Command structure, the first step is to have an initial Unified Command Meeting. This is critical to ensure effective implementation of the Unified Command and use of the Incident command System model when an incident occurs and plans need to be implemented. The establishment of a Unified Command must begin with an initial meeting of the agency level Incident Commanders and their respective Command and General Staffs from each of the involved jurisdictions and agencies.

During this meeting, which should be brief, the agency-level Incident Commanders must come to consensus on priorities, a collective set of incident objectives, an overall strategy, and selection of a Unified Command spokesperson before they can effectively work together to carry out the response.

The initial Unified Command Meeting also will provide an opportunity for the incident commanders to establish a Joint Information Center, as needed. In addition, if not established in preplanning activities, the agency level Incident Commanders must use the initial Unified Command Meeting as an opportunity to determine the appropriate roles and responsibilities of all representatives involved in the incident. This conversation will help establish the membership of the Unified Command.

Effective planning can facilitate assembly and conduct of the initial Unified Command Meeting. The responsibilities discussed above should be preplanned to the greatest extent possible. Although an initial meeting is critical for ensuring the effective integration of all responders into the Unified Command structure, the steps involved in the meeting may have to be revisited periodically as information on the incident or its demands changes. These meetings will provide a private opportunity for the agency Incident Commanders to speak openly and honestly about their priorities, considerations, and concerns. However, once participants in the Unified Command leave this meeting, they must speak with one voice. These are the meeting steps:

- Step 1: Set priorities and objectives
- Step 2: Present considerations
- Step 3: Develop a collective set of incident objectives
- Step 4: Adopt an overall strategy
- Step 5: Select a Unified Command spokesperson

Set Priorities and Objectives

For a Unified Command to work, each participant must be committed to working together to solve a common problem. Each responding agency will have individual objectives to carry out. In addition, the primary objectives of each responding agency are established under the national response priorities, which are to:

- Preserve the safety of human life
- Stabilize the situation to prevent the event from worsening
- Use all necessary containment and removal tactics in a coordinated manner to ensure a timely, effective response that minimizes adverse impacts to the environment and preserves property
- Where possible, address all three of these priorities concurrently

However, each responding entity will likely have other significant priorities requiring consideration, which might include the following factors:

- Maintaining business survival
- Minimizing response costs
- Maintaining or improving public image
- Minimizing economic or tourism impacts
- Minimizing environmental impacts
- Evaluating prospects of criminal prosecution
- Meeting certain reasonable stakeholder expectations

Understanding all the issues facing the Unified Command participants is important in any negotiation. Because consensus must be reached for the Unified Command to be effective, it is critical that the Unified Command engage in coordination whenever necessary.

Present Considerations

At the onset of the initial Unified Command Meeting, Unified Command members have an obligation to raise and discuss honestly what each response organization can provide in terms of authorities, equipment, skills, and experience, including their response capabilities. All agency-level Incident Commanders must be free to speak openly with the other members of the Unified Command about their constraints or limitations, whether practical or political in nature, because these constraints may have an impact on how the incident objectives can best be achieved.

Develop a Collective Set of Incident Objectives

The planning process for the Unified Command is similar to that used for a single jurisdiction or agency incident. However, because each agency will bring its own set of objectives and considerations to the response, the Unified Command must decide upon a collective set of incident-specific objectives to identify what the Unified Command as a whole needs to accomplish before an overall response strategy can be developed. To be effective, these objectives should be specific, measurable, assignable, reasonable, and time-related. The Unified Command must come to consensus on a set of general objectives that can then be documented to provide focus for the response organization. This process includes establishing and agreeing upon acceptable priorities.

Adopt an Overall Strategy

Strategy is the development of policies and plans to achieve the objectives of a response. If the Unified Command knows exactly how to accomplish an objective, it should specify the strategy. Because there are frequently multiple strategies that would accomplish the same objective, the Unified Command staff will often ask the plans section to recommend strategies for later Unified Command approval. This allows for better input and discussion from the responders and also reduces meeting time for the Incident Commanders.

Select a Unified Command Spokesperson

Frequently, the Unified Command will establish a Joint Information Center and designate a single spokesperson. The spokesperson is typically a member of the Unified Command and serves as a point of contact and a single voice for the members of the Incident Management Team at external and internal briefings. The spokesperson may change during the course of an incident as the situation develops. For example,

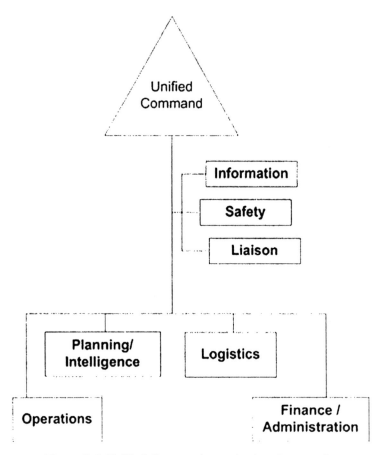

Figure 9.6 Unified Command organization chart overlay

a different agency may designate a spokesperson if it has more expertise in a particular area at a certain time. In addition, different departments within the same agency could designate a spokesperson at different times during the same incident, as appropriate.

CONCLUSION

The use of Incident Management Teams and Unified Command structures will make the job of Incident Commanders easier during complex, large-scale, or long-duration incidents. Overall, it will make the response to such potentially catastrophic events run more smoothly and the efforts more valuable in both the long and short run. Ultimately it will save lives.

Appendix A

INCIDENT COMMAND POST SYSTEMS POSITION DESCRIPTION CHECKLISTS

This series of checklists was originally created by Mr. David Nock, Program Manager at the National Emergency Rescue and Response Training Center's (NERRTC*) Weapons of Mass Destruction (WMD) Enhanced Incident Management/Unified Command (E IM/UC) Simulations Course, developed under a grant from the United States Department of Homeland Security (DHS).

The intended purpose of the checklists in that course for these checklists was to assist a person in that course that may not be familiar with the duties of a specific given position in the operation of an Incident Command Post (ICP) while they were expected to be well versed in the basic to concepts and tenets of Incident management System (IMS) and the operations of an fully staffed and integrated ICP. The simulation in that course was designed to create a virtual simulation of a significant WMD-type event and simulate a fully staffed and functioning Incident Command Post (ICP).

In the above use it became apparent to the author that this type of tool placed into a field usable format (padded, heavy card stock, laminated, etc.) could be helpful in larger scale incidents where a person who was familiar with the basic concepts and theories of the IMS and ICP, but not with all of the intended functions of a specific positions within the system, may be forced to work outside their own sphere of understanding.

They are intended to serve as illustrations of the concept of the idea of a field checklist as opposed to a "be all, end all" solution. These checklists, "or cheat sheets" can assist the responder and their supervisors in establishing an ICP even in less than

---

*The National Emergency Rescue and Response Training Center (NERRTC) is a division of the Texas Engineering Extension Service (TEEX) and is a part of the Texas A & M University Systems (TAMUS). Information on their incident management and other WMD training programs can be found at www.teex.com.
**Not all IMS positions are described here in these checklists as they are meant to be representative of the concept rather than the actual use.

ideal staffing scenarios. The checklists are designed to be flexible and can be modified to deal with specific situations and operational guidelines or procedures of any agency or jurisdiction. Agencies and jurisdictions are encouraged to use these checklists as a guide for the development of their own such tools more specific to their agency.

Incident Commander
Liaison Officer
Public Information Officer
Incident Safety Officer
Operations Section Chief
Staging Area Manager
Public Works Construction/Engineering and Debris Unit Leader
Public Health
Plans & Intelligence Section Chief
Situation Unit Leader
Resources Status Unit Leader
Documentation Unit Leader
Technical Specialists
Demobilization Unit Leader
Logistics Section Chief
Communications Officer
Security Officer
Administrate and Finance Section Chief

Incident Commander Checklist

Position Responsibility: The Incident Commander is responsible for all critical incident activities, including; operations, planning, logistics, finance, investigations, and staging.

Note: In the temporary absence or incapacitation of the Incident Commander, a deputy Incident Commander shall assume the responsibilities. In the absence of a Deputy Incident Command the transfer of command falls to the Operations Section Chief Functionary.

DUTY CHECKLIST: [Enter exact time in boxes below]

Task	Description	Completed
	Established the Command Post	
	• Strategically locate the command post near enough to command operations, far enough back to not be harmed if further problems develop	
	• Maintain centralized and systematized communication communication	
	• Logistically monitor and regulate personnel and equipment needs	
	• Acquire, analyze and properly distribute intelligence and other relevant information	
	• Coordinate activities with associated agencies	
	Assume command. Document the exact time and, if applicable, whom command was transferred from.	
	Command post location selection considerations	
	• Accessible to responders	
	• Defensible	
	• Sufficient space	
	• Restroom facilities	
	• Communications capabilities. Consider access to telephones, radio, television, microwave etc.	
	• Electricity and water	
	• Facility, structure for briefings, staff work space, protection from the weather	
	• Storage space, parking etc for vehicles and equipment, etc.	
	• Suggested Options, Schools, Universities, Parks, Large parking lots, water/power facilities, Churches, Military Armory	
	Appoint an Operations Officer	
	If the jurisdiction has established an EOC. Insure open and constant communications is available and operational	

Task	*Description*	*Completed*
	Contact City, County or Jurisdiction Office of Emergency Preparedness or the local Emergency Manager or Coordinator	
	Set up a Staging Area. Assign a Staging Area manager as required	
	Notify communication. Establish an uninterrupted communications and command link with responding units and ocal Headquarters or EOC	
	Assess available resources and request assistance as needed	
	Incident Response Considerations	
	Rescue Operations	Shelters	Public Information	
	Medical	Security	Public Warnings	
	Evacuations	Incident Perimeters	Material Hazards	
	Law Enforcement	Fire suppression	Staffing Requirements	
	Implement an action plan to include all necessary tactical strategies	
	Assign functions as necessary (Operations, Planning, Logistics, Admin and Finance, Public Information, Safety, Liaison, staging, communications, etc.	
	Assign a Communications Officer. Establish a dedicated communications link with HQ and or the EOC.	
	Assign a Security Officer for the Command Post	
	Assign a Field Media Representative, PIO. Authorize releases of information to the public.	
	Coordination with jurisdiction emergency management staff and elected officials for disaster declaration	
	Determine Plans, Reports, or Records to be maintained. Consider the following	
	• Incident Action Plan	
	• Status Reports required	
	• Incident site damage survey	
	• Routine Briefings	
	• Situational Staff Updates	
	• Shift Change Briefing Requirements	
	• Press Releases	
	• Shelter Requirements or plans	
	• Evacuation Requirements or plans	
	• State and Federal Information Requirements	

Task	Description			Completed
	• Jurisdictional Financial Reimbursement Data and Records Requirements			
	• Elected Officials Information Requirements			
	• Safety Alerts or messages			
	Monitor, coordinate and manage all incident activities			
	Keep a log of activities, to include events, times and personnel involved			
	Request and receive scheduled or periodic reports and briefings			
	Approve requests for additional personnel or resources			
	Approve requests for volunteers			
	Establish liaison with other responding agencies			
	Collect and maintain records on jurisdiction costs associated with the response and recovery efforts			
	Communicate with District Attorney on criminal matters			
	Keep Community, Jurisdiction elected leadership, Headquarters and or EOC Office Staff informed on all aspects of the incident			
	Approve plans for demobilization, release of personnel, equipment or resources			
	Collect all reports and logs			
	Prepare After Action Report			
	Other Points			
	• Staff to consider			
	Operations Section	Plans Section including an intelligence officer	Logistics Section	
	Admin and Finance Section	Public Information and Media Officer/ Advisor	Safety Officer	
	ICP Security staff	Communications Staff	Command Post Scribe, Recorder, Aide	
	Fire and Rescue Specialist	Law Enforcement	Public Works	
	Public Health	Mass Care including Evacuation and Shelter	Emergency Medical Services	
	Local Medical and Hospital Staff	Hazards Materials	Mutual Aid Jurisdictions	
	• Communications equipment and frequencies to use			
	• Consider requesting the recording of local TV and radio			
	• Maintaining enough phone line for the ICP staff as it grows to meet response requirements			
	• Establish demobilization criteria			

Liaison Officer Checklist

Position Responsibility: The Liaison Officer is the point of contact for any assisting agencies, to include fire, EMS, Red Cross, law enforcement, etc. Establishes and maintains a central location for all incoming agency representatives. Facilitates in the coordination of Private organizations and volunteer groups.

Note: Absent a delegated LNO the Incident Commander is responsible for this function. The Incident Commander is responsible for all critical incident activities, including the final authority on all aspects of operations, logistics, finance, investigations and staging.

DUTY CHECKLIST: [Enter exact time in boxes below]

Task	Description	Completed
	Report to the Incident Commander for briefing. Document exact time of arrival.	
	Activate, organize and brief any staff if needed	
	Keep a log of activities, including events and names of personnel involved	
	Identify assisting agency/jurisdictional representatives and contact them	
	Provide a point of contact or staging area for other agency representatives to meet	
	Integrate all responding agencies into compatible ICS positions	
	Establish uninterrupted communications links with other responding agencies	
	Maintain a log of all arrival and departure times for outside agencies	
	Coordinate with the Public Information Officer	
	Monitor the involvement of any outside agency assiting in the ncident	
	Confer with the Incident Commander on requests for mutual aid	
	Maintain a log of names and agencies responding to assist (This could be collected later)	
	Integrate assisting agencies and current jurisdiction into compatible functions if necessary	
	Maintain continual communications link with other agencies	
	Monitor the critical incident and the involvement of other agencies	
	Meet with the Incident Commander on all requests for mutual aid	
	Notify any relief personnel of the current status of the critical incident	
	Demobilize when ordered	
	Turn over all records and logs to the Incident Commander	
	Debrief any assistants	

Public Information Officer Checklist

Position Responsibility: The Public Information Officer (PIO) is the point of contact for any media information, news activities, media briefings and information pertaining to the incident that is offered to the citizens and public. The PIO drafts all news releases, obtains release authority of information and maintains media news related data of the incident. The PIO is part of the ICP command staff and reports to the incident commander.

Note: Absent a delegated PIO the Incident Commander is responsible for this function. The Incident Commander is responsible for all critical incident activities, including the final authority on all aspects of operations, logistics, finance, investigations and staging. In some jurisdictions this position is consolidated at the EOC.

DUTY CHECKLIST: [Enter exact time in boxes below]

Task	Description	Completed
	Report to the Incident Commander for briefing. Document exact time of arrival.	
	Activate, organize and brief any staff if needed	
	Keep a log of activities, including events and names of personnel involved	
	Obtain media or news constraints (policy guidance) on the release of information by the incident commander	
	Obtain approval for release of information from the IC. The PIO may receive blanket release authority from the IC.	
	Establish a single incident information center whenever possible. Establish a Media Information Center or Joint Information Office/Center providing the necessary space, materials, telephones, etc.	
	Contact and Coordinate with EOC and State level PIOs (all other PIOs)	
	Review the Incident Action Plan	
	Obtain a copy of all Situation Status Reports	
	Develop content for Emergency Alert Systems (EAS) releases if available. Monitor EAS releases as necessary.	
	Identify assisting, responding or supporting agencies and jurisdictional PIO representatives	
	Establish a media briefing area (conference space, etc.) or information center where it will not interfere with command post operations	
	Ensure that the public within the affected area receives complete accurate and consistent information about life safety procedures, public health advisories, relief and assistance programs and other vital information. In coordination with all effected or responding PIOs provide advisories and public instructions for life safety, health, and assistance	

Task	*Description*	*Completed*
	Act as the rumor control officer to correct false and erroneous information	
	Provide appropriate staffing and telephones to efficiently handle incoming media and public calls	
	Consider a 1-800 number or local number to provide a recorded message to the public (Disaster Hotline)	
	Arrange necessary work space, materials telephones and staffing	
	Prepare, update and distribute to the public a Disaster Assistance Information Directory which contains locations to obtain food, shelter, supplies, health services, etc.	
	Consider announcements and emergency information translation to other local languages and the hearing impaired	
	Monitor news broadcasts	
	Obtain all required forms (maintain and update as needed)	
	Release news to media and post information in command post and other appropriate locations	
	Attend meetings to update news releases and public notification activities	
	Arrange/Coordinate meetings between media and incident personnel	
	Provide escort service to media and VIPs	
	Provide protective clothing to visitors as required	
	Respond to special requests for information	
	Develop the format for press conferences or media briefings	
	Develop a media briefing schedule (press conferences) to keep the media and the public informed. Schedule them at a regular interval. Don't conduct press conferences with subject matter experts (SMEs) being available to discuss technical issues	
	Ensure that file copies of all news releases are maintained	
	Maintain a positive relationship with media representatives	
	Things to have in place prior to the incident:	
	• Designate a PIO in each emergency relief organization. The PIO duties could be part of a person's regular emergency relief duties. Having a designated PIO on-staff provides reporters with a central spokesperson in everyday and emergency situations. • Organize a "what if" brainstorming session with others in your office. Come up with "what if" scenarios about potential crisis and disaster situations. Determine steps on how you would respond to the "what if" crises. • Have a crisis communication/emergency communication plan before a disaster strikes. With an emergency communication plan in place, PIOs will be able to respond and perform in a	

:	*Description*	*Completed*
	• proactive stance, as opposed to a reactive mode, thus better controlling the information and news coverage in a disaster.	
	• Select disaster/crisis communication teams. Who is responsible for communicating with the media during a crisis? Who fields telephone calls? Who makes decisions about what to say to the media? Everyone in your office should know who are on the crisis communication and crisis management teams.	
	• Provide all PIOs with communications-related training opportunities. Emphasize topics PIOs believe to be important when communicating with the media. It is not enough to have a designated PIO on staff; that PIO should be properly trained in communication methods. (Many PIOs are volunteers.)	
	• Initiate World Wide Web page development training for PIOs or a designated person on staff. Reporters and the general public are becoming more adamant about getting almost immediate, online information. Firefighter PIOs did not recognize the need for online information to be strong; however, reporters in the study said otherwise. As much emergency information as possible should be made available on the Web.	
	During the Incident	
	• Gather and classify information into categories, such as facts and rumors. Facts should be routinely updated; rumors should be verified or exposed as myths.	
	• Cater to local media before national media. Local reporters will provide immediate, important information to area constituents.	
	• Remember newspaper reporters have information needs. The immediacy of television and radio coverage may have caused PIOs to provide more resources to television reporters and video photographers. However, newspaper reporters' information and photography needs also should be provided.	
	• Consider "media pool coverage," especially of video footage, and/or media tours to disaster-damaged areas. This should be a standard feature at all emergency command center sites and not change from site to site.	
	• Be accessible or designate someone to be accessible to the media at all times. Reporters should have a contact person's telephone number, cellular phone number, fax number and electronic mail address for around-the-clock contact.	
	• Provide necessary resources (cellular phones, laptop computers) to PIOs in the field.	
	• Provide other automated services, such as a 24-hour telephone hotline, for the public to use for emergency updates.	

Task	*Description*	*Completed*
	• Get the facts. Miscommunication heightens during a crisis and can be exaggerated by half-truths, distortions, or negative perceptions. Get to the heart of the real story and tell it.	
	• Take the offensive when a serious matter occurs. Be active, not reactive. Tell it all and tell it fast.	
	• Deal with rumors swiftly. Tell only the truth about what you know to be fact. Do not repeat others opinions, hearsay, or possibilities.	
	• Centralize information. Designate one spokesperson. A central spokesperson provides a singular "face" for the reporters. Viewers begin to become familiar with a central spokesperson, so this is one way to begin building credibility with the organization, if the person comes across as trustworthy. Centralized information also will minimize miscommunication.	
	• Don't get mad. Don't get mad. Don't get mad. Keep your cool in an interview or news conference with reporters. Some of their questions may be hostile, and some questions and comments may seem to be a personal attack to you, but remember that they are trying to get information on a crisis-oriented story that may have widespread impact to their audiences. So don't get mad when you are asked the "hard" questions.	
	• Stay "on the record" in all interviews. Do not go "off the record." Any comment worth saying should be said "on th record." If you go "off the record," be ready to read it in print the next day. Is this unethical for reporters to report "off the record" comments? Sure, but anything can, may, and will be done to advance a story. You should not be lured into going "off the record" under any circumstance.	
	• No "no comments." Try to have an answer for reporters' questions. But if you don't have an answer, don't be afraid to say, "I don't know, but I'll find out." Saying "no comment" instead, appears to television news viewers and newspaper readers that you have something to hide.	
	• Write everything down. Maintain a crisis communication inventory of what was said by whom and at what time. This way, you will have a record of the event and how it was communicated. You can evaluate your responses so you will be better prepared if another crisis happens in the future.	
	Ensure that the IC obtains a copy of news releases. Post a copy of all news releases in the command post.	
	Develop a shift change briefing format for the PIO	
	Demobilize when ordered	
	Turn over all records and logs to the Incident Commander	

Task	Description	Completed
	Debrief any assistants	
	After the Incident Don't just sit back and do nothing; you won't be ready for the next disaster or crisis! It is time to evaluate how you handled the crisis. Your review should include the following: • A review of why the crisis or disaster occurred. Could you have done anything to prevent it? • An evaluation of how the crisis was handled and communicated You may want to use the crisis communication inventory you maintained to evaluate how communication was handled. Was information disseminated through one spokesperson? Did miscommunication occur? • An examination of similar scenarios. What would you do in a similar situation in the future? What did others do in similar situations?	

Safety Officer Checklist

Position Responsibility: The Incident Safety officer is the point of contact for any safety related issue that pertains to the emergency responders within the incident site or under the control of the Incident Commander. The Incident Safety Officer keeps the IC aware of any safety issues that are developing. The Incident Safety Officer is part of the ICP command staff and reports to the incident commander.

Note: Absent a delegated ISO the Incident Commander is responsible for this function. The Incident Commander is responsible for all critical incident activities, including the final authority on all aspects of operations, logistics, finance, investigations and staging. In some jurisdictions this position is consolidated at the EOC.

DUTY CHECKLIST: [Enter exact time in boxes below]

Task	Description	Completed
	Obtain Briefing from IC	
	Identify hazardous situations associated with the incident	
	Participate in planning meetings	
	Review Incident Actions Plans	
	Exercise emergency authority to stop and prevent unsafe acts	
	Investigate accidents that have occurred within incident areas	
	Tour all response facilities and locations	
	Determine any fire response requirements at command post facilities such as fire extinguishers evacuation routes, etc.	
	Prepare and present safety briefings for command and general staff	
	Consider the effects of the incident, how it started, etc and help prepare any possible follow on effects (as an example: earthquake – aftershock, explosive device – any secondary device, chemical or biological agent – required protective suits for responders in the hot zone)	
	Assist with the development of hot, warm and cold zone marking and procedures	
	Maintain log	
	Demobilize when ordered	
	Turn over all records and logs to the Incident Commander	
	Debrief any assistants	

Operations Section Chief Checklist

Position Responsibility: The Operations Office is responsible for the management of all uniformed operations applicable to the critical incident. Coordinates plans, requests resources, changes plans as necessary and reports to the Incident Commander.

Note: In the absence of an assigned Operations Officer, the duties shall be completed by the Incident Commander. The Incident Commander is responsible for all critical incident activities, including the final authority on all aspects of operations, logistics, finance, investigations and staging

DUTY CHECKLIST: [Enter exact time in boxes below]

Task	Description		Completed
	Report to the Incident Commander and get a briefing		
	Activate and brief your operational teams and uniformed personnel as necessary		
	Activate Branches, Divisions, Teams as required		
	Consider the following Branches:		
	Fire & Rescue Branch	Mass Care Branch	
	Law Enforcement Branch		
	Consider establishing the following units or groups:		
	Staging Area Manager	Coroner	
	EMS & Medical	HAZMAT & Decon	
	Fire & Rescue	Shelter	
	Law Enforcement	Public Works, Construction/ Engineer& Debris Removal	
	Public Health	Evacuation	
	Determine need for mutual aid:		
	Fire	Law Enforcement	Hazmat & Decon	
	Shelter	Public Works	Public Health	
	Determine need for 24 hour operations		
	Determine need and requirements for shift change briefing		
	Determine operational periods in consultation with the IC, and Chiefs of Logistics and Admin/Finance		
	Determine need for outside agencies representation with liaison officer		
	In coordination with Logistics, consider communications needs		
	Be Proactive, Think ahead		
	Keep IC and EOC (if activated) advised of your location and status of the incident		
	Keep command staff and section chiefs advised of status of the incident		

Task	*Description*	*Completed*
	Develop format and facilitate the shift change informational exchange within the ICP staff	
	Supervise operations	
	Ensure coordination sessions are maintained with the Planning Section	
	Keep a log of activities and observations	
	Advise the Incident Commander on operational strategy and tactics, Report information, special activities and events as needed	
	Draft the Incident Action Plan (IAP), develop the operations of the IAP and ensure it is presented to the command and general ICP staff after approval by the IC	
	Review tactical plans from special team leaders	
	Ensure tactical intervention is carried out after approval is given by the IC	
	Approve requests for response resources and requests for release of resources	
	Maintain contact with resources in the staging area via Staging Area Manager	
	Establish a traffic management plan	
	Fire and Rescue Considerations	
	• Prepare the objectives of the Fire & Rescue Branch • Coordinate mobilization and transportation of resources through plans and logistics sections	
	Activate units to consider:	

	Fire & Rescue	Hazmat & Decon	Disaster Medical	
	Search & Rescue			

	• Keep Operations Section Chief and Plans/Intelligence Section up to date with overall summary of operations period • Ensure fiscal and administrative requirements are forwarded to admin/finance section • Coordinate with LE unit for perimeter maintenance/security, investigation and evacuation issues • Coordinate with Mass Care for shelter and evacuation operations • Determine need for Fire and Rescue mutual aid • Ensure communications with Responding Emergency Disciplines in coordination with the Logistics Section	
	Law Enforcement Considerations	
	• Prepare the objectives of the Law Enforcement Branch	

Task	Description			Completed
	Consider activating :			
	Law Enforcement	Perimeter Security	Incident Investigations	
	Evacuation	Traffic		
	• Determine need for Law Enforcement mutual aid • Ensure fiscal and administrative needs are provided to admin/finance section • Ensure communications with all responding disciplines • In conjunction with Plans/Intel Section determine impact of current and forecasted weather on LE Operations • Coordinate with Mass Care Branch for shelter and evacuations operations • Help ensure proper media coordination with the ICP PIO • Provide relief shift change brief • With Incident Investigations Unit, gather crime scene intelligence, incident evidence information • Establish and maintain contact with DA • Ensure evidence preservation • Establish and maintain contact with FBI, ATF and INS as required			
	Public Health Considerations			
	• Prepare the objectives of the Public Health Unit			
	Consider activating:			
	Fatality Management	Medical Surveillance	Medical Diagnosis, Prophylaxis	
	Epidemiological Investigation	Regional Health Organizations Liaison	Medical Hazard Assessment and Mitigation	
	Critical Incident Stress Management (CISM)	Family Care Center	Coroner or Medical Examiner Support	
	• Alert hospitals and confirm actions taken by hospitals • Advise EMS of situation and advisories • Issue guidance to hospitals and clinics including patient care guidelines, signs and symptoms • Issue public health alert, through the PIO, to warn the public of signs, symptoms and self-care, including shelter guidance (shelter-in-place) • Facilitate request of additional public health supplies including the National Pharmaceutical Stockpile (NPS) if needed.			

Task	Description	Completed
	• Initiate case finding and tracking • Initiate CISM for patients, staff, responders, families, etc. • Patient decon and triage support for hospitals	
	Maintain log	
	Demobilize on command from the Incident Commander	
	Debrief all assigned personnel	
	Forward all logs and reports to the Incident Commander	

Staging Area Manager, Operations Section Checklist

Position Responsibility: The Staging Officer is responsible for managing and organizing all staging areas for all disciplines of response to a critical incident. Staging areas are maintained for collection of all resources until they are allocated to a given assignment.

Note: The Incident Commander or Operations Chief performs this function in the absence of an assigned Staging Functionary

DUTY CHECKLIST: [Enter exact time in boxes below]

Task	Description	Completed
	Report to the Incident Commander and or the Operations Chief for briefing and assignment	
	Document exact time of arrival	
	Activate and brief any staff	
	Establish an uninterrupted communications link with the Incident Commander, Operations Section and EOC.	
	Identify locations and establish staging areas: (Keep the ICP EOC Advised)	
	Staging Area is located away from any threats	
	Readily identifiable location	
	Arranged so that staged resources can be easily dispersed.	
	Determine any support needs for equipment, feeding, sanitation and security	
	Set up uniformed security at all staging areas	
	Establish check-in and check-out procedures as required	
	Establish traffic patterns into and away from staging areas.	
	Advise the ICP and EOC (if established) when staging areas are ready	
	Maintain a log of activities to include arrival times of resources	
	Organize the release of resources on demand from any functionary	
	Determine any support needs for staging areas, to include, sanitation and food.	
	Obtain and issues receipts for equipment issued from the staging area to responding resources	
	Request maintenance service for equipment in staging as required	
	Establish a media staging point and coordinate with the PIO as needed	
	Establish a staging point for volunteers	
	Establish a Reception Center for victims and families	
	Establish a Relocation Center for evacuated persons.	

Task	Description	Completed
	Establish a field mortuary in conjunction with the Coroner/Medical Examiner	
	Maintain log	
	Demobilize on command from the Incident Commander	
	Debrief all assigned personnel	
	Forward all logs and reports to the Incident Commander	

Public Works Construction/Engineering and Debris Unit, Operations Section Checklist

Position Responsibility: Coordinate all public works operations, survey utility systems and infrastructure including roads and bridges, restore disrupted systems, assess damage to facilities, according to the Incident Action Plan (IAP).

Note: The Incident Commander or Operations Chief performs this function in the absence of an assigned Public Works Functionary.

DUTY CHECKLIST: [Enter exact time in boxes below]

Task	Description	Completed
	Report to the Incident Commander and or the Operations Chief for briefing and assignment	
	Document exact time of arrival	
	Activate and brief any staff	
	Prepare the objectives of the Public Works Unit	
	Consider and advise on the need for Public Works mutual aid. Determine response needs	
	Support Law Enforcement with barrier material, fencing, road block, etc. to create manage the incident perimeter	
	Assist in the current recovery operations and development of future operational period plans	
	Provide jurisdictional public works response including electrical, water, waste water, drainage/ runoff issues, etc.	
	Assist the safety officer in engineering, public works issues	
	Maintain log	
	Demobilize on command from the Incident Commander	
	Debrief all assigned personnel	
	Forward all logs and reports to the Incident Commander	

Public Health Group Supervisor Checklist

Position Responsibility: Coordinate all health operations that impact the public health, according to the Incident Action Plan (IAP).

Note: The Incident Commander or Operations Chief performs this function in the absence of an assigned Public Health Functionary.

DUTY CHECKLIST: [Enter exact time in boxes below]

Task	Description			Completed
	Report to the Incident Commander and or the Operations Chief for briefing and assignment			
	Document exact time of arrival			
	Activate and brief any staff			
	Prepare the objectives of the Public Health Unit			
	Consider activating:			
	Fatality Management	Medical Surveillance	Medical Diagnosis, Prophylaxis	
	Epidemiological Investigation	Regional Health Organizations Liaison	Medical Hazard Assessment and Mitigation	
	Critical Incident Stress Management (CISM)	Family Care Center	Coroner or Medical Examiner Support	
	Alert hospitals and confirm actions taken by hospitals			
	Advise EMS			
	Issue guidance to hospitals and clinics including patient care guidelines, signs and symptoms			
	Issue public health alert, through the PIO, to warn the public of signs, symptoms and self-care, including shelter guidance (shelter-in-place)			
	Facilitate request of additional public health supplies including the National Pharmaceutical Stockpile (NPS) if needed.			
	Initiate case finding and tracking			
	Initiate CISM for patients, staff, responders, families etc			
	Patient decon and triage support for hospitals			
	Maintain log			
	Demobilize on command from the Incident Commander			
	Debrief all assigned personnel			
	Forward all logs and reports to the Incident Commander			

Plans & Intelligence Section Chief Checklist

Position Responsibility: Establish the planning function for incident management as required plans including: collecting, analyzing, displaying situational information; preparing situation reports; preparing and distributing the Incident Action Plan (IAP) and facilitating the plan development; conduct advance incident response planning activities; provide technical support services to all Command Post sections, branches, etc; and documentation of incident actions and maintenance of files (historical record) on ICP activities.

Note: The Incident Commander performs this function in the absence of an assigned Planning Functionary.

DUTY CHECKLIST: [Enter exact time in boxes below]

Task	Description		Completed
	Report to the Incident Commander for briefing and assignment		
	Document exact time of arrival		
	Activate and brief any staff		
	Prepare the objectives of the Plans Section		
	Establish and supervise the appropriate level of organization for the Planning & Intelligence Section		
	Activate branches and units as needed		
	Exercise overall responsibility for the coordination of branch/ unit activities within the section		
	Keep the IC and Operations Section Chief informed of significant issues affecting planning and intelligence		
	In coordination with all sections and branches within the Command Post ensure that status reports are completed, updated as required and used to complete the IAP		
	Ensure the section is properly set up and equipped with supplies, maps status boards, etc.		
	Establish information requirements and reporting schedules for all organizational elements for use in preparing planning documents and incident records		
	Consider activating: (as required)		
	Situation Unit	Demobilization Unit	
	Documentation Unit	Technical Specialists (as required)	
		Resources Status Unit	
	Request staffing for 24 hour operations		
	Establish contact with the Operations Section, meet with the Operations Section Chief, review any incident reports, coordinate status reports		
	Review responsibilities of branches in section; develop plans procedures, etc. to ensure all responsibilities are addressed		

Task	Description	Completed
	Perform operational planning for the Planning Section	
	Maintain a list of key issues (the ICs incident response priorities)	
	Anticipate situations, plan ahead	
	Provide periodic predictions on incident potential to the staff	
	Ensure section logs and other necessary incident files are maintained	
	Ensure the Situation Unit is maintaining current information for the status report. The Situation Unit maintains the current status boards for the command post. Ensure the section is planning the next operational period and maintaining/updating the IAP. Ensure the section is planning, developing and distributing a report which highlights forecasted events or conditions likely to occur beyond the forthcoming operational period; particularly those situations which may influence the overall strategic objectives of the jurisdiction EOC.	
	Ensure the Documentation Unit is collecting and maintaining information pertaining to the incident for auditing purposes, jurisdiction reimbursement and capturing of lessons learned. Ensure that the Documentation Unit maintains files on all ICP activities and provides reproduction and archiving services for the EOC, as required.	
	Ensure the Technical Specialists are providing the subject matter expert guidance to all branches and sections within the Command Post. The Specialists also support the specifics technical aspects of the IAP. The Technical Specialists should include all required expertise (SMEs) from all emergency disciplines required to respond to the incident. Identify the need for specialized response or recovery resources.	
	Ensure the Resources Status Unit is accounting for all personnel and equipment responding to the incident including accounting for names and locations of assigned personnel.	
	Ensure that major incident reports, section and branch status reports are completed by each section and are accessible by the Plans Section for planning requirements and incident documentation	
	Establish a weather data collection system as necessary	
	Long Range Planning Considerations: • Establish contact with all ICP Section Chiefs to determine the best estimates of future direction and outcomes of the incident. • Develop an advanced plan identifying future policy related issues, social and economic impacts, significant response or recovery resources needs, and any other key issues likely to affect the incident or the jurisdiction within the next 36-72 hours.	

Task	Description	Completed
	• Maintain a list of key issues (the ICs incident response priorities) • Anticipate situations, plan ahead • Submit advance plans to the Plans Section Chief for review and approval prior to conducting briefings for the staff • Review action planning objectives submitted by each section for the upcoming operational period. I coordination with the general staff and the section chiefs recommend a transition strategy to the IC for long term response (if needed),recovery operations and jurisdictional recovery operations at the EOC level	
	Provide incident traffic plan	
	Ensure that a situation report is produced and distributed to all command post sections, all other active command posts and the jurisdiction EOC, at least once per operational period.	
	Assemble and disassemble strike teams, task forces etc based on the next operational period plan in coordination with the Logistics Section	
	Ensure that all status boards and other displays are current and posted information is neat and legible.	
	Ensure that the Public Information Office (PIO) has immediate and unlimited access to all status reports and displays	
	Conduct periodic briefings with section staff and work to reach consensus among staff on section objectives for up coming operational periods	
	Facilitate the ICs incident action planning meetings prior to the end of the operational period	
	Ensure that objectives for each section are completed , collected and posted in preparation for the next Action Planning meeting.	
	Ensure that the ICP Action Plan is completed and distributed prior to the start of the next operational period.	
	Work closely with each branch/unit within the Planning/ Intelligence Section to ensure the section objectives, as defined in the current IAP are being addressed.	
	Provide technical specialists, such as energy, chemical, health, etc. advisors and other technical specialists to all ICP sections as required.	
	Ensure that fiscal and administrative requirements are coordinated through the Finance/Administration Section.	
	Maintain the log for the section	
	Demobilize on command from the Incident Commander	
	Debrief all assigned personnel	
	Forward all logs and reports to the Incident Commander	

Situation Unit Leader Checklist

Position Responsibility: Establish the Situational Analysis Unit or Branch as part of the planning function for incident management as required including: collecting, organization and analyzing of disaster situation information. Display situational information, prepare situation reports as required, prepare and distribute the Incident Action Plan (IAP) and facilitate the development of the plan. Conduct advance incident response planning activities as required.

Note: The Incident Commander and or the Plans/Intelligence Section Chief perform this function in the absence of an assigned Situational Analysis Functionary.

DUTY CHECKLIST: [Enter exact time in boxes below]

Task	Description	Completed
	Report to the Planning and Intelligence Section Chief for briefing and assignment	
	Document exact time of arrival	
	Activate and brief any staff	
	Prepare the objectives of the Situational Analysis Unit	
	Establish and supervise the appropriate level of organization for the Unit	
	Activate subunits as needed	
	Exercise overall responsibility for the coordination of branch/ unit activities within the section	
	Keep the IC and Operations Section Chief informed of significant issues affecting planning and intelligence	
	In coordination with all sections and branches within the Command Post ensure that status reports are completed, updated as required and used to complete the IAP	
	Ensure the section is properly set up and equipped with supplies, maps status boards, etc.	
	Request staffing for 24 hour operations	
	Establish contact with the Operations Section, meet with the Operations Section Chief, review any incident reports, coordinate status reports	
	Oversee the preparation and distribution of situation statue reports. Coordination with the Documentation Unit as required.	
	Maintain a list of key issues (the ICs incident response priorities)	
	Anticipate situations, plan ahead	
	Ensure unit/branch logs and other necessary incident files are maintained	
	Ensure the Situation Unit is maintaining current information for the status report. The Situation Unit maintains the current incident status board within the command post	

Task	Description	Completed
	Assist the Documentation Unit in collecting and maintaining information pertaining to the incident for auditing purposes, jurisdiction reimbursement and capturing of lessons learned.	
	Ensure that major incident reports, section and branch status reports are completed by each section and are accessible by the Plans Section for planning requirements and incident documentation	
	Ensure that a situation report is produced and distributed to all command post sections, all other active command posts and the jurisdiction EOC, at least once per operational period.	
	Ensure that all status boards and other displays are current and posted information is neat and legible. Ensure that adequate staff is assigned to maintain maps, status boards and other displays.	
	Ensure that the Public Information Office has immediate and unlimited access to all status reports and displays. Meet with the Public Information Office (PIO) to determine the best method for ensuring access to current information.	
	Conduct periodic briefings with section staff and work to reach consensus among staff on section objectives for up coming operational periods	
	Facilitate the ICs incident action planning meetings prior to the end of the operational period. Prepare a situation summary for the ICs Incident Action Planning meeting. Ensure each ICP section provides their objectives at least 30 minutes prior to each IAP planning meeting. Ensure that objectives for each section are completed, collected and posted in preparation for the next IAP planning meeting. In preparation for the IAP planning meeting, ensure that all current ICP objectives are clearly posted, and that the meeting room is set up with appropriate equipment and materials (easels, markers, reports, etc.)	
	Ensure that the Documentation Unit publishes and distributes the IAP prior to the beginning of the next operational period.	
	Ensure that all collected information is validated prior to posting	
	Work closely with each branch/unit within the Planning/Intelligence Section to ensure the section objectives, as defined in the current IAP are being addressed.	
	Assist the advance planning unit to develop and distribute a report which highlights forecasted events or conditions likely to occur beyond the forthcoming operational period; particularly those situations which may influence the overall strategic objectives of the jurisdiction EOC.	
	Ensure that each ICP Section provides the Situation Unit with Section, Branch, Unit Status Reports, as required.	

Task	Description	Completed
	Ensure that fiscal and administrative requirements are coordinated through the Finance/Administration Section.	
	Maintain the log for the unit / branch	
	Demobilize on command from the Incident Commander and Planning Intelligence Section Chief	
	Debrief all assigned personnel	
	Forward all logs and reports to the Incident Commander	

Resources Status Unit Leader Checklist

Position Responsibility: Establish the Resources Unit or Branch as part of the planning function for incident management as required including: establishing all incident check-in activities, preparation and processing of resource status change information, maintenance of all command post display charts, lists and incident organization structure charts, incident status charts etc. Maintain the master check in list of resources assigned to the incident.

Note: The Incident Commander and or the Plans/Intelligence Section Chief perform this function in the absence of an assigned Resource Functionary.

DUTY CHECKLIST: [Enter exact time in boxes below]

Task	Description	Completed
	Report to the Planning and Intelligence Section Chief for briefing and assignment	
	Document exact time of arrival	
	Activate and brief any staff	
	Prepare the objectives of the Resource Unit	
	Establish and supervise the appropriate level of organization for the Unit	
	Activate subunits as needed	
	Exercise overall responsibility for the coordination of branch/ unit activities within the unit	
	Maintain master roster for all resources checked in at the incident	
	In coordination with all sections, branches and units within the Command Post ensure that status reports are completed, updated as required and used to complete the IAP	
	Ensure the section is properly set up and equipped with supplies, maps status boards, etc.	
	Request staffing for 24 hour operations	
	Using the incident briefing (ICS 201) prepare and maintain the command post display including the organizational chart, resources allocation and deployment sections	
	Provide resources summary information to situation unit as requested	
	Maintain a list of key issues (the ICs incident response priorities)	
	Anticipate situations, plan ahead	
	Ensure unit/branch logs and other necessary incident files are maintained	
	Assist the Documentation Unit in collecting and maintaining information pertaining to the incident for auditing purposes, jurisdiction reimbursement and capturing of lessons learned.	
	Ensure that all status boards and other displays are current and posted information is neat and legible. Ensure that adequate staff is assigned to maintain maps, status boards and other displays.	

Task	Description	Completed
	Ensure that all collected information is validated prior to posting	
	Work closely with each branch/unit within the Planning/ Intelligence Section to ensure the section objectives, as defined in the current IAP are being addressed.	
	Ensure that fiscal and administrative requirements are coordinated through the Finance/Administration Section.	
	Maintain the log for the unit/branch	
	Demobilize on command from the Incident Commander and Planning Intelligence Section Chief	
	Debrief all assigned personnel	
	Forward all logs and reports to the Incident Commander	

Documentation Unit Leader Checklist

Position Responsibility: Establish the Documentation Unit or Branch as part of the planning function for incident management as required including: collecting, organization and filing all completed event and disaster related forms, logs, situation status reports, IAPs and any other relater incident information.

Note: The Incident Commander and or the Plans/Intelligence Section Chief performs this function in the absence of an assigned Documentation Functionary.

DUTY CHECKLIST: [Enter exact time in boxes below]

Task	Description	Completed
	Report to the Planning and Intelligence Section Chief for briefing and assignment	
	Document exact time of arrival	
	Activate and brief any staff	
	Prepare the objectives of the Documentation Unit	
	Establish and supervise the appropriate level of organization for the Unit	
	Activate subunits as needed	
	Exercise overall responsibility for the coordination of branch/ unit activities within the section	
	Keep the IC and Section Chief and all ICP Section Chiefs informed of significant issues affecting incident documentation	
	In coordination with all sections and branches within the Command Post ensure that status reports are completed, updated as required	
	Ensure the section is properly set up and equipped	
	Request staffing for 24 hour operations	
	Establish contact with all ICP section Chiefs to ensure the unit receives section documentation in a timely fashion	
	Oversee the preparation and distribution of situation statue reports. Coordination with the Situational Analysis Unit as required.	
	Maintain a list of key issues (the ICs incident response priorities)	
	Anticipate situations, plan ahead	
	Ensure ICP section/unit/branch logs and other necessary incident files are maintained as official records	
	Meet with the Recovery and Resources Units leaders to determine what ICP materials and documentation are necessary to provide accurate records and documentation for jurisdictional financial audit and recovery purposes.	
	Initiate and maintain a roster of all activated ICP staff positions to ensure that positions logs are accounted for and submitted to the documentation unit at the end of the operational period or shift	

Task	Description	Completed
	Assist the Situational Analysis Unit in collecting and maintaining information pertaining to the incident for auditing purposes, jurisdiction reimbursement and capturing of lessons learned.	
	Ensure that major incident reports, section and branch status reports are completed by each section and are accessible by the Plans Section for planning requirements and incident documentation	
	Ensure that a situation report is produced and distributed to all command post sections, all other active command posts and the jurisdiction EOC, at least once per operational period. Keep extra copies of reports, plans etc for special distribution as required. Set up and maintain documentation reproduction services for the ICP	
	Ensure that the Documentation Unit publishes and distributes the IAP prior to the beginning of the next operational period.	
	Work closely with each branch/unit within the Planning/ Intelligence Section to ensure the section objectives, as defined in the current IAP are being collected as required.	
	Assist the advance planning unit to develop and distribute a report which highlights forecasted events or conditions likely to occur beyond the forthcoming operational period; particularly those situations which may influence the overall strategic objectives of the jurisdiction EOC.	
	Ensure that fiscal and administrative requirements are coordinated through the Finance/Administration Section.	
	Maintain the log for the unit / branch	
	Demobilize on command from the Incident Commander and Planning Intelligence Section Chief	
	Debrief all assigned personnel	
	Forward all logs and reports to the Incident Commander	

Technical Specialists Checklist

Position Responsibility: Establish the Technical Specialists as part of the planning function for incident management as required including: provide technical observations and recommendations to the IC and all sections chiefs in specialized areas required during the incident response.

Note: The Incident Commander and or the Plans/Intelligence Section Chief performs this function in the absence of an assigned Technical Services Functionary.

DUTY CHECKLIST: [Enter exact time in boxes below]

Task	Description			Completed
	Report to the Planning and Intelligence Section Chief for briefing and assignment			
	Document exact time of arrival			
	Activate and brief any staff			
	Prepare the objectives of the Technical Services Unit			
	Establish and supervise the appropriate level of organization for the Unit			
	Activate specialists as needed (Consider:)			
	Fire	Rescue	Law Enforcement	
	Public Health	Health Care (Medical)	Public Works	
	Hazmat	EMS	Emergency Management	
	Public Information	Communications	WMD	
	Shelter	Evacuation	Any other as required	
	Exercise overall responsibility for the coordination of branch/ unit activities within the section			
	Keep the IC, General staff and Section Chief and all ICP Section Chiefs informed addressing developing technical issues			
	Coordinate with the Logistics section to ensure that the technical staff is located and mobilized			
	Ensure the unit is properly set up and equipped			
	Request staffing for 24 hour operations			
	Establish contact with all ICP section Chiefs to determine the best estimates of future direction and outcomes of the incident.			
	Assign technical staff to assist ICP staff sections in coordinating specialized areas of response or recovery			
	Assign technical staff to assist the Logistics section with interpreting specialized resources capabilities and requests			
	Maintain a list of key issues (the ICs incident response priorities)			

Task	Description	Completed
	Anticipate situations, plan ahead	
	Review action planning objectives submitted by each section for the up coming operational period. Ensure the technical staff is available to support upcoming specialized response or recovery issues based on the current plan or future actions	
	Ensure that fiscal and administrative requirements are coordinated through the Finance/Administration Section.	
	Maintain the log for the unit / branch	
	Demobilize on command from the Incident Commander and Planning Intelligence Section Chief	
	Debrief all assigned personnel	
	Forward all logs and reports to the Incident Commander	

Demobilization Unit Leader Checklist

Position Responsibility: Establish the Demobilization Unit or Branch as part of the planning function for incident management as required including: provide technical observations and recommendations to the IC and all sections chiefs in specialized areas required during the incident response.

Note: The Incident Commander and or the Plans/Intelligence Section Chief performs this function in the absence of an assigned Demobilization Functionary.

DUTY CHECKLIST: [Enter exact time in boxes below]

Task	Description	Completed
	Report to the Planning and Intelligence Section Chief for briefing and assignment	
	Document exact time of arrival	
	Activate and brief any staff	
	Prepare the objectives of the Technical Services Unit	
	Establish and supervise the appropriate level of organization for the Unit	
	Exercise overall responsibility for the coordination of branch/ unit activities within the section	
	Keep the IC, General staff and Section Chief and all ICP Section Chiefs informed addressing any demobilization issues	
	Develop the incident demobilization plan for review by Plans Section Chief and approval by IC	
	Ensure that the demobilization plan is coordinated with the Logistics Section	
	Ensure the unit is properly set up and equipped	
	Request staffing for 24 hour operations	
	Maintain a list of key issues (the ICs incident response priorities)	
	Anticipate situations, plan ahead	
	Advise all activated sections that demobilized staff must complete al reports, time sheets, and documentation requirement	
	Ensure that fiscal and administrative requirements are coordinated through the Finance/Administration Section.	
	Maintain the log for the unit/branch	
	Demobilize on command from the Incident Commander and Planning Intelligence Section Chief	
	Debrief all assigned personnel	
	Forward all logs and reports to the Incident Commander	

Situation Unit Leader Checklist

Position Responsibility: The Logistics Section is responsible for providing facilities, services, and material in support of the incident. The section chief participates in development and implementation of the Incident Action Plan and activates and supervises the branches and units within the logistics section.

Note: The Incident Commander performs this function in the absence of an assigned Logistics Functionary.

DUTY CHECKLIST: [Enter exact time in boxes below]

Task	*Description*			*Completed*
	Report to the IC for briefing and assignment			
	Document exact time of arrival			
	Activate and brief any staff			
	Prepare the objectives of the Logistics Section			
	Establish and supervise the appropriate level of organization for the section. Exercise overall responsibility for the coordination of branch/unit activities within the section			
	Keep the IC, General staff and Section Chief and all ICP Section Chiefs informed addressing developing Logistical issues			
	Activate subunits as needed (Consider:)			
	Ground Support	Facilities	Security	
	Supply	Medical	Communications	
	Consider activating branches (Consider:)			
	Support	Services		
	Request staffing for 24 hour operations as required			
	Maintain a list of key issues (the ICs incident response priorities)			
	Be proactive, Think ahead, Anticipate situations			
	Ensure the unit is properly set up and equipped			
	Notify resources unit of the Logistics Section of units activated including names and locations of assigned personnel			
	Participate in the preparation of the Incident Action Plan			
	Identify service and support requirements for planned and expected operations			
	Provide input to and review the communications plan, the medical plan, and the traffic plan			
	Coordinate and process requests for additional resources			
	Review the Incident Action Plan and estimate section needsfor the next operational period			
	Prepare service and support elements of the Incident Action Plan			
	Advise on current and estimate future service and support requirements			
	Recommend release of units within the demobilization plan			

Task	Description	Completed
	Medical Considerations: • Consider activation of medical unit • Prepare the medical emergency plan, including procedures for major medical emergency • Declare major medical emergency as appropriate • Respond to requests for responder medical aid • Respond to requests for responder medical transportation • Respond to requests for responder medical supplies Supply Considerations: • Provide kits to Planning, Logistics, and Administrative/ Financial Sections • Determine the type and amount of supplies enroute • Arrange for receiving ordered supplies • Review the Incident Action Plan for information on operations of the supply unit • Order, receive, distribute and store supplies and equipment • Process requests for personnel, supplies and equipment • Maintain inventory of supplies and equipment as required	
	Facilities Considerations: • Determine requirements for each facility to be established, such as the staging area, family support, etc. • Prepare layouts of incident facilities • Activate incident facilities • Obtain personnel to operate facilities • Provide sleeping facilities, security services, maintenance, sanitation, lighting and clean up	
	Ground Support Considerations: • Review the incident action plan • Implement the traffic plan as defined by the Plans Section • Support out of service resources • Notify resources units of all status changes on support and transportation • Arrange for and activate fuel, maintenance and repair of ground resources • Maintain inventory of support and transportation vehicles • Provide transportation services • Collect use information on rented and contracted equipment • Support debris removal operations	
	Ensure that fiscal and administrative requirements are coordinated through the Finance/Administration Section.	
	Maintain the log for the unit / branch	

Task	Description	Completed
	Demobilize on command from the Incident Commander and Planning Intelligence Section Chief	
	Debrief all assigned personnel	
	Forward all logs and reports to the Incident Commander	

Communications Unit Leader Checklist

Position Responsibility: The Communications Officer is responsible for developing plans for the effective use of telephone and radio communications during an incident. The Communications Officer insures prompt distribution of communications equipment to personnel and the maintenance of equipment.

Note: The Incident Commander and Logistics Section Chief are responsible for this function. The Incident Commander is responsible for all critical incident activities, including the final authority on all aspects of operations, logistics, finance, investigations and staging.

DUTY CHECKLIST: [Enter exact time in boxes below]

Task	Description	Completed
	Check in with the Incident Commander and or Logistic Section Chief for a briefing and assignment	
	Document exact time on duty	
	Maintain a log of activities	
	Appoint, organize, brief staff as appropriate	
	Manage incident communications operations	
	Contact Amateur Radio operators if necessary and manage their involvement	
	Ensure that emergency communications links are established between field functions, the ICP and jurisdiction EOC. Maintain Communications liaison between County/City/State EOC, if activated	
	Advise the Incident Commander, Command and General Section Chiefs of any anticipated problems regarding communications equipment	
	Manage the distribution of radio equipment as available and necessary	
	Recover communications equipment from relieved, released or demobilized units	
	Prepare and implement the Incident Radio Communication Plan	
	Establish the Incident Communication Center and Message Center	
	Manage relief personnel for 24hour communications requirements	
	Set up the telephone and public address systems. Ensure communications are installed and tested	
	Establish appropriate communications distribution and maintenance locations	
	Ensure accountability of all communication equipment	
	Provide technical communications specialist information to the incident command post	

Task	Description	Completed
	Establish, support and maintain dispatching operations as required	
	Maintain the section log and forward logs and reports the IC	
	Demobilize on command from the Incident Commander	
	Debrief all assigned personnel	

Security Unit Leader Checklist

Position Responsibility: The responsibility of the ICP Security Staff is to provide uniformed personnel to maintain the safety and security of personnel working in the Command Post and supporting facilities. Once assignments are made, the security officers should report to the functionaries in charge of the duty area.

Note: The Incident Commander and Logistics Section Chief are responsible for this function. The Incident Commander is responsible for all critical incident activities, including the final authority on all aspects of operations, logistics, finance, investigations, staging and Headquarters EOC

Note: In the temporary absence or incapacitation of the Incident Commander, a Deputy Incident Commander shall assume the responsibilities. In the absence of a Deputy Incident Command the transfer of command falls to the Operations Functionary.

DUTY CHECKLIST: [Enter exact time in boxes below]

Task	Description	Completed
	Report to the Logistics Functionary for briefing and assignment (Command Post Security Officers will subsequently report to the Incident Commander	
	Document exact time of arrival	
	Maintain a log of activities	
	Control access to Command Post facilities and other response facilities	
	Determine Command post Security requirements	
	Keep track of personnel at all security areas	
	Keep Logistics Functionary informed of deployment of uniformed personnel	
	Develop security plans, maintain a defensive posture	
	Keep the peace, prevent assaults, unwarranted intrusion into protected areas,	
	Document complaints from citizens and suspicious activities and forward them to Logistics Functionary	
	Prevent theft of property	
	Forward all logs of activities and associated documents to the Logistics Functionary	
	If applicable, ensure the following areas are provided with uniformed security:	
	Command Post	Staging Areas	Relocation Centers	
	Field Mortuary	Hospitals	Triage Areas	
	Media Centers	Victim Assembly Areas	Headquarters EOC	
	Headquarters	Government Buildings	Helipads	

Task	Description	Completed
	If assistance is necessary, organize and brief your staff.	
	Identify any threat to the command post or command post occupants	
	Prepare plans to assure the safety of the command post or its occupants	
	In cases regarding hazardous materials, contact an expert to get advice on safety	
	Exercise authority to stop all unsafe acts or uninvited entry from anyone into the command post.	
	Investigate and prepare reports on any injuries or accidents relative to the command post or its occupants.	
	Maintain a log of activities, to include significant events and names of security personnel	
	Review and understand EMS plan if needed at the command post	
	Set up a staging area near the command post for EMS personnel and ambulance(s) and provide security.	
	Plan to deal with large numbers of Volunteers	
	Confer with public Information Officer to establish areas for media personnel	
	Provide vehicular and pedestrian traffic control	
	Secure food, water, medical and health/first Aid resources	
	Demobilize on orders from the Logistics Staff section	
	Debrief personnel	

Administrative and Finance Section Chief Checklist

Position Responsibility: Is responsible for all financial and cost analysis aspects of the incident and for supervising members of the finance section.

Note: The Incident Commander performs this function in the absence of an assigned Administrative and Finance Functionary.

DUTY CHECKLIST: [Enter exact time in boxes below]

Task	Description	Completed
	Report to the IC for briefing and assignment. Report to the IC when the sectional is fully operational.	
	Document exact time of arrival	
	˙Activate and brief any staff	
	Prepare the objectives of the Administrative and Financial Section	
	Establish and supervise the appropriate level of organization	
	Activate subunits as needed (Consider:)	
	Time · Procurement · Cost	
	Financial Documentation	
	Exercise overall responsibility for the coordination of section activities	
	Keep the IC, General staff and Section Chief and all ICP Section Chiefs informed addressing developing administrative and financial issues	
	Ensure the unit is properly set up and equipped	
	Request staffing for 24 hour operations	
	Develop an operating plan for financial actions	
	Determine a need for commissary operations	
	Meet with assisting and cooperating agency representatives as required to discuss incident financial and personnel coordination	
	Provide input in all planning sessions on personnel, financial and incident cost matters	
	Maintain contact with agency administrative headquarters on financial matters	
	Insure that all personnel time records are transmitted to home agencies according to policy	
	Participate in all demobilization planning	
	Insure that all obligation documents initiated at the incident are properly prepared and completed	
	Brief agency administration personnel on all incident related business management issues needing attention, and follow-up prior to leaving the incident	

Task	*Description*	*Completed*
	Maintain a list of key financial and administrative issues (the ICs incident response priorities)	
	Anticipate situations, plan ahead	
	Ensure that incident fiscal and administrative requirements are coordinated through the Finance/Administration Section.	
	Time Unit Considerations: • Establish a process to account for personnel time and equipment usage. • Consider establishing a commissary for long term incident response • Determine incident requirements for the time recording function • Establish contact with appropriate supporting/responding agency personnel representatives • Ensure that daily personnel time recording documents are prepared and in compliance to time policy • Submit cost estimates to Cost Unit as required • Maintain records security • Ensure that al records are current and complete prior to demobilization • Time reports from assisting agencies should be released to respective agency representatives prior to demobilization • Set up the equipment time recording function • Advise the Ground Support Unit of the requirement to do a daily record of equipment use/time • Maintain current postings on all charges for fuel, parts, services and commissary • Verify all time data with owner/ operators of equipment • Establish and maintain employee time records within the first operational period • Initiate, gather and update a time record from all applicable personnel assigned to the incident for each operational period • Ensure that all employee identification information is verified to be correct	
	Procurement Unit Considerations: • Coordinate with local supply sources for procurement requirements, Establish contracts with supply vendors as required • Obtain the Incident Procurement Plan • Prepare and sign contracts and land use agreements as required • Interpret contracts and resolve claims within delegated authority	

Task	Description	Completed
	• Coordinate with Compensation / Claims Unit on procedures for handling claims • Complete processing of procurement requests and forward documents for payment • Coordinate cost data in contracts with Cost Unit Leader	
	Compensation and Claims Unit Considerations: • Establish contact with incident safety officer and liaison officer • Determine the need for compensation claims specialists • Obtain a copy of the Incident Medical Plan • Maintain a compensation and claim log	
	Cost Unit Considerations: • Coordinate with jurisdiction on cost reporting procedures • Obtain and record all cost data • Prepare incident cost summaries • Prepare resources-use cost estimates for planning • Maintain cumulative incident cost records	
	Maintain the section log and forward logs and reports the IC	
	Demobilize on command from the Incident Commander	
	Debrief all assigned personnel	

Appendix B

INCIDENT COMMAND SYSTEM FORMS

THE CONCEPT BEHIND THE USE OF ICS/IMS FORMS

The concept behind the use of forms associated with incident command is relatively simple to failitate, more effective information management. It becomes very obvious, very quickly in any emergency event that the Incident Commander along with all of the persons that will work in any Incident Command Post (ICP) will be come overwhelmed with data flowing into them. Effective management of that data will be key to the effective management of the incident both strategically and tactically and later when administrative things such as After Action Reports and such need to be generated the proper use of forms to document the actions taken by whom and when will be invaluable to those responsible for the assembly of such reports.

THE DILEMMA WITH FORMS

The "forms dilemma" is created by a set of seemingly minor and irrelevant points that are relatively simple in nature yet, cause considerable consternation to the concept of a truly integrated and unified incident management and command system. With the advent of the National Incident Management System (NIMS) a concerted effort at the federal level has begun to deal with this dilemma.

There have been several sets of "standardized" ICS/IMS forms created by at least three "significant players" in the world of IMS systems development, those entities are:

- The California FIRESCOPE (FIrefighting REsources of California Organized for Potential Emergencies) Program (http://www.firescope.org/)
- The National Wildfire Coordinating Group (http://www.nwcg.gov)
- The National Oceanic and Atmospheric Administration (http://www.noaa.gov)

While all of these agencies have used similar form styles and even names to their forms (to some extent) no one set of forms exists that is considered the "Universal ICS/IMS Forms set".

The real problem is not so much in the forms content but rather the format that the form takes on in its final generation, also some movement towards an electronic capture of data has been made as well with the advent of newer technology.

There is also the problem of "ICS Forms Mindset", which being that only the set of forms that a given agency or entity uses are the "right" ones to be used for that purpose. The importance of the DATA must be maintained not the format per se so all of the available forms from the above entities are presented in this appendix.

CALIFORNIA FIRESCOPE

The California FIRESCOPE Program forms can be found at http://www.firescope.org/.

Note 1) In practice many of these forms have instructions for that form on the backside of the form, they were intentionally omitted from this section, as they are basic and routine instructions and are often redundant from form to form.

Note 2) Not all forms that are commonly referred to in various FIRESCOPE literature are currently available at this site, the currently available FIRESCOPE forms on are:

FORM NAME	FORM TITLE
ICS 201	Incident Briefing
ICS 202	Objectives
ICS 203	Organization Assignment List
ICS 204	Assignment List
ICS 205	Incident Radio Communications Plan
ICS 206	Medical Plan
ICS 207	Incident Organization Chart (24″ × 15″)
ICS 208-HM	Site Safety and Control Plan
ICS 209	National ICS 209 Form
ICS 211	Check in List
ICS 212	Incident Demobilization Vehicle Safety Inspection
ICS 214	Unit/Activity Log
ICS 215-A	Incident Safety Analysis LCES
ICS 215-G	Operational Planning Worksheet-Generic
ICS 215-W	Operational Planning Worksheet-Wildland
ICS 216	Radio Requirements Worksheet
ICS 218	Support Vehicle Inventory
ICS 220	Air Operations Summary
ICS 221	Demobilization checkout
ICS 222	Incident Weather Forecast Request
ICS 223	Tentative Release list

FORM NAME	FORM TITLE	
ICS 223	Tentative Release list	
ICS 225	Incident Personnel Performance Rating	
ICS 226	Compensation for Injury Log	
ICS 227	Claims Log	Page B-33
ICS 229	Incident Cost Work Summary	Page B-34

THE NATIONAL WILDFIRE COORDINATING GROUP

The National Wildfire Coordinating Group (NWCG) can be found at www.nwcg.gov.

Note 1) In practice many of these forms have instructions for that form on the backside of the form, they were intentionally omitted from this section, as they are basic and routine instructions and are often redundant from form to form.

Note 2) Not all forms that are commonly referred to in various NWCG literature are currently available at this site, the currently available NWCG forms on are:

FORM NAME	FORM TITLE
ICS-201	Incident Briefing
ICS-202	Incident Objectives List
ICS-203	Organization Assignment List
ICS-204	Division Assignment List
ICS-205	Incident Radio Communications Plan
ICS-206	Medical Plan
ICS-207	Organizational Chart
ICS-209	Incident Status Summary Report
ICS-210	Status Change Card
ICS-211	Incident Check-In Lists
ICS-213	General Message Form
ICS-214	Unit Log Form
ICS-215	Operational Planning Work Sheet
ICS-215a	Incident Action Plan Safety Analysis
ICS-216	Radio Requirements Worksheet
ICS-217	Radio Frequency Assignment Worksheet
ICS-218	Support Vehicle Inventory Form
ICS-219-2	Card Stock—Crew (GREEN)
ICS-219-4	Card Stock—Helicopter (BLUE)
ICS-219-6	Card Stock—Aircraft (ORANGE)
ICS-219-7	Card Stock—Dozer (YELLOW)
ICS-220	Air Operations Summary
ICS-221	Demobilization Checkout
ICS-224	Crew Performance Rating Form
ICS-225	Incident Personnel Performance Rating Form

FORM NAME	FORM TITLE
ICS-226	Individual Personnel Rating
IMT1	OSHA Abatement Form
	(Incident Management Team Checklist)
IRSS01	IRSS Check-in Form
RES ORDER A	Resource Order
RES ORDER B	Resource Order (Continuation)

THE NATIONAL OCEANIC AND ATMOSPHERIC ADMINISTRATION

The National Oceanic and Atmospheric Administration (NOAA) developed IMS forms for the specific use response to oil spill and/or other pollution events on the ocean at the behest of the United States Coast Guard. Many of the additions are very oil-spill-specific and have limited use outside of that realm. NOAA did follow for the most part the fire-based forms numbering sequence and their ICS structure does mimic the "other" forms. This proves that the forms themselves can be modified to meet a specific need but that the integrity of the system can be maintained regardless of what the forms' context is. These forms may be found on www.noaa.gov.

Note 1) In practice many of these forms have instructions on the backside of the form. They were intentionally omitted from this section as they are basic and routine instructions and are often redundant from form to form.

Note 2) Not all forms that are commonly referred to in various NWCG literature are currently available at this site, the currently available NWCG forms are:

FORM NAME	FORM TITLE
ICS 201-OS	Incident Briefing
ICS 202-OS	Incident Objectives
ICS 203-OS	Organization Assignment List
ICS 204-OS	Assignment List
ICS 204a-OS	Assignment List Attachment
ICS 205-OS	Incident Radio Communications Plan
ICS 205a-OS	Communications List
ICS 206-OS	Medical Plan
ICS 207-OS	Incident Organization Chart
ICS 208-OS	Site Safety and Control Plan
ICS 209-OS	Incident Status Summary
ICS 210-OS	Status Change
ICS 211-OS	Check-In List
ICS 211e-OS	Check-In List (Equipment)
ICS 211p-OS	Check-In List (Personnel)
ICS 213-OS	General Message
ICS 214-OS	Unit Log

FORM NAME	**FORM TITLE**
ICS 214a-OS	Individual Log
ICS 215-OS	Operational Planning Worksheet
ICS 215a-OS	Incident Action Plan Safety Analysis
ICS 216-OS	Radio Requirements Worksheet
ICS 217-OS	Radio Frequency Assignment Worksheet
ICS 218-OS	Support Vehicle Inventory
ICS 219-OS	Resource Status Card—Crew
ICS 219-4-OS	Resource Status Card—Helicopter
ICS 219-6-OS	Resource Status Card—Aircraft
ICS 219-7-OS	Resource Status Card—Dozer
ICS 220-OS	Air Operations Summary
ICS 221-OS	Demobilization Checkout
ICS 230-OS	Daily Meting Schedule
ICS 231-OS	Meeting Summary
ICS 232-OS	Resources At Risk Summary
ICS 232a-OS	ACP Site Index
	Executive Summary
	General Plan
	Version A IAP Cover Sheet
	Version B IAP Cover Sheet
	Initial Incident Information

INCIDENT BRIEFING	1. INCIDENT NAME	2. DATE PREPARED	3. TIME PREPARED

4. MAP SKETCH

ICS 201 5-94	PAGE 1	8. PREPARED BY (NAME AND POSITION)

OBJECTIVES **ICS 202**	1. INCIDENT NAME	2. DATE PREPARED	3. TIME PREPARED

4. OPERATIONAL PERIOD (Date/Time)

5. OVERALL INCIDENT OBJECTIVE:

6. OBJECTIVES FOR THIS OPERATIONAL PERIOD:

7. WEATHER FORECAST FOR OPERATIONAL PERIOD

8. GENERAL/SAFETY MESSAGE

9. ATTACHMENTS (✔ IF ATTACHED)

☐ ORGANIZATION LIST (ICS 203)   ☐ MEDICAL PLAN (ICS 206)   ☐ _____

☐ ASSIGNMENT LISTS (ICS 204)   ☐ INCIDENT MAP   ☐ _____

☐ COMMUNICATIONS PLAN (ICS 205)   ☐ TRAFFIC PLAN   ☐ _____

ICS 202 5-94	10. PREPARED BY (Planning Section Chief)	11. APPROVED BY (Incident Commander)

ORGANIZATION ASSIGNMENT LIST **ICS 203**	1. INCIDENT NAME	2. DATE PREPARED	3. TIME PREPARED

POSITION	NAME	4. OPERATIONAL PERIOD (DATE/TIME)

5. INCIDENT COMMANDER AND STAFF

INCIDENT COMMANDER	
DEPUTY	
SAFETY OFFICER	
INFORMATION OFFICER	
LIAISON OFFICER	

6. AGENCY REPRESENTATIVES

AGENCY	NAME

7. PLANNING SECTION

CHIEF	
DEPUTY	
RESOURCES UNIT	
SITUATION UNIT	
DOCUMENTATION UNIT	
DEMOBILIZATION UNIT	
TECHNICAL SPECIALISTS	

8. LOGISTICS SECTION

CHIEF	
DEPUTY	

a. SUPPORT BRANCH

DIRECTOR	
DEPUTY	
SUPPLY UNIT	
FACILITIES UNIT	
GROUND SUPPORT UNIT	

b. SERVICE BRANCH

DIRECTOR	
DEPUTY	
COMMUNICATIONS UNIT	
MEDICAL UNIT	
FOOD UNIT	

9. OPERATIONS SECTION

CHIEF	
DEPUTY	

a. BRANCH I - DIVISIONS/GROUPS

BRANCH DIRECTOR	
DEPUTY	
DIVISION/GROUP	
DIVISION/GROUP	
DIVISION/GROUP	
DIVISION/GROUP	
DIVISION/GROUP	

b. BRANCH II - DIVISIONS/GROUPS

BRANCH DIRECTOR	
DEPUTY	
DIVISION/GROUP	
DIVISION/GROUP	
DIVISION/GROUP	
DIVISION/GROUP	
DIVISION/GROUP	

c. BRANCH III - DIVISIONS/GROUPS

BRANCH DIRECTOR	
DEPUTY	
DIVISION/GROUP	
DIVISION/GROUP	
DIVISION/GROUP	
DIVISION/GROUP	
DIVISION/GROUP	

d. AIR OPERATIONS BRANCH

AIR OPERATIONS BR. DIR.	
DEPUTY	
AIR TACTICAL SUPERVISOR	
AIR SUPPORT SUPERVISOR	
HELICOPTER COORDINATOR	
AIR TANKER/ FIXED WING COORDINATOR	

10. FINANCE/ADMINISTRATION SECTION

CHIEF	
DEPUTY	
TIME UNIT	
PROCUREMENT UNIT	
COMPENSATION/CLAIMS UNIT	
COST UNIT	

PREPARED BY (RESOURCES UNIT)

ICS 203   **5/94**

1. BRANCH	2. DIVISION/GROUP	**ASSIGNMENT LIST**		ICS 204 (5-94)

3. INCIDENT NAME	4. OPERATIONAL PERIOD
	DATE
	TIME

5. OPERATIONS PERSONNEL

OPERATIONS CHIEF		DIVISION/GROUP SUPERVISOR	
BRANCH DIRECTOR		AIR TACTICAL SUPERVISOR	

6. RESOURCES ASSIGNED THIS PERIOD

RESOURCE DESIGNATOR	LEADER	NUMBER PERSONS	TRANS. NEEDED	DROP OFF PT./TIME	PICK UP PT./TIME

7. CONTROL ASSIGNMENT (S)

8. SPECIAL INSTRUCTIONS/SAFETY MESSAGE

9. DIVISION/GROUP COMMUNICATION SUMMARY

FUNCTION		FREQ.	SYSTEM	CHAN.	FUNCTION		FREQ.	SYSTEM	CHAN.
COMMAND	LOCAL				SUPPORT	LOCAL			
	REPEAT					REPEAT			
DIV/GROUP TACTICAL					GROUND TO AIR				

PREPARED BY (RESOURCE UNIT LEADER)	APPROVED BY (PLANNING SECTION CHIEF)	DATE	TIME

INCIDENT RADIO COMMUNICATIONS PLAN				1. INCIDENT NAME	2. DATE/TIME PREPARED	3. OPERATIONAL PERIOD DATE/TIME
SYSTEM/CACHE	CHANNEL	FUNCTION	FREQUENCY	ASSIGNMENT		REMARKS

4. PREPARED BY (COMMUNICATIONS UNIT)

ICS 205   2-95

MEDICAL PLAN	1. INCIDENT NAME	2. DATE PREPARED	3. TIME PREPARED	4. OPERATIONAL PERIOD

5. INCIDENT MEDICAL AID STATIONS

MEDICAL AID STATIONS	LOCATION	PARAMEDICS YES	NO

6. TRANSPORTATION

A. AMBULANCE SERVICES

NAME	ADDRESS	PHONE	PARAMEDICS YES	NO

B. INCIDENT AMBULANCES

NAME	LOCATION	PARAMEDICS YES	NO

7. HOSPITALS

NAME	ADDRESS	TRAVEL TIME AIR	GRND	PHONE	HELIPAD YES	NO	BURN CENTER YES	NO

8. MEDICAL EMERGENCY PROCEDURES

ICS 206 **5-94**	9. PREPARED BY (MEDICAL UNIT LEADER)	10. REVIEWED BY (SAFETY OFFICER)

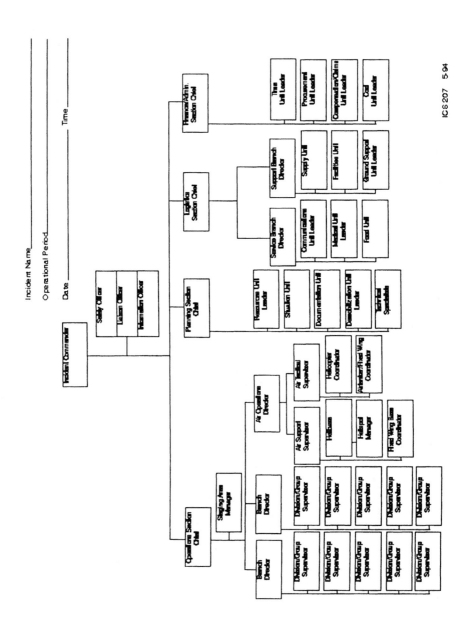

Incident Name

Operational Period

Date _____ Time _____

ICS 207   5 94

SITE SAFETY AND CONTROL PLAN ICS 208 HM	1. Incident Name:	2. Date Prepared:	3. Operational Period: Time:

Section I. Site Information			
4. Incident Location:			

Section II. Organization

5. Incident Commander:	6. HM Group Supervisor:	7. Tech. Specialist - HM Reference:
8. Safety Officer:	9. Entry Leader:	10. Site Access Control Leader:
11. Asst. Safety Officer - HM:	12. Decontamination Leader:	13. Safe Refuge Area Mgr:
14. Environmental Health:	15.	16.

17. Entry Team: (Buddy System)		18. Decontamination Element:	
Name:	PPE Level	Name:	PPE Level
Entry 1		Decon 1	
Entry 2		Decon 2	
Entry 3		Decon 3	
Entry 4		Decon 4	

Section III. Hazard/Risk Analysis

19. Material:	Container type	Qty.	Phys. State	pH	IDLH	F.P.	I.T.	V.P.	V.D.	S.G.	LEL	UEL

Comment:

Section IV. Hazard Monitoring

20. LEL Instrument(s):	21. O$_2$ Instrument(s):
22. Toxicity/PPM Instrument(s):	23. Radiological Instrument(s):

Comment:

Section V. Decontamination Procedures

24. Standard Decontamination Procedures:	YES:	NO:

Comment:

Section VI. Site Communications

25. Command Frequency:	26. Tactical Frequency:	27. Entry Frequency:

Section VII. Medical Assistance

28. Medical Monitoring:	YES:	NO:	29. Medical Treatment and Transport In-place:	YES:	NO:

Comment:

Incident Status Summary (ICS - 209)

1: Date	2: Time	3: Initial Update Final	4: Incident Number			5: Incident Name
			State	Unit	Number	

6: Incident Kind	7: Start Date/Time	8: Cause	9: Incident Commander	10: IMT Type	11: State-Unit

12: County	13: Latitude and Longitude	14: Short Location Description (in reference to nearest town):
	Lat: Long:	

Current Situation

15:Size/Area Involved	16: % Contained or MMA	17: Expected Containment Date: Time:	18: Line to Build	19: Costs to Date	20: Declared Controlled Date: Time:

21: Injuries this reporting Period	22: Injuries to Date	23: Fatalities	24: Structure Information			
			Type of Structure	# Threatened	# Damaged	# Destroyed

25: Threat to Human Life/Safety:		Residence			
Evacuation(s) in progress _____		Commercial			
No evacuation(s) imminent _____		Property			
Potential future threat _____		Outbuilding/Other			
No likely threat					

26: Communities/Critical Infrastructure Threatened (in 12, 24, 48 and 72 hour time frames):

12 hours:

24 hours:

48 hours:

72 hours:

27: Critical Resource Needs (kind and amount, in priority order):
1
2
3

28: Major Problems and Concerns (control problems, social/political/economic concerns or impacts, etc.). Relate critical resource needs identified above to the Incident Action Plan.

29: Resources Threatened (kind(s) and value/significance):

CHECK-IN LIST ICS-211	1. INCIDENT NAME	2. CHECK-IN LOCATION			3. DATE/TIME
5-94		BASE	CAMP	STAGING AREA	
		ICP RESOURCES UNIT		HELIBASE	

CHECK-IN INFORMATION

4. LIST PERSONNEL (OVERHEAD) BY AGENCY & NAME
-OR-
LIST EQUIPMENT BY THE FOLLOWING FORMAT

AGENCY	4. SINGLE TYPE S/T	KIND	TYPE	5. I.D. NO./NAME	5. ORDER/ REQUEST NUMBER	6. DATE/TIME CHECK-IN	7. LEADER'S NAME	8. TOTAL NO PERSONNEL	9. MANIFEST YES	NO	10. CREW WEIGHT OR INDIVIDUALS WEIGHT	11. HOME BASE	12. DEPARTURE POINT	13. METHOD OF TRAVEL	14. INCIDENT ASSIGNMENT	15. OTHER QUALIFICATIONS	16. SENT TO RESTAT TIME/INT.

18. Prepared by (Name and Position)   Use back for remarks or comments

17.

Page _____ of _____

Incident Demobilization Vehicle Safety Inspection

Vehicle Operator: Complete items above double lines prior to inspection

Incident Name		Order No.	
Vehicle: License No.	Agency		Reg/Unit
Type (Eng., Bus., Sedan)	Odometer Reading		Veh. ID No.

Inspection Items	Pass	Fail	Comments
1. Gauges and lights. See back*			
2. Seat belts. See back *			
3. Glass and mirrors. See back*			
4. Wipers and horn. See back *			
5. Engine compartment. See back			
6. Fuel system. See back *			
7. Steering. See back *			
8. Brakes. See back *			
9. Drive line U-joints. Check play			
10. Springs and shocks. See back			
11. Exhaust system. See back *			
12. Frame. See back *			
13. Tire and wheels. See back *			
14. Coupling devices. * Emergency exit (Buses)			
15. Pump Operation			
16. Damage on Incident			
17. Other			
* Safety Item - Do not Release Until Repaired			

Additional Comments:

	HOLD FOR REPAIRS			RELEASE	
Date		Time		Date		Time	

Inspector Name (Print)	Operator Name (Print)
Inspector Signature	Operator Signature

This form may be photocopied, but three copies must be completed.
Distribution: Original to Inspector, copy to vehicle operator, copy to Incident Documentation Unit

ICS 212   2/96

	1. INCIDENT NAME		2. DATE	3. TIME
UNIT/ACTIVITY LOG **ICS 214 5-94**			PREPARED	PREPARED
4. ORGANIZATION POSITION	5. LEADER NAME		6. OPERATIONAL PERIOD	

7.   PERSONNEL ROSTER ASSIGNED

NAME	ICS POSITION	HOME BASE

8.   ACTIVITY LOG (CONTINUE ON REVERSE)

TIME	MAJOR EVENTS

OTHER RISK ANALYSIS

Other Risk Mitigations

Hazardous Materials

Transportation / Hr.*

Communication

Structure Protection

Multi-aircraft

***LCES ANALYSIS OF TACTICAL APPLICATIONS**

*LCES Mitigations

Snags

Unburned Area

Spotting, wind-driven

Extreme Conditions

Anchor Points

Frontal Assault

Mid-slope Fireline

Indirect Fireline

Underslung Fireline

Downhill Fireline

Division /Group

| ICS 215A 11/93 | **INCIDENT SAFETY ANALYSIS (*LCES)** *Lookouts, Communications, Escape routes, Safety zones | 1. Incident Name | 2. Date Prepared / Time Prepared | 3. Operational Period (Date/Time) |

OPERATIONAL PLANNING WORKSHEET
(Generic)

1. Incident Name	2. Date Prepared	3. Operational Period (Date/Time)
	Time Prepared	

5. Division/ Group/ Staging or Other Location	6. Work Assignments	Resources By Type (Show Strike Teams as ST)	7. Overhead	8. Special Equipment	9. Supplies	10. Reporting Location	11. Requested Arrival Time
		Req.					
		Have					
		Need					
		Req.					
		Have					
		Need					
		Req.					
		Have					
		Need					
		Req.					
		Have					
		Need					
		Req.					
		Have					
		Need					
		Req.					
		Have					
		Need					
		Req.					
		Have					
		Need					
		Req.					
		Have					
		Need					
		Req.					
		Have					
		Need					
		Req.					
		Have					
		Need					

12.			
Total Resources Required	Single Resources	Strike Teams	
Total Resources on Hand	Single Resources	Strike Teams	
Total Resources Needed	Single Resources	Strike Teams	

13. Prepared by (Name and Position)

2-98

ICS 215-G

OPERATIONAL PLANNING WORKSHEET DRAFT

1. Incident Name	2 Date Prepared	Time Prepared	3. Operational Period (Date/Time)

4. Division/Group or Other Location	5. Work Assignments		Resources By Type (Show Strike Teams as ST) Engine 1	Engine 2	Engine 3	Engine 4	Water Tenders 1	Water Tenders 2	Hand Crews 1	Hand Crews 2	Dozers 1	Dozers 2	Dozers 3	Helicopters 1	Helicopters 2	Helicopters 3	Air Tankers 1	Air Tankers 2	Air Tankers 3	6. Overhead	7. Special Equipment	8. Other	9. Reporting Location	10. Requested Arrival Time
		Req																						
		Have																						
		Need																						
		Req.																						
		Have																						
		Need																						
		Req																						
		Have																						
		Need																						
		Req.																						
		Have																						
		Need																						
		Req.																						
		Have																						
		Need																						
		Req																						
		Have																						
		Need																						
		Req.																						
		Have																						
		Need																						
		Req																						
		Have																						
		Need																						
		Req.																						
		Have																						
		Need																						
		Req.																						
		Have																						
		Need																						
ICS 6-92 215-W	Total Resources Required	Single Resources / Strike Teams																					Prepared by: (Name and Position)	
	Total Resources on Hand																							
	Total Resources Needed																							

RADIO REQUIREMENTS WORKSHEET

1. INCIDENT NAME	2. JURISDICTION	3. OPERATIONAL PERIOD

4. BRANCH	5. PAGE	6. PREPARED BY (NAME AND POSITION)

7. DIVISION / GROUP		DIVISION / GROUP		DIVISION / GROUP		DIVISION/ GROUP	
FREQUENCY		FREQUENCY		FREQUENCY		FREQUENCY	

8. AGENCY	ID NO.	RADIO RQMTS	AGENCY	ID NO.	RADIO RQMTS	AGENCY	ID NO.	RADIO RQMTS	AGENCY	ID NO.	RADIO RQMTS

ICS 216 (2/95)

SUPPORT VEHICLE INVENTORY

(USE SEPARATE SHEET FOR EACH VEHICLE CATEGORY)

1. INCIDENT NAME	2. DATE PREPARED	3. TIME PREPARED

4.

VEHICLE INFORMATION

a. TYPE	b. MAKE	c. CAPACITY / SIZE	d. AGENCY / OWNER	e. I.D. NO.	f. LOCATION	g. RELEASE TIME

ICS 218	8-78	PAGE	5. PREPARED BY (GROUND SUPPORT UNIT)

AIR OPERATIONS SUMMARY

1. INCIDENT NAME		2. OPERATIONAL PERIOD		3. DISTRIBUTION		
		DATE	TIME	HELIBASES	FIXED WING BASES	

4.	NAME	AIR / AIR FREQUENCY	AIR / GROUND FREQUENCY	5. REMARKS (Specific Instructions, Safety Notes, Hazards, Priorities)
PERSONNEL & COMMUNICATIONS				
AIR OPERATIONS DIRECTOR				
AIR TACTICAL SUPERVISOR				
HELICOPTER COORDINATOR				
AIR TANKER/FIXED WING COORDINATOR				

6.	7.	8. FIXED WING		9. HELICOPTERS		10. TIME		11.	12.
LOCATION / FUNCTION	ASSIGNMENT	NO.	TYPE	NO.	TYPE	AVAILABLE	COMMENCE	AIRCRAFT ASSIGNED	OPERATING BASE
	13. TOTALS								

14. AIR OPERATIONS SUPPORT EQUIPMENT	15. PREPARED BY	DATE	TIME

ICS 220 (5/94)

DEMOBILIZATION CHECKOUT		ICS-221
1. INCIDENT NAME / NUMBER	2. DATE / TIME	3. DEMOB. NO.

4. UNIT / PERSONNEL RELEASED

5. TRANSPORTATION TYPE / NO.

6. ACTUAL RELEASE DATE / TIME	7. MANIFEST YES NO
	NUMBER_____
8. DESTINATION	**9. AGENCY / REGION / AREA NOTIFIED**
_____	NAME_____
	DATE____

10. UNIT LEADER RESPONSIBLE FOR COLLECTING PERFORMANCE RATING

11. UNIT / PERSONNEL YOU AND YOUR RESOURCES HAVE BEEN RELEASED SUBJECT TO SIGNOFF FROM THE FOLLOWING:

(DEMOB. UNIT LEADER CHECK ✔ APPROPRIATE BOX)

<u>LOGISTICS SECTION</u>

☐ SUPPLY UNIT_____

☐ COMMUNICATIONS UNIT_____

☐ FACILITIES UNIT _____

☐ GROUND SUPPORT UNIT_____

<u>PLANNING SECTION</u>

☐ DOCUMENTATION UNIT_____

<u>FINANCE/ADMINISTRATION SECTION</u>

☐ TIME UNIT_____

<u>OTHER</u>

☐ _____   _____

☐ _____   _____

12. REMARKS

_____

_____

ICS 221	5-94	

INSTRUCTIONS ON BACK

INCIDENT WEATHER FORECAST REQUEST

REQUEST TIME/DATE	REQUESTING AGENCY	INCIDENT OR PROJECT NAME

LOCATION (LAT/LONG AND OR TOWNSHIP/SECTION/RANGE)

ASPECT OR EXPOSURE	ELEVATION (TOP/BOTTOM)	DRAINAGE NAME

PROJECT SIZE	FUEL TYPE	CONTACT PERSON OR INCIDENT COMMANDER

SEND FORECAST BY (FAX, E-MAIL, BOTH)	PHONE NUMBER(S) (OFFICE,CELL,PGR)

FAX NUMBER(S)	E-MAIL ADDRESS

ON SITE WEATHER OBSERVATIONS

TIME	ELEV	TEMP DRY	TEMP WET	WIND DIR/SPD 20 FT	WIND DIR/SPD EYE LVL	RH	DP	REMARKS

FORECAST ELEMENTS (CHECK THE FOLLOWING)

WEATHER DISCUSSION:		SKY/WEATHER	
TEMPERATURE		TRANSPORT WINDS	
RELATIVE HUMIDITY		MIXING HEIGHTS	
RIDGETOP WINDS		INVERSIONS	
EYE LEVEL WINDS		HAINS INDEX	
20 FOOT WINDS		PROBABILITY OF PRECIPITATION	

FORECAST PERIODS (CHECK THE FOLLOWING)

0-12 HOURS		3 TO 5 DAY	
0-24 HOURS		6 TO 10 DAY	
0-48 HOURS		OTHER	

COMMENTS OR REMARKS

Rev. 06/01

ICS 222

TENTATIVE RELEASE LIST

ICS 223

1. Function _____

2. The following resources are surplus to my needs as of _____ hours on _____
   At that time, these resources are available for release processing.

3.

Name of Individual /Crew or Equipment	Position on Incident
1.	
2.	
3.	
4.	
5.	
6.	
7.	
8.	
9.	
10.	
11.	
12.	
13.	
14.	
15.	
16.	
17.	
18.	
19.	
20.	
21.	

4. _____ _____ _____
   Signature of Section Chief   Date   Time

<table>
<tr><td colspan="2">

**INCIDENT PERSONNEL
PERFORMANCE RATING**

</td><td>

INSTRUCTIONS: The immediate job supervisor will prepare this form for each subordinate. It will be delivered to the planning section before the rater leaves the fire. Rating will be reviewed with employee who will sign at the bottom.

</td></tr>
</table>

THIS RATING IS TO BE USED ONLY FOR DETERMINING AN INDIVIDUAL'S PERFORMANCE

1. Name	2. Fire Name and Number

3. Home Unit (address)	4. Location of Fire (address)

5. Fire Position	6. Date of Assignment From: To:	7. Acres Burned	8. Fuel Type(s)

9. Evaluation

Enter X under appropriate rating number and under proper heading for each category listed. Definition for each rating number follows:

0--Deficient. Does not meet minimum requirements of the individual element.
    DEFICIENCIES MUST BE IDENTIFIED IN REMARKS.

1--Needs to improve. Meets some or most of the requirements of the individual element.
    IDENTIFY IMPROVEMENT NEEDED IN REMARKS.

2--Satisfactory. Employee meets all requirements of the individual element.

3--Superior. Employee consistently exceeds the performance requirements.

Rating Factors	Hot Line				Mop-Up				Camp				Other (Specify)			
	0	1	2	3	0	1	2	3	0	1	2	3	0	1	2	3
Knowledge of the job																
Ability to obtain performance																
Attitude																
Decisions under stress																
Initiative																
Consideration for personnel welfare																
Obtain necessary equipment and supplies																
Physical ability for the job																
Safety																
Other (specify)																

10. Remarks

11. Employee (signature) This rating has been discussed with me		12. Date
13. Rated By (signature)	14. Home Unit	15. Position on Fire	16. Date

NFES 1576   9/86   ICS Form 225

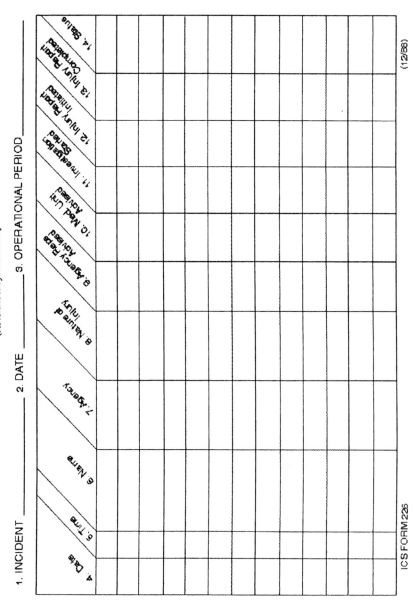

COMPENSATION FOR INJURY LOG
(See reverse side for instructions)

1. INCIDENT _____ 2. DATE _____ 3. OPERATIONAL PERIOD _____

4. Date	5. Time	6. Name	7. Agency	8. Nature of Injury	9. Agency Rep Advised	10. Med Unit Advised	11. Investigation Started	12. Injury Report Started	13. Injury Report Completed	14. Status

ICS FORM 226   (12/88)

CLAIMS LOG
(See reverse side for instructions)

1. INCIDENT _____   2. DATE _____   3. OPERATIONAL PERIOD _____

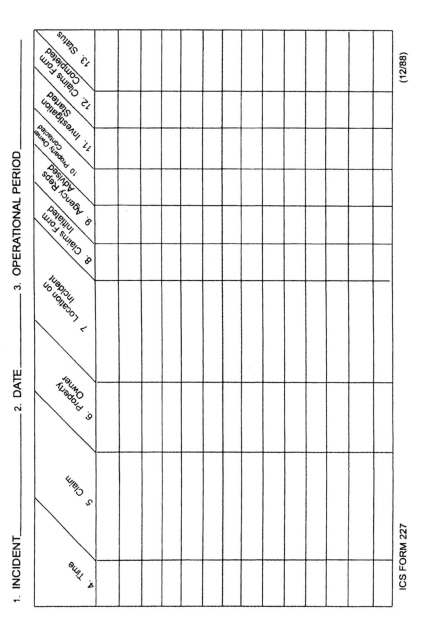

4. Time	5. Claim	6. Property Owner	7. Location on Incident	8. Claims Form Initiated	9. Agency Reps Advised	10. Property Owner Contacted	11. Investigation Started	12. Claims Form Completed	13. Status

ICS FORM 227   (12/88)

INCIDENT COST SUMMARY
(See reverse side for instructions)

Incident Name_____

Date_____Operational Period_____

I.   Engine Costs   _____

II. Hand Crew Costs   _____

III. Dozer Costs   _____

IV. Aircraft Costs (incl. retardant)   _____

V.   Overhead Costs   _____

VI. Miscellaneous Costs   _____

         Est. Oper. Period Total   _____

         Est. Incident Total   _____

Prepared by_____
             Cost Unit Leader

INCIDENT BRIEFING	1. INCIDENT NAME	2. DATE PREPARED	3. TIME PREPARED

4. MAP SKETCH

ICS 201 (12/93) NFES 1325	PAGE 1	5. PREPARED BY (NAME AND POSITION)

6. SUMMARY OF CURRENT ACTIONS

ICS 201 (12/93) NFES 1325	PAGE 2	

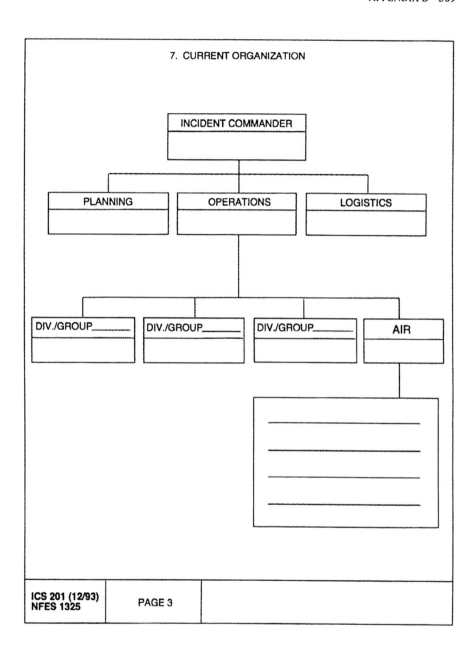

7. CURRENT ORGANIZATION

INCIDENT COMMANDER

PLANNING

OPERATIONS

LOGISTICS

DIV./GROUP_____

DIV./GROUP_____

DIV./GROUP_____

AIR

ICS 201 (12/93)
NFES 1325

PAGE 3

8. RESOURCES SUMMARY				
RESOURCES ORDERED	RESOURCES IDENTIFICATION	ETA	ON SCENE √	LOCATION/ASSIGNMENT
ICS 201 (12/93) **NFES 1325**	PAGE 4			

INCIDENT OBJECTIVES	1. INCIDENT NAME	2. DATE PREPARED	3. TIME PREPARED

4. OPERATIONAL PERIOD (DATE/TIME)

5. GENERAL CONTROL OBJECTIVES FOR THE INCIDENT (INCLUDE ALTERNATIVES)

6. WEATHER FORECAST FOR OPERATIONAL PERIOD

7. GENERAL SAFETY MESSAGE

8. ATTACHMENTS (✓ IF ATTACHED)

☐ ORGANIZATION LIST (ICS 203) ☐ MEDICAL PLAN (ICS 206) ☐ _____

☐ ASSIGNMENT LIST (ICS 204) ☐ INCIDENT MAP ☐ _____

☐ COMMUNICATIONS PLAN (ICS 205) ☐ TRAFFIC PLAN ☐ _____

9. PREPARED BY (PLANNING SECTION CHIEF)	10. APPROVED BY (INCIDENT COMMANDER)

202 ICS (1/99)   NFES 1326

ORGANIZATION ASSIGNMENT LIST	1. INCIDENT NAME	2. DATE PREPARED	3. TIME PREPARED

POSITION	NAME	4. OPERATIONAL PERIOD (DATE/TIME)		

5. INCIDENT COMMANDER AND STAFF		**9. OPERATIONS SECTION**	
INCIDENT COMMANDER		CHIEF	
DEPUTY		DEPUTY	
SAFTEY OFFICER		a BRANCH I- DIVISION/GROUPS	
INFORMATION OFFICER		BRANCH DIRECTOR	
LIAISON OFFICER		DEPUTY	
		DIVISION/GROUP	
6. AGENCY REPRESENTATIVES		DIVISION/GROUP	
AGENCY	NAME	DIVISION/GROUP	
		DIVISION/GROUP	
		DIVISION/GROUP	
		b BRANCH II- DIVISION/GROUPS	
		BRANCH DIRECTOR	
		DEPUTY	
		DIVISION/GROUP	
7. PLANNING SECTION		DIVISION/GROUP	
CHIEF		DIVISION/GROUP	
DEPUTY		DIVISION/GROUP	
RESOURCES UNIT		DIVISION/GROUP	
SITUATION UNIT			
DOCUMENTATION UNIT		c. BRANCH III- DIVISION/GROUPS	
DEMOBILIZATION UNIT		BRANCH DIRECTOR	
TECHNICAL SPECIALISTS		DEPUTY	
		DIVISION/GROUP	
		DIVISION/GROUP	
		DIVISION/GROUP	
		DIVISION/GROUP	
		DIVISION/GROUP	
8. LOGISTICS SECTION		d. AIR OPERATIONS BRANCH	
CHIEF		AIR OPERATIONS BR. DIR.	
DEPUTY		AIR TACTICAL GROUP SUP.	
		AIR SUPPORT GROUP SUP	
a SUPPORT BRANCH		HELICOPTER COORDINATOR	
DIRECTOR		AIR TANKER/FIXED WING CRD.	
SUPPLY UNIT			
FACILITIES UNIT		**10. FINANCE/ADMINISTRATION SECTION**	
GROUND SUPPORT UNIT		CHIEF	
		DEPUTY	
b. SERVICE BRANCH		TIME UNIT	
DIRECTOR		PROCUREMENT UNIT	
COMMUNICATIONS UNIT		COMPENSATION/CLAIMS UNIT	
MEDICAL UNIT		COST UNIT	
FOOD UNIT			

PREPARED BY(RESOURCES UNIT)

203 ICS (1/99)   NFES 1327

1. BRANCH	2. DIVISION/GROUP	**ASSIGNMENT LIST**

3. INCIDENT NAME	4. OPERATIONAL PERIOD
	DATE _____ TIME _____

5. OPERATIONAL PERSONNEL

OPERATIONS CHIEF _____   DIVISION/GROUP SUPERVISOR _____

BRANCH DIRECTOR _____   AIR TACTICAL GROUP SUPERVISOR _____

6. RESOURCES ASSIGNED THIS PERIOD

STRIKE TEAM/TASK FORCE/ RESOURCE DESIGNATOR	EMT	LEADER	NUMBER PERSONS	TRANS. NEEDED	PICKUP PT./TIME	DROP OFF PT./TIME

7. CONTROL OPERATIONS

8. SPECIAL INSTRUCTIONS

9. DIVISION/GROUP COMMUNICATIONS SUMMARY

FUNCTION		FREQ.	SYSTEM	CHAN.	FUNCTION		FREQ.	SYSTEM	CHAN.
COMMAND	LOCAL				SUPPORT	LOCAL			
	REPEAT					REPEAT			
DIV./GROUP TACTICAL					GROUND TO AIR				

PREPARED BY (RESOURCE UNIT LEADER)	APPROVED BY (PLANNING SECT. CH.)	DATE	TIME

204 ICS (1/99)   NFES 1328

INCIDENT RADIO COMMUNICATIONS PLAN

1. INCIDENT NAME	2. DATE/TIME PREPARED	3. OPERATIONAL PERIOD DATE/TIME

4. BASE RADIO CHANNEL UTILIZATION

SYSTEM/CACHE	CHANNEL	FUNCTION	FREQUENCY/TONE	ASSIGNMENT	REMARKS

5. PREPARED BY (COMMUNICATIONS UNIT)

205 ICS (9/66)

NFES 1330

MEDICAL PLAN	1. INCIDENT NAME	2. DATE PREPARED	3. TIME PREPARED	4. OPERATIONAL PERIOD

5. INCIDENT MEDICAL AID STATIONS

MEDICAL AID STATIONS	LOCATION	PARAMEDICS	
		YES	NO

6. TRANSPORTATION

A. AMBULANCE SERVICES

NAME	ADDRESS	PHONE	PARAMEDICS	
			YES	NO

B. INCIDENT AMBULANCES

NAME	LOCATION	PARAMEDICS	
		YES	NO

7. HOSPITALS

NAME	ADDRESS	TRAVEL TIME		PHONE	HELIPAD		BURN CENTER	
		AIR	GRND		YES	NO	YES	NO

8. MEDICAL EMERGENCY PROCEDURES

206 ICS 8/78	9. PREPARED BY (MEDICAL UNIT LEADER)	10 REVIEWED BY (SAFETY OFFICER)

NFES 1331

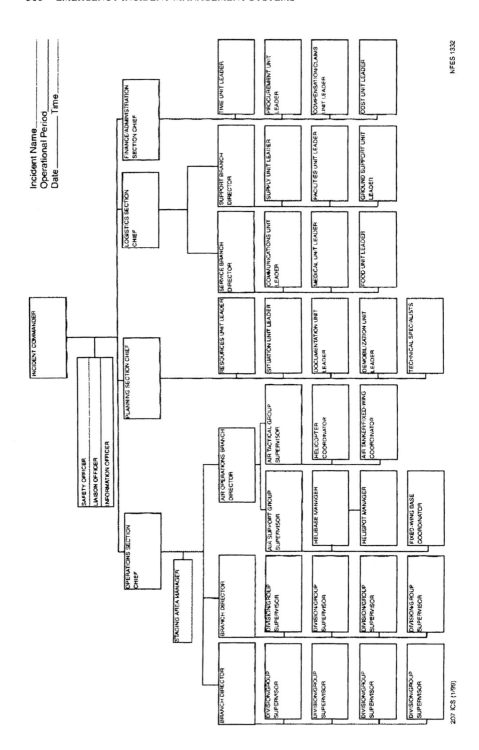

Incident Status Summary (ICS-209)

1: Date 2: Time 3: Initial | Update | Final 4: Incident Number 5: Incident Name

6: Incident Kind 7: Start Date Time 8: Cause 9: Incident Commander 10: IMT Type 11: State-Unit

12: County 13: Latitude and Longitude Lat: Long: 14: Short Location Description (in reference to nearest town):

Current Situation

15: Size/Area Involved 16: % Contained or MMA 17: Expected Containment Date: Time: 18: Line to Build 19: Costs to Date 20: Declared Controlled Date: Time:

21: Injuries this Reporting Period: 22: Injuries to Date: 23: Fatalities 24: Structure Information

Type of Structure	# Threatened	# Damaged	# Destroyed
Residence			
Commercial			
Property			
Outbuilding/Other			

25: Threat to Human Life/Safety:
Evacuation(s) in progress ----
No evacuation(s) imminent --
Potential future threat --------
No likely threat --------------

26: Communities/Critical Infrastructure Threatened (in 12, 24, 48 and 72 hour time frames):

12 hours:

24 hours:

48 hours:

72 hours:

27: Critical Resource Needs (kind & amount, in priority order):
1.
2.
3.

28: Major problems and concerns (control problems, social/political/economic concerns or impacts, etc.) Relate critical resources needs identified above to the Incident Action Plan.

29: Resources threatened (kind(s) and value/significance):

DESIGNATOR
NAME/ ID. NO. _____

_____

STATUS

☐ ASSIGNED   ☐ AVAILABLE   ☐ O/S REST
☐ O/S MECHANICAL   ☐ O/S MANNING
_____ ETR (O/S= Out of Service)

FROM	LOCATION	TO
	DIVISION/GROUP	
	STAGING AREA	
	BASE/ICP	
	CAMP	
	ENROUTE	ETA
	HOME AGENCY	

MESSAGES

RESTAT

TIME _____   PROCESS ☐

ICS   STATUS CHANGE CARD
FORM
210   6/83   NFES 1334

CHECK-IN LIST

1. INCIDENT NAME	2. CHECK-IN LOCATION	3. DATE/TIME
	☐ BASE ☐ CAMP ☐ STAGING AREA ☐ ICP RESOURCES ☐ HELIBASE	

CHECK-IN INFORMATION

4. PERSONNEL (OVERHEAD) BY AGENCY & NAME -OR-

LIST EQUIPMENT BY THE FOLLOWING FORMAT

AGENCY	SINGLE RESOURCE KIND	TYPE	I.D. NO./NAME	5. ORDER/ REQUEST NUMBER	6. DATE/TIME CHECK-IN	7. LEADER'S NAME	8. TOTAL NO. PERSONNEL	9. MANIFEST YES | NO	10. CREW WEIGHT INDIVIDUAL WEIGHT	11. HOME BASE	12. DEPARTURE POINT	13. METHOD OF TRAVEL	14. INCIDENT ASSIGNMENT	15. OTHER QUALIFICATION	16. SENT TO RESOURCES TIME/INT.

17. PAGE ___ OF ___	18. PREPARED BY (NAME AND POSITION)	USE BACK FOR REMARKS OR COMMENTS

211 ICS (1/99)

NFES 1335

*U.S. GPO: 1009-793-975

GENERAL MESSAGE

TO: _____ : POSITION _____

FROM _____ : POSITION _____

SUBJECT _____ │DATE _____

MESSAGE:

DATE   TIME   SIGNATURE/POSITION

213 ICS 1/79
NFES 1336

PERSON RECEIVING GENERAL MESSAGE KEEP THIS COPY

SENDER REMOVE THIS COPY FOR YOUR FILES

*U.S. GPO: 1009-793-975

GENERAL MESSAGE

TO: _____ | POSITION _____

FROM | POSITION

SUBJECT | DATE

MESSAGE:

DATE   TIME   SIGNATURE/POSITION

213 ICS 1/79
NFES 1336

PERSON RECEIVING GENERAL MESSAGE KEEP THIS COPY

SENDER REMOVE THIS COPY FOR YOUR FILES

UNIT LOG	1. INCIDENT NAME	2 DATE PREPARED	3. TIME PREPARED
4. UNIT NAME/DESIGNATORS.	5. UNIT LEADER (NAME AND POSITION)	6. OPERATIONAL PERIOD	

7. PERSONNEL ROSTER ASSIGNED		
NAME	ICS POSITION	HOME BASE

8. ACTIVITY LOG (CONTINUE ON REVERSE)	
TIME	MAJOR EVENTS

NFES 1337

OPERATIONAL PLANNING WORKSHEET

1. INCIDENT NAME	
2. DATE PREPARED	TIME PREPARED
3. OPERATIONAL PERIOD (DATE/TIME)	

4. DIVISION OR OTHER LOCATION	5. WORK ASSIGNMENTS	6. RESOURCE	RESOURCES BY TYPE (SHOW STRIKE TEAM AS ST)	7. REPORTING LOCATION	8. REQUESTED ARRIVAL TIME
		TYPE			
		REQ			
		HAVE			
		NEED			
		REQ			
		HAVE			
		NEED			
		REQ			
		HAVE			
		NEED			
		REQ			
		HAVE			
		NEED			
		REQ			
		HAVE			
		NEED			
		REQ			
		HAVE			
		NEED			
		REQ			
		HAVE			
		NEED			

9. TOTAL RESOURCES REQUIRED	SINGLE RESOURCES	STRIKE TEAMS
TOTAL RESOURCES ON HAND		
TOTAL RESOURCES NEEDED		

10. PREPARED BY (NAME AND POSITION)

215 ICS 8-86

NFES 1338

INCIDENT ACTION PLAN SAFETY ANALYSIS	1. Incident Name	2. Date	3. Time

LCES* Analysis of Tactical Applications
Lookouts Communications Escape routes Safety zones

		Dynamic map	Tactical Area	Downhill Fire	Underslung Fire-ae	Escape Routes	Portal Areas	Anchor Points	Extreme Conditions (Spotting, Wind driven)	Reburn Potential	LCES Mitigations											Hazard Material	Transportation, 1 Hr +	Communications	Structure Protection	Other Risk Analysis	Other Risk Mitigations

Prepared by (Name and Position)

ICS 215A

RADIO REQUIREMENTS WORKSHEET

1. INCIDENT NAME

2. DATE

3. TIME

4. BRANCH

5. AGENCY

6. OPERATIONAL PERIOD

7. TACTICAL FREQUENCY

8. DIVISION/GROUP

AGENCY

9. AGENCY

DIVISION/GROUP		
AGENCY		
AGENCY	ID NO.	RADIO RQMTS

DIVISION/GROUP		
AGENCY		
AGENCY	ID NO.	RADIO RQMTS

DIVISION/GROUP		
AGENCY		
AGENCY	ID NO.	RADIO RQMTS

AGENCY	ID NO.	RADIO RQMTS

PAGE

5. PREPARED BY (COMMUNICATIONS UNIT)

216 ICS 3-82

NFES 1339

*GPO 1985-0-593-005/14,028

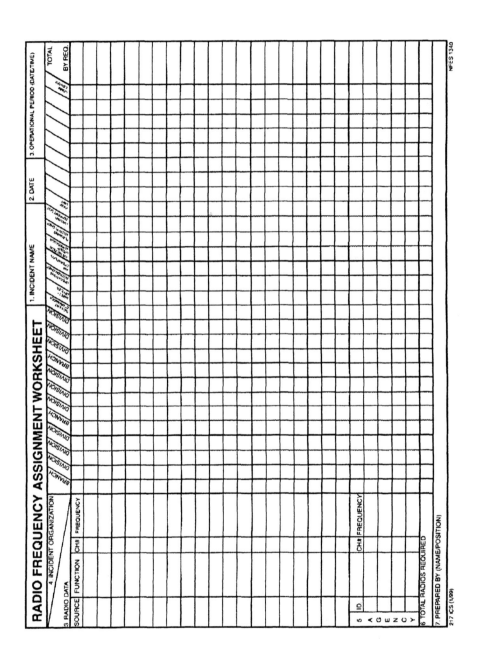

SUPPORT VEHICLE INVENTORY
(USE SEPARATE SHEET FOR EACH VEHICLE CATEGORY)

1. INCIDENT NAME	2. DATE PREPARED	3. TIME PREPARED

VEHICLE INFORMATION

a. TYPE	b. MAKE	c. CAPACITY/SIZE	d. AGENCY/OWNER	e. I.D. NO.	f. LOCATION	g. RELEASE TIME

PAGE	5. PREPARED BY (GROUND SUPPORT UNIT)

218 ICS 8-78

NFES 1341

GREEN CARD STOCK (CREW)

AGENCY	ST	KIND	TYPE	I.D. NO

ORDER/REQUEST NO   DATE/TIME CHECK IN

HOME BASE

DEPARTURE POINT

LEADER NAME

CREW ID NO./NAME (FOR STRIKE TEAMS)

NO PERSONNEL	MANIFEST	WEIGHT
	YES NO	

METHOD OF TRAVEL
☐ OWN   ☐ BUS   ☐ AIR

OTHER

DESTINATION POINT   ETA

TRANSPORTATION NEEDS
☐ OWN   ☐ BUS   ☐ AIR

OTHER

ORDERED DATE/TIME   CONFIRMED DATE/TIME

REMARKS

ICS 219-2 (Rev. 4/82) CREW NFES 1344

AGENCY	TF	KIND	TYPE	I.D. NO /NAME

INCIDENT LOCATION   TIME

STATUS
☐ ASSIGNED   ☐ O/S REST   ☐ O/S PERS
☐ AVAILABLE   ☐ O/S MECH   ☐ ETR

NOTE

INCIDENT LOCATION   TIME

STATUS
☐ ASSIGNED   ☐ O/S REST   ☐ O/S PERS.
☐ AVAILABLE   ☐ O/S MECH   ☐ ETR

NOTE

INCIDENT LOCATION   TIME

STATUS
☐ ASSIGNED   ☐ O/S REST   ☐ O/S PERS.
☐ AVAILABLE   ☐ O/S MECH   ☐ ETR

NOTE

INCIDENT LOCATION   TIME

STATUS
☐ ASSIGNED   ☐ O/S REST   ☐ O/S PERS
☐ AVAILABLE   ☐ O/S MECH   ☐ ETR

NOTE

*U.S. GPO 1990-794-001

BLUE CARD STOCK (HELICOPTER)

AGENCY	ST	KIND	TYPE	I.D. NO.

ORDER/REQUEST NO   DATE/TIME CHECK IN

HOME BASE

DEPARTURE POINT

PILOT NAME

DESTINATION POINT   ETA

REMARKS

INCIDENT LOCATION

STATUS
[ ] ASSIGNED   [ ] O/S REST   [ ] O/S PERS
[ ] AVAILABLE   [ ] O/S MECH   [ ] ETR
NOTE

INCIDENT LOCATION   TIME

STATUS
[ ] ASSIGNED   [ ] O/S REST   [ ] O/S PERS.
[ ] AVAILABLE   [ ] O/S MECH   [ ] ETR
NOTE

ICS 219-4 (Rev. 4/82) HELICOPTER NFES 1346

AGENCY	TYPE	MANUFACTURER	I.D. NO.

INCIDENT LOCATION   TIME

STATUS
[ ] ASSIGNED   [ ] O/S REST   [ ] O/S PERS
[ ] AVAILABLE   [ ] O/S MECH   [ ] ETR
NOTE

INCIDENT LOCATION   TIME

STATUS
[ ] ASSIGNED   [ ] O/S REST   [ ] O/S PERS
[ ] AVAILABLE   [ ] O/S MECH   [ ] ETR
NOTE

INCIDENT LOCATION   TIME

STATUS
[ ] ASSIGNED   [ ] O/S REST   [ ] O/S PERS
[ ] AVAILABLE   [ ] O/S MECH   [ ] ETR
NOTE

INCIDENT LOCATION   TIME

STATUS
[ ] ASSIGNED   [ ] O/S REST   [ ] O/S PERS.
[ ] AVAILABLE   [ ] O/S MECH   [ ] ETR
NOTE

*U.S. GPO: 1988-594-771   NFES 1346

ORANGE CARD STOCK (AIRCRAFT)

AGENCY	TYPE	MANUFACTURER	I.D. NO.

ORDER/REQUEST NO.   DATE/TIME CHECK IN

HOME BASE

DATE TIME RELEASED

INCIDENT LOCATION		TIME

STATUS
- [ ] ASSIGNED   [ ] O/S REST   [ ] O/S PERS
- [ ] AVAILABLE   [ ] O/S MECH   [ ] ETR

NOTE

INCIDENT LOCATION		TIME

STATUS
- [ ] ASSIGNED   [ ] O/S REST   [ ] O/S PERS
- [ ] AVAILABLE   [ ] O/S MECH   [ ] ETR

NOTE

INCIDENT LOCATION		TIME

STATUS
- [ ] ASSIGNED   [ ] O/S REST   [ ] O/S PERS
- [ ] AVAILABLE   [ ] O/S MECH   [ ] ETR

NOTE

ICS 219-6 (4/82) AIRCRAFT

AGENCY	TYPE	MANUFACTURER NAME/NO	I.D. NO.

INCIDENT LOCATION		TIME

STATUS
- [ ] ASSIGNED   [ ] O/S REST   [ ] O/S PERS.
- [ ] AVAILABLE   [ ] O/S MECH   [ ] ETR

NOTE

INCIDENT LOCATION		TIME

STATUS
- [ ] ASSIGNED   [ ] O/S REST   [ ] O/S PERS.
- [ ] AVAILABLE   [ ] O/S MECH   [ ] ETR

NOTE

INCIDENT LOCATION		TIME

STATUS
- [ ] ASSIGNED   [ ] O/S REST   [ ] O/S PERS.
- [ ] AVAILABLE   [ ] O/S MECH   [ ] ETR

NOTE

INCIDENT LOCATION		TIME

STATUS
- [ ] ASSIGNED   [ ] O/S REST   [ ] O/S PERS.
- [ ] AVAILABLE   [ ] O/S MECH   [ ] ETR

NOTE

*U.S. GPO: 695-182-1988   NFES 1348

YELLOW CARD STOCK (DOZERS)

AGENCY	ST	TF	KIND	TYPE	I.D. NO.

ORDER/REQUEST NO.   DATE/TIME CHECK IN

HOME BASE

DEPARTURE POINT

LEADER NAME

RESOURCE ID. NO S/NAMES

DESTINATION POINT   ETA

REMARKS

INCIDENT LOCATION   TIME

STATUS
☐ ASSIGNED   ☐ O/S REST   ☐ O/S PERS.
☐ AVAILABLE   ☐ O/S MECH   ☐ ETR
NOTE

ICS 219-7 (Rev. 4/82) DOZERS NFES 1349

AGENCY	ST	TF	KIND	TYPE	I.D. NO.

INCIDENT LOCATION   TIME

STATUS
☐ ASSIGNED   ☐ O/S REST   ☐ O/S PERS.
☐ AVAILABLE   ☐ O/S MECH   ☐ ETR
NOTE

INCIDENT LOCATION   TIME

STATUS
☐ ASSIGNED   ☐ O/S REST   ☐ O/S PERS.
☐ AVAILABLE   ☐ O/S MECH   ☐ ETR
NOTE

INCIDENT LOCATION   TIME

STATUS
☐ ASSIGNED   ☐ O/S REST   ☐ O/S PERS.
☐ AVAILABLE   ☐ O/S MECH   ☐ ETR
NOTE

INCIDENT LOCATION   TIME

STATUS
☐ ASSIGNED   ☐ O/S REST   ☐ O/S PERS
☐ AVAILABLE   ☐ O/S MECH   ☐ ETR
NOTE

*U.S. GPO. 1990-794-006

AIR OPERATIONS SUMMARY

PREPARED BY:		PREPARED DATE/TIME:

1. INCIDENT NAME

2. OPERATIONAL PERIOD DATE: | START TIME: | END TIME: | SUNRISE: | SUNSET:

3. REMARKS (Safety Notes, Hazards, Air Operations Special Equipment, etc.)

4. MEDEVAC A/C:

5. TFR:
Radius: _____ NM
Altitude: _____ ' MSL
Centerpoint: Lat: _____
Long: _____

6. PERSONNEL	Phone	7. FREQUENCIES	AM	FM	8. FIXED-WING	#Available/Type/ Make-Model/ FAA N#/ Bases
AOBD:		AIR/AIR FW			Airtankers	
ATGS:		AIR/AIR RW:				
HLCO:		AIR/GROUND:				
ASGS:		COMMAND: (Simplex)			Leadplanes	
HEBM:		COMMAND RPT	Rx:	Tx:	Base FAX#	
ATB MGR		DECK FREQ.:			ATGS Aircraft	
		TOLC FREQ.:				
					Other	

9. HELICOPTERS (Use Additional Sheets As Necessary)

FAA N#	TY	MAKE/MODEL	BASE	AVAIL	START	REMARKS	FAA N#	TY	MAKE/MODEL	BASE	AVAIL	START	REMARKS

220 ICS (2/99)

NFES 1351

DEMOBILIZATION CHECKOUT

ICS-221

1. INCIDENT NAME/NUMBER	2. DATE/TIME	3. DEMOB NO.

4. UNIT/PERSONNEL RELEASED

5. TRANSPORTATION TYPE/NO.

6. ACTUAL RELEASE DATE/TIME	7. MANIFEST YES NO
	NUMBER

8. DESTINATION	9. AREA/AGENCY/REGION NOTIFIED
	NAME
	DATE

10. UNIT LEADER RESPONSIBLE FOR COLLECTING PERFORMANCE RATING

11. UNIT/PERSONNEL   YOU AND YOUR RESOURCES HAVE BEEN RELEASED SUBJECT TO SIGNOFF FROM THE FOLLOWING:

(DEMOB. UNIT LEADER CHECK ✓ APPROPRIATE BOX)

LOGISTICS SECTION

SUPPLY UNIT

COMMUNICATIONS UNIT

FACILITIES UNIT

GROUND SUPPORT UNIT LEADER

PLANNING SECTION

DOCUMENTATION UNIT

FINANCE/ADMINISTRATION SECTION

TIME UNIT

OTHER

12. REMARKS

221 ICS 1/83

INSTRUCTIONS ON BACK

CREW PERFORMANCE RATING	Instructions: This rating is to be used only for determining an individual's fire fighting qualifications. All blocks must be completed. Crew will be rated by the immediate supervisor, not crew representative. If deficiencies are indicated for items 9 and 10, explain in item 1.

1. Crew Name and Number	2. Fire Name and Number	3. Crew Boss (name)

4. Crew Home Unit and Address	5. Location of Fire (complete address)

6. Crew Representative	7. Dates on Fire	8. Number of Shifts Worked

9. Crew Evaluation

Rating Factors	Excellent	Satisfactory	Deficient	Special Improve
Physical Condition				
Hot Line Construction				
Mop Up				
Off Line Conduct				
Use of Safe Practices				
Crew Organization and Equipment				
Other (specify)				

10. Supervisory Performances

Crew Boss				
Squad Bosses				
Crew Representative				

11. Areas Needing Improvement

12. Names of Outstanding Workers (comment)	13. Names of Individuals Needing Improvement (indicate area(s))

14. Remarks

15. Crew Boss (signature) This rating has been discussed with me.	16. Date

17. Rated By (signature)	18. Home Unit (address)	19. Position of Fire	20. Date

ICS 224   NFES 1577

INCIDENT PERSONNEL PERFORMANCE RATING	*INSTRUCTIONS:* The immediate job supervisor will prepare this form for each subordinate. It will be delivered to the planning section before the rater leaves the fire. Rating will be reviewed with employee who will sign at the bottom.

THIS RATING IS TO BE USED ONLY FOR DETERMINING AN INDIVIDUAL'S PERFORMANCE

1. Name	2. Fire Name and Number
3. Home Unit *(address)*	4. Location of Fire *(address)*

5. Fire Position	6. Date of Assignment From: To:	7. Acres Burned	8. Fuel Type(s)

9. Evaluation

Enter X under appropriate rating number and under proper heading for each category listed. Definition for each rating number follows:

0 - Deficient. Does not meet minimum requirements of the individual element.

   DEFICIENCIES MUST BE IDENTIFIED IN REMARKS.

1 - Needs to improve. Meets some or most of the requirements of the individual element.

   IDENTIFY IMPROVEMENT NEEDED IN REMARKS.

2 - Satisfactory. Employee meets all requirements of the individual element.

3. - Superior. Employee consistently exceeds the performance requirements.

Rating Factors	Hot Line				Mop-Up				Camp				Other specify)			
	0	1	2	3	0	1	2	3	0	1	2	3	0	1	2	3
Knowledge of the job																
Ability to obtain performance																
Attitude																
Decisions under stress																
Initiative																
Consideration for personnel welfare																
Obtain necessary equipment and supplies																
Physical ability for the job																
Safety																
Other *(specify)*																

10. Remarks

11. Employee *(signature)* This rating has been discussed with me		12. Date	
13. Rated By *(signature)*	14. Home Unit *(address)*	15. Position of Fire	16. Date

ICS 225

INDIVIDUAL PERFORMANCE RATING	INSTRUCTIONS: The immediate supervisor will prepare this form for a subordinate person. Rating will be reviewed with the individual who will sign and date the form. The completed rating will be given to the Planning Section Chief before the rater leaves the incident.

1. NAME	2. INCIDENT NAME AND NUMBER	START DATE OF INCIDENT

3. HOME UNIT ADDRESS	4. INCIDENT AGENCY AND ADDRESS

5. POSITION HELD ON INCIDENT	6. TRAINEE POSITION ☐ YES ☐ NO	7. INCIDENT COMPLEXITY ☐ I ☐ II ☐ III	8. DATE OF ASSIGNMENT FROM: TO:

9. List the main duties from the Position Checklist, on which the position will be rated. Enter X under the appropriate column indicating the individuals level of performance for each duty listed.	Did not apply on this incident	Unacceptable	Need to Improve	Fully Successful	Exceeds Successful
		EXPLAIN IN REMARKS			

PERFORMANCE LEVEL

10. REMARKS

11. THIS RATING HAS BEEN DISCUSSED WITH ME (Signature of individual being rated.)	12. DATE

13. RATED BY (Signature)	14. HOME UNIT	15. POSITION HELD ON THIS INCIDENT	16. DATE

	OSHA Abatement Plan
	Incident Management Team Checklist
☐	Direction has been given and IMT will ensure a briefing be given to all arriving fire line personnel.
☐	IMT should monitor performance of firefighters and fire line supervisors and ensure that they understand and exercise their responsibilities.
☐	Incident Management Teams should utilize Incident Safety Debriefing forms.
☐	IMT should implement procedures requiring appropriate briefing for personnel assigned to fires. Briefing should include the following: • Incident Organization • Objectives/Operational Plan • Safety Information • Weather • Fire Danger/Fire Behavior • Fuels • Topography
☐	Incident Management Team will provide a briefing to fire line personnel prior to each operational period.
☐	Fire weather forecasts should be communicated twice daily and include NFDRS outputs one a day.
☐	Crew supervisors and firefighters know they must receive a briefing prior to beginning their assignment and I.C. is responsible for ensuring that all resources assigned to an incident are briefed prior to fire line duty.
☐	Fire line supervisors need to ensure that lookouts are posted in potentially dangerous situations.
☐	IMT needs to take immediate corrective actions when unsafe practices and processes are identified. It is the responsibility of every person to take immediate action when violations are found.
☐	Local Agency Administrators should monitor and evaluate suppression operations and individual performance on all fires.
☐	Policy of "The Fire Orders are Firm; We Don't Bend Them, We Don't Break Them" should be communicated to all fire line personnel.
☐	Fire line supervisors should ensure that downhill/indirect fire line will not begin until the elements of the downhill/indirect fire line construction guidelines have been evaluated and the decision has been made that the operations can be implemented safely.
☐	IMT will monitor fire suppression operations to ensure that the guidelines for downhill/indirect fire line construction are adhered to and take immediate corrective action when violations are found.
☐	Assigned Safety Personnel, I.C.s, and fire line supervisors have authority to correct any and all safety violations.
☐	FBA should obtain and distribute fire weather information in a timely and consistent manner by radio transmission or other appropriate methods. This has high priority over all other transmissions except life threatening situations.
☐	IMT should receive a hard copy of both morning and afternoon weather forecasts. In addition, these forecasts should be read over the radio in morning and afternoon. Forecasts should be relayed to incident and incoming personnel.
☐	IMT should monitor fire situations and implement appropriate suppression organizations commensurate with the fire threat and complexity of situation.
☐	IMT has a role in ensuring that the appropriate suppression organizations are in place and are commensurate with complexity of the situation.
☐	Special weather updates should be evaluated and adjustments made to operating plans. These updates will be communicated to personnel assigned to fires.
☐	IMT should monitor fire activity and ensure fire management systems and processes are in place and functioning and that safety requirements are not compromised.
☐	IMT should evaluate individual fire suppression operations and perform post fire critiques to ensure safe practices have been implemented. Immediate corrective actions should be taken when unsafe practices or processes are identified.

IRSS Check-In Form

Check-in Location _____ Date/Time _____

Agency	S T/F S/T	Date/Time Check in	Leader's Name	Home Unit	City State	Airport	Method of Travel	Transport ID	Incident Assignment:	Other Qualifications Put a (T) for Training Quals	Last R&R Dt

ck for comments

INCIDENT/PROJECT ORDER NUMBER

RESOURCE ORDER

| INITIAL DATE/TIME | 2. INCIDENT/PROJECT NAME | 3. INCIDENT /PROJECT ORDER NUMBER | 4. OFFICE REFERENCE NUMBER |

5. DESCRIPTIVE LOCATION/RESPONSE AREA | 6. SEC. | TWN | RNG | Base MDM | 8. INCIDENT BASE/PHONE NUMBER | 9. JURISDICTION/AGENCY

7. MAP REFERENCE | 10. ORDERING OFFICE

11. AIRCRAFT INFORMATION | LAT. | | LONG.

| BEARING | DISTANCE | BASE OR OMNI | AIR CONTACT | FREQUENCY | | Ground Contact | FREQUENCY | RELOAD BASE | OTHER AIRCRAFT/HAZARDS |

12.
| Request Number | Ordered Date/Time | From | To | QTY | RESOURCE REQUESTED | Needed Date/Time | Deliver To | To | From | Time | Agency ID | RESOURCE ASSIGNED | ETD | ETA | RELEASED Date | To | ETA Time |

ACTION TAKEN

13.
| Req. No. | Date | Time | ORDER RELAYED To/From | | ACTION TAKEN | Req. No. | Date | Time | ORDER RELAYED To/From | | ACTION TAKEN |

ICS 259-3 (7/87)   NFES 2202

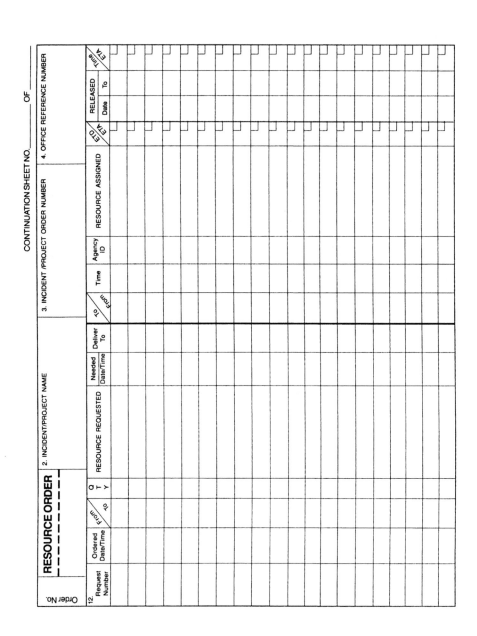

AIR OPERATIONS SUMMARY

PREPARED BY:	PREPARED DATE/TIME:

1. INCIDENT NAME

2. OPERATIONAL PERIOD DATE:

START TIME:	END TIME:	SUNRISE:	SUNSET:

3. REMARKS (Safety Notes, Hazards, Air Operations Special Equipment, etc.).

4. MEDEVAC A/C:

5. TFR:
Radius _____ NM
Altitude: _____ ' MSL
Centerpoint: Lat: _____
Long: _____

6. PERSONNEL

	Phone
AOBD:	
ATGS:	
HLCO:	
ASGS:	
HEBM:	
ATB MGR	

7. FREQUENCIES

	AM	FM
AIR/AIR FW:		
AIR/AIR RW:		
AIR/GROUND:		
COMMAND: (Simplex)		
COMMAND RPT	Rx:	Tx:
DECK FREQ.:		
TOLC FREQ.:		

8. FIXED-WING #Available/Type/ Make-Model/ FAA N#/ Bases

Airtankers

Leadplanes

Base FAX#

ATGS Aircraft

Other

9. HELICOPTERS (Use Additional Sheets As Necessary)

FAA N#	TY	MAKE/MODEL	BASE	AVAIL	START	REMARKS

FAA N#	TY	MAKE/MODEL	BASE	AVAIL	START	REMARKS

220 ICS (2/99)

PAGE 1 OF 2

NFES 1351

1. Incident Name	2. Operational Period (Date / Time)	EXECUTIVE SUMMARY
	From: To:	

3. Operations

4. Environmental

5. Planning

6. Other

7. Prepared by	Date / Time

EXECUTIVE SUMMARY   June 2000

Electronic version: NOAA 1.0 June 1, 2000

1. Incident Name			GENERAL PLAN
2. Prepared By	Date / Time Prepared	3. Operational Period (Date / Time) From: To:	
4. Notification (Date and time completed)		5. Response Initiation (Date and time completed)	

6. Plan Item	Timeframe ⟹ (Enter days or weeks)
Site Characterization, Forecasts, and Analysis	
Site Safety	
Site Security	
Source Stabilization, Salvage, and Lightering	
Surveillance	
On Water Containment and Recovery	
Sensitive Areas / Resources at Risk	
Alternative Response Technology	
Shoreline Protection and Recovery	
Wildlife Protection and Rehabilitation	
Logistics Support	
Response Organization	
Communications	
Public Information	
Financial Management and Cost Documentation	
NRDA and Claims	
Training	
Information Management	
Restoration / Mitigation	
Waste Management	
Demobilization	

June 2000

GENERAL PLAN

Electronic version: NOAA 1.0 June 1, 2000

1. Incident Name	2. Operational Period to be covered by IAP (Date / Time) From: To:	**IAP COVER SHEET**

3. Approved by:

FOSC _____

SOSC _____

RPIC _____

_____   _____

_____   _____

INCIDENT ACTION PLAN

The items checked below are included in this Incident Action Plan:

☐ ICS 202-OS (Response Objectives)

☐ ICS 203-OS (Organization List) - OR - ICS 207-OS (Organization Chart)

☐ ICS 204-OSs (Assignment Lists)
One Copy each of any ICS 204-OS attachments:

- ☐ Map
- ☐ Weather forecast
- ☐ Tides
- ☐ Shoreline Cleanup Assessment Team Report for location
- ☐ Previous day's progress, problems for location

☐ ICS 205-OS (Communications List)

☐ ICS 206-OS (Medical Plan)

☐ _____

☐ _____

☐ _____

☐ _____

☐ _____

☐ _____

4. Prepared by:	Date / Time

IAP COVER SHEET   June 2000

1. Incident Name	2. Operational Period to be covered by IAP (Date / Time)	IAP COVER SHEET
	From: To:	

3. Approved by:

FOSC _____

SOSC _____

RPIC _____

_____ _____

_____ _____

INCIDENT ACTION PLAN

4. Prepared by: Date / Time

IAP COVER SHEET	June 2000	

1. Incident Name	2. Prepared by: (name)	INCIDENT BRIEFING ICS
	Date: Time:	201-OS (pg 1 of 4)

3. Map / Sketch   (Include maps drawn here or attached, showing the total area of operations, the incident site/area, overflight results, trajectories, impacted shorelines, or other graphics depicting situational and response status)

INCIDENT BRIEFING	June 2000	ICS 201-OS (pg 1 of 4)

1. Incident Name	2. Operational Period (Date / Time) From:	**INCIDENT OBJECTIVES** ICS 202-OS

3. Overall Incident Objective(s)

Ensure the Safety of Citizens and Response Personnel
Control the Source of the Spill
Manage a Coordinated Response Effort
Maximize Protection of Environmentally-Sensitive Areas
Contain and Recover Spilled Material
Recover and Rehabilitate Injured Wildlife
Remove Oil from Impacted Areas
Minimize Economic Impacts
Keep Stakeholders and Public Informed of Response Activities

4. Objectives for specified Operational Period

5. Safety Message for specified Operational Period

Approved Site Safety Plan Located at:

6. Weather	**See Attached Weather Sheet**
7. Tides / Currents	**See Attached Tide / Current Data**
8. Time of Sunrise	**Time of Sunset**

9. Attachments (mark "X" if attached)

☐ Organization List (ICS 203-OS)	☐ Medical Plan (ICS 206-OS)	☐ Resource at Risk Summary (ICS 232-OS)
☐ Assignment List (ICS 204-OS)	☐ Incident Map(s)	☐ ....
☐ Communications List (ICS 205-OS)	☐ Traffic Plan	☐ ....

10. Prepared by: (Planning Section Chief)	Date / Time

INCIDENT OBJECTIVES	June 2000	ICS 202-OS

Electronic version. NOAA 1.0 June 1, 2000

1. Incident Name	2. Operational Period (Date / Time) From:	**ORGANIZATION ASSIGNMENT LIST** **ICS 203-OS**

3. Incident Commander and Staff	7. OPERATION SECTION

3. Incident Commander and Staff

	Primary	Deputy
Federal:		
State:		
RP(s):		
Safety Officer:		
Information Officer:		
Liaison Officer:		

4. Agency Representatives

Agency	Name	

5. PLANNING SECTION

Chief	
Deputy	
Resources Unit	
Situation Unit	
Environmental Unit	
Documentation Unit	
Demobilization Unit	
Technical Specialists	

6. LOGISTICS SECTION

Chief	
Deputy	
a. Support Branch Director	
Supply Unit	
Facilities Unit	
Transportation Unit	
Vessel Support Unit	
Ground Support Unit	
b. Service Branch Director	
Communications Unit	
Medical Unit	
Food Unit	

7. OPERATION SECTION

Chief	
Deputy	

a. Branch I - Division/Groups

Branch Director	
Deputy	
Division / Group	
Division / Group	
Division / Group	
Division / Group	
Division / Group	

b. Branch II - Division/Groups

Branch Director	
Deputy	
Division / Group	
Division / Group	
Division / Group	
Division / Group	
Division / Group	

c. Branch III - Division/Groups

Branch Director	
Deputy	
Division / Group	
Division / Group	
Division / Group	
Division / Group	
Division / Group	

d. Air Operations Branch

Air Operations Br. Dir	
Air Tactical Supervisor	
Air Support Supervisor	
Helicopter Coordinator	
Fixed Wing Coordinator	

8. FINANCE / ADMINISTRATION SECTION

Chief	
Deputy	
Time Unit	
Procurement Unit	
Compensation/Claims Unit	
Cost Unit	

9. Prepared By: (Resources Unit)	Date / Time

ORGANIZATION ASSIGNMENT LIST	June 2000	ICS 203-OS

Electronic version: NOAA 1.0 June 1. 2000

1. Incident Name	2. Operational Period (Date / Time)	ASSIGNMENT LIST
	From: To:	ICS 204-OS

3. Branch	4. Division/Group

5. Operations Personnel Name Affiliation Contact # (s)

Operations Section Chief: _____

Branch Director: _____

Division/Group Supervisor: _____

6. Resources Assigned This Period   "X" Indicates 204a attachment with special instructions

Strike Team / Task Force / Resource Identifier	Leader	Contact Info. #	# of Persons	Notes / Remarks	
					☐
					☐
					☐
					☐
					☐
					☐
					☐

7. Assignments

8. Special Instructions for Division / Group

9. Communications (radio and / or phone contact numbers needed for this assignment)

Name / Function   Radio: Freq. / System / Channel   Phone   Pager

_____   _____   _____   _____
_____   _____   _____   _____
_____   _____   _____   _____

Emergency Communications

Medical _____ Evacuation _____ Other _____

10. Prepared By (Resource Unit Leader)	Date / Time	11. Approved By (Planning Section Chief)	Date / Time

ASSIGNMENT LIST	June 2000	ICS 204-OS

Electronic version: NOAA 1.0 June 1, 2000

1. Incident Name	2. Operational Period (Date / Time)		**ASSIGNMENT LIST ATTACHMENT**
	From:	To:	**ICS 204a-OS**

3. Branch	4. Division / Group

5. Strike Team / Task Force / Resource Identifier	6. Leader	7. Assignment Location

8. Work Assignment Special Instructions (if any)   [Ops]

9. Special Equipment / Supplies Needed for Assignment (if any)   [Ops]

10. Special Environmental Considerations (if any)   [P.S.C.]

11. Special Site-Specific Safety Considerations (if any)   [S.O.]

Approved Site Safety Plan Located at:

12. Other Attachments (as needed)

☐ Map   ☐ Shoreline Cleanup Assessment Team Report   ☐ _____

☐ Weather Forecast   ☐ Tides   ☐ _____

13. Prepared by: (Resources Unit Leader)   Date / Time

ASSIGNMENT LIST ATTACHMENT   June 2000   ICS 204a-OS

1. Incident Name	2. Operational Period (Date / Time) From: To:	INCIDENT RADIO COMMUNICATIONS PLAN ICS 205-OS

3. BASIC RADIO CHANNEL USE

SYSTEM / CACHE	CHANNEL	FUNCTION	FREQUENCY	ASSIGNMENT	REMARKS

4. Prepared by: (Communications Unit)	Date / Time

INCIDENT RADIO COMMUNICATIONS PLAN	June 2000	ICS 205-OS

Electronic version: NOAA 1 0 June 1, 2000

1. Incident Name	2. Operational Period (Date / Time)		**COMMUNICATIONS LIST**
	From:	To:	**ICS 205A-OS**

3. Basic Local Communications Information

Assignment	Name	Method(s) of contact (radio frequency, phone, pager, cell #(s), etc.)

4. Prepared by: (Communications Unit)	Date / Time

COMMUNICATIONS LIST	June 2000	ICS 205a-OS

Electronic version: NOAA 1.0 June 1, 2000

1. Incident Name	2. Operational Period (Date / Time)	MEDICAL PLAN
	From: To:	ICS 206-OS

3. Medical Aid Stations

Name	Location	Contact #	Paramedics On site (Y/N)

4. Transportation

Ambulance Service	Address	Contact #	Paramedics On board (Y/N)

5. Hospitals

Hospital Name	Address	Contact #	Travel Time		Burn Ctr?	Heli-Pad?
			Air	Ground		

6. Special Medical Emergency Procedures

7. Prepared by: (Medical Unit Leader) Date / Time	8. Reviewed by: (Safety Officer) Date / Time

MEDICAL PLAN	June 2000	ICS 206-OS

Electronic version: NOAA 1.0 June 1, 2000

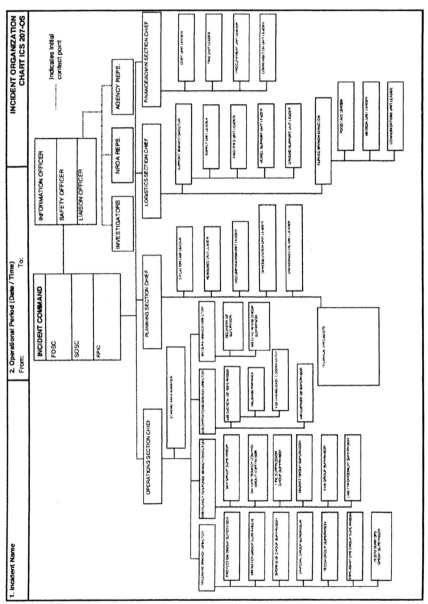

SITE SAFETY AND CONTROL PLAN ICS 208 HM	1. Incident Name:	2. Date Prepared:	3. Operational Period: Time:

Section I. Site Information

4. Incident Location:

Section II. Organization

5. Incident Commander:	6. HM Group Supervisor:	7. Tech. Specialist - HM Reference:
8. Safety Officer:	9. Entry Leader:	10. Site Access Control Leader:
11. Asst. Safety Officer - HM:	12. Decontamination Leader:	13. Safe Refuge Area Mgr:
14. Environmental Health:	15.	16.

17. Entry Team: (Buddy System)		18. Decontamination Element:	
Name:	PPE Level	Name:	PPE Level
Entry 1		Decon 1	
Entry 2		Decon 2	
Entry 3		Decon 3	
Entry 4		Decon 4	

Section III. Hazard/Risk Analysis

19. Material:	Container type	Qty.	Phys. State	pH	IDLH	F.P.	I.T.	V.P.	V.D.	S.G.	LEL	UEL

Comment:

Section IV. Hazard Monitoring

20. LEL Instrument(s):	21. O₂ Instrument(s):
22. Toxicity/PPM Instrument(s):	23. Radiological Instrument(s):

Comment:

Section V. Decontamination Procedures

24. Standard Decontamination Procedures:		YES:	NO:

Comment:

Section VI. Site Communications

25. Command Frequency:	26. Tactical Frequency:	27. Entry Frequency:

Section VII. Medical Assistance

28. Medical Monitoring:	YES:	NO:	29. Medical Treatment and Transport In-place:	YES:	NO:

Comment:

1. Incident Name	2. Operational Period (Date / Time)		Time of Report	**INCIDENT STATUS SUMMARY ICS 209-OS**
	From:	To:		

3. Spill Status (Estimated, in Barrels) [Ops & EUL/SSC]

Source Status: Remaining Potential (bbl):

☐ Secured   Rate of Spillage (bbl/hr):

☐ Unsecured

	Since Last Report	Total
Volume Spilled		

Mass Balance / Oil Budget

Recovered Oil		
Evaporation		
Natural Dispersion		
Chemical Dispersion		
Burned		
Floating, Contained		
Floating, Uncontained		
Onshore		
Total spilled oil accounted for:		

4. Waste Management (Estimated) [Ops / Disposal]

	Recovered	Stored	Disposed
Oil (bbl)			
Oily Liquids (bbl)			
Liquids (bbl)			
Oily Solids (tons)			
Solids (tons)			

5. Shoreline Impacts (Estimated, in miles) [PSC / EUL / SSC]

Degree of Oiling	Affected	Cleaned	To Be Cleaned
Light			
Medium			
Heavy			
Total			

6. Wildlife Impacts [Ops / Wildlife Br.]

Numbers in ( ) indicate subtotal that are threatened / endangered species.

	Captured	Cleaned	Released	DOA	Died in Facility Euth.	Other
Birds						
Mammals						
Reptiles						
Fish						
Total						

7. Safety Status [Safety Officer]

	Since Last Report	Total
Responder Injury		
Public Injury		

8. Equipment Resources [RUL]

Description	Ordered	Available / Staged	Assigned	Out of Service
Spill Resp. Vsls				
Fishing Vessels				
Tugs				
Barges				
Other Vessels				
Skimmers				
Boom (ft.)				
Sbnt/Snr Bm. (ft.)				
Vacuum Trucks				
Helicopters				
Fixed Wing				

9. Personnel Resources [RUL]

Description	People in Cmd. Post	People in the Field	Total People On Scene
Federal			
State			
Local			
RP			
Contract Personnel			
Volunteers			
Total Response Personnel from all Organizations:			

10. Special Notes

11. Prepared by: (Situation Unit Leader)

INCIDENT STATUS SUMMARY   June 2000   **ICS 209-OS**

1. Incident Name	2. Operational Period (Date / Time)		STATUS CHANGE
	From: To:		ICS 210-OS
3. Personnel / Resource Name or I.D.			

4. New Status

   . .   Available / Staged   . . .   Assigned . .   ..   : . .   Out of Service

5. FROM Location or Status	6. TO Location or Status

7. Time of Location / Status Change

8. Comments

| 9. Prepared by: | Date / Time |
| 10. Processed by: (Resource Unit) | Date / Time |

| STATUS CHANGE | June 2000 | ICS 210-OS |

Electronic version: NOAA 1.0 June 1, 2000

CHECK-IN LIST

1. INCIDENT NAME	2. CHECK-IN LOCATION	3. DATE/TIME

2. CHECK-IN LOCATION: BASE / CAMP / STAGING AREA / ICP / RESOURCES / HELIBASE

CHECK-IN INFORMATION

4. LIST PERSONNEL (OVERHEAD) BY AGENCY & NAME, OR LIST EQUIPMENT BY THE FOLLOWING FORMAT:

5. ORDER/ REQUEST NUMBER	6. DATE/TIME CHECK IN	7. LEADER'S NAME	8. TOTAL NO. PERSONNEL	9. MANIFEST YES/NO	10. CREW WEIGHT OR INDIVIDUALS WEIGHT	11. HOME BASE	12. DEPARTURE POINT	13. METHOD OF TRAVEL	14. INCIDENT ASSIGNMENT	15. OTHER QUALIFICATION	16. SENT TO RESOURCES UNIT/CNT.

SINGLE / KIND / TYPE / ID. NO./NAME

17. PREPARED BY (NAME AND POSITION)   USE BACK FOR REMARKS OR COMMENTS

ICS 211   PAGE ___ of ___

Electronic version: NOAA 1.0 June 1, 2009

1. Incident Name	2. Operational Period (Date / Time)		3. Check-in Location	CHECK-IN LIST (Equipment) ICS 211e-OS		
	From:	To:	☐ Command Post ☐ Staging Area ____ ☐ Other ____			
Equipment Check-in Information					9. Initial Incident Check-in?	10. Time
4. Equipment Description	5. Equipment Identifier	6. Supplier/Owner	7. Assignment	8. Contact Information	(X)	In	Out
					☐		
					☐		
					☐		
					☐		
					☐		
					☐		
					☐		
					☐		
					☐		
					☐		
					☐		
					☐		
					☐		
					☐		
					☐		
					☐		
					☐		
					☐		
					☐		
					☐		
					☐		
					☐		

11. Prepared by:	Date / Time	12. Date / Time Sent to Resource Unit

CHECK-IN LIST (Equipment)   June 2000   ICS 211e-OS

Electronic version: NOAA 1.0 June 1, 2000

CHECK-IN LIST (Personnel)
ICS 211p-OS

1. Incident Name	2. Operational Period (Date / Time)		3. Check-in Location	
	From:	To:	☐ Command Post ☐ Other ☐ Staging Area ____	

Personnel Check-in Information

8. Initial Incident Check-In?

4. Name	5. Company / Agency	6. ICS Section / Assignment / Quals.	7. Contact Information	(X)	9. Time	
					In	Out
				☐		
				☐		
				☐		
				☐		
				☐		
				☐		
				☐		
				☐		
				☐		
				☐		
				☐		
				☐		
				☐		
				☐		
				☐		
				☐		
				☐		
				☐		
				☐		
				☐		

10. Prepared by:	Date / Time	11. Date / Time Sent to Resources Unit

CHECK-IN LIST (Personnel)   June 2000   ICS 211p-OS

Electronic version: NOAA 1.0 June 1, 2000

1. Incident Name	2. Date and Time of Message	**GENERAL MESSAGE** **ICS 213-OS**
3. TO:	ICS Position	
4. FROM:	ICS Position	
5. Subject:		

6. Message

7. Reply

8. Signature / Position (person replying)	Date / Time of reply

GENERAL MESSAGE	June 2000	ICS 213-OS

1. Incident Name	2. Operational Period (Date / Time)		**UNIT LOG**
	From: To:		**ICS 214-OS**

3. Unit Name / Designators	4. Unit Leader (Name and ICS Position)

5. Personnel Assigned

NAME	ICS POSITION	HOME BASE

6. Activity Log (Continue on Reverse)

TIME	MAJOR EVENTS

7. Prepared by:	Date / Time

UNIT LOG	June 2000	ICS 214-OS

Electronic version: NOAA 1.0 June 1, 2000

1. Incident Name		2. Operational Period (Date / Time)		INDIVIDUAL LOG
		From: To:		ICS 214a-OS
3. Individual Name		4. ICS Section	5. Assignment / Location	

6. Activity Log	Page of
Time	**Major Events**

7. Prepared by:	Date / Time

INDIVIDUAL LOG	June 2000	ICS 214a-OS

Electronic version: NOAA 1.0 June 1, 2000

OPERATIONAL PLANNING WORKSHEET
ICS 215-OS

1. Incident Name	2. Operational Period (Date / Time) From: To:	

3. Division / Group or Location	4. Work Assignments	Resource		5. Resource / Equipment	6. Notes / Remarks	7. Reporting Location	8. Requested Arrival Time	9. "X" here if 204s Needed
		Req. / Have / Need						☐
		Req. / Have / Need						☐
		Req. / Have / Need						☐
		Req. / Have / Need						☐
		Req. / Have / Need						☐
		Req. / Have / Need						☐

10. Total Resources Required

11. Total Resources On Hand

12. Total Resources Needed

13. Prepared by:

Date   Time

OPERATIONAL PLANNING WORKSHEET   June 2000   ICS 215-OS

Electronic version NOAA 1.0 June 1, 2000

1. Incident Name	2. Operational Period (Date / Time) From:	3. Date	4. Time

INCIDENT ACTION PLAN SAFETY ANALYSIS
ICS 215A-OS

RISKS

RISK MITIGATION

Division/Group	Weather	Biohazard	Hazardous Materials	Communications	River/Water Hazard	SHA	Fatigue	Diving Hazards/Bends	Dehydration	CISM

Prepared by (Name and Position)

November 2003

Electronic version: NOAA 1.0 November 1, 2003

RADIO REQUIREMENTS WORKSHEET

1. INCIDENT NAME		2. DATE	3. TIME

4. BRANCH	5. AGENCY	6. OPERATIONAL PERIOD	7. TACTICAL FREQUENCY

8. DIVISION/GROUP	DIVISION/ GROUP	DIVISION/ GROUP	DIVISION/ GROUP

AGENCY		AGENCY	AGENCY	AGENCY

9. AGENCY	ID NO.	RADIO RQMTS	AGENCY	ID NO.	RADIO RQMTS	AGENCY	ID NO.	RADIO RQMTS	AGENCY	ID NO.	RADIO RQMTS

PAGE	10. PREPARED BY (COMMUNICATIONS UNIT)

216 KS 3-82

Electronic version: NOAA 1.0 June 1, 2000

NFES 1339

RADIO FREQUENCY ASSIGNMENT WORKSHEET

1. INCIDENT NAME:

2. DATE:

3. OPERATIONAL PERIOD (DATE/TIME):
From:

4. INCIDENT ORGANIZATION

5. RADIO DATA

SOURCE	FUNCTION	CH#	FREQUENCY																				TOTAL BY REQ

6.

ID	AGENCY	CH#	FREQUENCY

7. TOTAL RADIOS REQUIRED

8. PREPARED BY (NAME/POSITION)

212 ICS

Electronic version NOAA 1.0 June 1, 2000

SUPPORT VEHICLE INVENTORY
(USE SEPARATE SHEET FOR EACH VEHICLE CATEGORY)

1. INCIDENT NAME	2. DATE PREPARED	3. TIME PREPARED

VEHICLE INFORMATION

a. TYPE	b. MAKE	c. CAPACITY/SIZE	d. AGENCY/OWNER	e. I.D. NO.	f. LOCATION	g. RELEASE TIME

5. PREPARED BY (GROUND SUPPORT UNIT)
PAGE

218 ICS 8-78

NFES 1341

Electronic version NOAA 1.0 June 1 2000

GREEN CARD STOCK (CREW)

AGENCY	ST	KIND	TYPE	I.D. NO./NAME

ORDER/REQUEST NO.	DATE/TIME CHECK IN

HOME BASE

DEPARTURE POINT

LEADER NAME

CREW ID NO./NAME (FOR STRIKE TEAMS)

NO. PERSONNEL	MANIFEST	WEIGHT

METHOD OF TRAVEL

OTHER

DESTINATION POINT	ETA

TRANSPORTATION NEEDS

OTHER

ORDERED DATE/TIME	CONFIRMED DATE/TIME

REMARKS

ICS 219-2 (Rev. 4/82) CREW NFES 1344

AGENCY	ST	TF	KIND	TYPE	I.D. NO./NAME

INCIDENT LOCATION	TIME

STATUS

ETR

NOTE

INCIDENT LOCATION	TIME

STATUS

ETR

NOTE

INCIDENT LOCATION	TIME

STATUS

ETR

NOTE

INCIDENT LOCATION	TIME

STATUS

ETR

NOTE

BLUE CARD STOCK (HELICOPTER)

AGENCY	TYPE	MANUFACTURER NAME/NO	I.D. NO.

ORDER/REQUEST NO	DATE/TIME CHECK IN

HOME BASE

DEPARTURE POINT

PILOT NAME

DESTINATION POINT	ETA

REMARKS

INCIDENT LOCATION	TIME

STATUS

ETR
NOTE

INCIDENT LOCATION	TIME

STATUS

ETR
NOTE

ICS 219-4 (Rev. 4/82) HELICOPTER NFES 1346

AGENCY	TYPE	MANUFACTURER	I.D. NO.

INCIDENT LOCATION	TIME

STATUS

ETR
NOTE

INCIDENT LOCATION	TIME

STATUS

ETR
NOTE

INCIDENT LOCATION	TIME

STATUS

ETR
NOTE

INCIDENT LOCATION	TIME

STATUS

ETR
NOTE

NFES 1346

Electronic version: NOAA 1.0 June 1, 2000

ORANGE CARD STOCK (AIRCRAFT)

AGENCY	TYPE	MANUFACTURER NAME/NO	I.D. NO.

ORDER/REQUEST NO	DATE/TIME CHECK IN

HOME BASE

DATE TIME RELEASED

INCIDENT LOCATION	TIME

STATUS

ETR

NOTE

INCIDENT LOCATION	TIME

STATUS

ETR

NOTE

INCIDENT LOCATION	TIME

STATUS

ETR

NOTE

ICS 219-6 (4/82) AIRCRAFT

AGENCY	TYPE	MANUFACTURER	I.D. NO.

INCIDENT LOCATION	TIME

STATUS

ETR

NOTE

INCIDENT LOCATION	TIME

STATUS

ETR

NOTE

INCIDENT LOCATION	TIME

STATUS

ETR

NOTE

INCIDENT LOCATION	TIME

STATUS

ETR

NOTE

NFES 1348

Electronic version: NOAA 1.0 June 1, 2000

YELLOW CARD STOCK (DOZERS)

AGENCY	ST	TF	KIND	TYPE	I.D. NO.

ORDER/REQUEST NO.	DATE/TIME CHECK IN

HOME BASE

DEPARTURE POINT

LEADER NAME

RESOURCE I.D. NO.S/NAMES

DESTINATION POINT	ETA

REMARKS

INCIDENT LOCATION	TIME

STATUS

ETR

NOTE

ICS 219-7 (Rev. 4/82) DOZERS NFES 1349

AGENCY	ST	TF	KIND	TYPE	I.D. NO./NAME

INCIDENT LOCATION	TIME

STATUS

ETR

NOTE

INCIDENT LOCATION	TIME

STATUS

ETR

NOTE

INCIDENT LOCATION	TIME

STATUS

ETR

NOTE

INCIDENT LOCATION	TIME

STATUS

ETR

NOTE

Electronic version: NOAA 1 0 June 1, 2000

AIR OPERATIONS SUMMARY
ICS 220-OS

1. Incident Name	2. Operational Period (Date / Time) From: To:

3. Distribution ☐ Fixed-Wing Bases ☐ Helibase

4. Personnel and Communications

	Air Operations Director	Air / Air Frequency	Air / Ground Frequency
Air Operations Director			
Air Tactical Supervisor			
Air Support Supervisor			
Helicopter Coordinator			
Fixed-Wing Coordinator			

5. Remarks (Spec. Instructions, Safety Notes, Hazards, Priorities)

6. Location / Function	7. Assignment	8. Fixed-Wing		9. Helicopter		10. Time		11. Aircraft Assigned	12. Operating Base
		NO.	TYPE	NO.	TYPE	Available	Commence		
13. TOTALS									

14. Air Operation Support Equipment

15. Prepared by	Date / Time

AIR OPERATIONS SUMMARY   June 2000   ICS 220-OS

Electronic version. NOAA 1.0 June 1, 2000

1. Incident Name	2. Operational Period (Date / Time)		**DEMOB. CHECK-OUT**
	From:	To:	**ICS 221-OS**
3. Unit / Personnel Released		4. Release Date / Time	

5. Unit / Personnel

You and your resources have been released, subject to signoff from the following:
(Demob. Unit Leader "X" appropriate box(es))

Logistics Section

☐ Supply Unit _____

☐ Communications Unit _____

☐ Facilities Unit _____

☐ Ground Unit _____

Planning Section

☐ Documentation Unit _____

Finance / Admin. Section

☐ Time Unit _____

Other

☐ _____   _____

☐ _____   _____

☐ _____   _____

6. Remarks

_____

_____

_____

_____

7. Prepared by:	Date / Time
DEMOB. CHECK-OUT	June 2000

ICS 221-OS

| 1. Incident Name | 2. Operational Period (Date / Time) | | **DAILY MEETING SCHEDULE** |
| | From: To: | | **ICS 230-OS** |

3. Meeting Schedule (Commonly-held meetings are included)

Date / Time	Meeting Name	Purpose	Attendees	Location
	Tactics Meeting	Develop primary and alternate Strategies to meet Incident Objectives for the next Operational Period.	PSC, OPS, LSC, EUL, RUL & SUL	
	Planning Meeting	Review status and finalize strategies and assignments to meet Incident Objectives for the next Operational Period.	Determined by the IC/UC	
	Operations Briefing	Present IAP and assignments to the Supervisors / Leaders for the next Operational Period.	IC/UC, Command Staff, General Staff, Branch Directors, Div. Sups., Task Force/Strike Team Leaders and Unit Leaders	
	Unified Command Objectives Meeting	Review / identify objectives for the next operational period.	Unified Command members	

4. Prepared by: (Situation Unit Leader)   Date / Time

| DAILY MEETING SCHEDULE | June 2000 | ICS 230-OS |

Electronic version. NOAA 1.0 June 1, 2000

1. Incident Name	2. Meeting Date / Time	**MEETING SUMMARY** **ICS 231-OS**
3. Meeting Name		
4. Meeting Location		
5. Facilitator		
6. Attendees		
7. Notes (with summary of decisions and action items)		
8. Prepared by:	Date / Time	

MEETING SUMMARY	June 2000	ICS 231-OS

Electronic version: NOAA 1.0 June 1, 2000

1. Incident Name	2. Operational Period (Date / Time)		RESOURCES AT RISK SUMMARY
	From:	To:	ICS 232-OS

3. Environmentally-Sensitive Areas and Wildlife Issues

Site #	Priority	Site Name and/or Physical Location	Site Issues

Narrative

4. Archaeo-cultural and Socio-economic Issues

Site #	Priority	Site Name and/or Physical Location	Site Issues

Narrative

5. Prepared by: (Environmental Unit Leader)   Date / Time

RESOURCES AT RISK SUMMARY	June 2000	ICS 232-OS

Electronic version: NOAA 1.0 June 1, 2000

| 1. Incident Name | 2. Operational Period (Date / Time) | | ACP Site Index |
| | From: To: | | ICS 232a-OS |

3. Index to ACP/GRP sites shown on Situation Map

Site #	Priority	Site Name and/or Physical Location	Action	Status

Note: This form is designed to be posted next to the situation map. Use additional sheets, as needed.

| 4. Prepared by: | Date / Time |

| ACP Site Index | June 2000 | ICS 232a-OS |

1. Incident Name	2. Operational Period (Date / Time)	EXECUTIVE
	From: To:	SUMMARY

3. Operations

4. Environmental

5. Planning

6. Other

7. Prepared by	Date / Time

EXECUTIVE SUMMARY	June 2000

GENERAL PLAN

1. Incident Name		
2. Prepared By	Date / Time Prepared	3. Operational Period (Date / Time) From: To:
4. Notification (Date and time completed)		5. Response Initiation (Date and time completed)

6. **Plan Item** Timeframe ⟹ (Enter days or weeks)

- Site Characterization, Forecasts, and Analysis
- Site Safety
- Site Security
- Source Stabilization, Salvage, and Lightering
- Surveillance
- On Water Containment and Recovery
- Sensitive Areas / Resources at Risk
- Alternative Response Technology
- Shoreline Protection and Recovery
- Wildlife Protection and Rehabilitation
- Logistics Support
- Response Organization
- Communications
- Public Information
- Financial Management and Cost Documentation
- NRDA and Claims
- Training
- Information Management
- Restoration / Mitigation
- Waste Management
- Demobilization

June 2000

GENERAL PLAN

Electronic version: NOAA 1.0 June 1, 2000

1. Incident Name	2. Operational Period to be covered by IAP (Date / Time)	**IAP COVER SHEET**
	From: To:	

3. Approved by:

FOSC _____

SOSC _____

RPIC _____

_____   _____

_____   _____

INCIDENT ACTION PLAN

The items checked below are included in this Incident Action Plan:

☐ ICS 202-OS (Response Objectives)

☐ ICS 203-OS (Organization List) - OR - ICS 207-OS (Organization Chart)

☐ ICS 204-OSs (Assignment Lists)
One Copy each of any ICS 204-OS attachments:

  ☐ Map
  ☐ Weather forecast
  ☐ Tides
  ☐ Shoreline Cleanup Assessment Team Report for location
  ☐ Previous day's progress, problems for location

☐ ICS 205-OS (Communications List)

☐ ICS 206-OS (Medical Plan)

☐ _____

☐ _____

☐ _____

☐ _____

☐ _____

☐ _____

4. Prepared by:	Date / Time

IAP COVER SHEET	June 2000

Electronic version: NOAA 1.0 June 1, 2000

1. Incident Name	2. Operational Period to be covered by IAP (Date / Time)	IAP COVER SHEET
	From: To:	

3. Approved by:

FOSC _____

SOSC _____

RPIC _____

_____   _____

_____   _____

INCIDENT ACTION PLAN

4. Prepared by:	Date / Time

IAP COVER SHEET	June 2000

INITIAL INCIDENT INFORMATION	INCIDENT NAME	Information as of:	
		Date	Time

NAME OF PERSON REPORTING THE INCIDENT

Call-Back Number(s) of person reporting the incident:

VESSEL/FACILITY INFORMATION AND POINTS OF CONTACT

Vessel / Facility Name:	Number of people onboard/on site:
Location:	
Type of Vessel / Facility:	
Contact / Agent:	Phone:
Owner:	Phone:
Operator / Charterer:	Phone:

VESSEL SPECIFIC INFORMATION

Last Port of Call:	Destination:		Flag:
Particulars: Length: Ft. Tonnage (Gross/Net/DWT):	Draft Fwd:	Aft:	Year Built:

Type of Hull: ☐ Single ☐ Double ☐ Double-Bottom ☐ Double-Sided

Hull Material:

Type of Propulsion: ☐ Diesel ☐ Steam ☐ Gas Turbine ☐ Nuclear ☐ Other

Petroleum Products or Crude Oil   ☐ Yes ☐ No

Type of Cargo:	Total Number of Tanks on Vessel:
Total Quantity: Barrels x 42=	Gallons Total Capacity Barrels
Type of Fuel:	Quantity on Board: Barrels

INCIDENT INFORMATION

Location:	Lat/Long:

Type of Casualty: ☐ Grounding ☐ Collision ☐ Allision ☐ Explosion ☐ Fire ☐ Other

Number of Tanks Impacted:	Total Capacity of Affected Tanks:
Material(s) Spilled:	Viscosity:
Estimated Quantity Spilled:	(☐ Gallons / ☐ Barrels) Classification: ☐ Minor ☐ Medium ☐ Major
Source Secured?: ☐ Yes ☐ No	If Not, Estimated Spill Rate: ☐ Barrels ☐ Gallons / Hour

Notes:

INCIDENT STATUS

Injuries/Casualties:	☐ SAR Underway

Vessel Status: ☐ Sunk ☐ Aground ☐ Dead in Water   Set and Drift:

☐ Anchored ☐ Berthed ☐ Under Tow   Estimated Time to Dock / Anchor:

☐ Enroute to Anchorage / Berth Under Own Power   Estimated Time of Arrival:

☐ Holed: ☐ Above Waterline ☐ Below Waterline ☐ At Waterline   Approximate Size of Hole:

☐ Fire: ☐ Extinguished ☐ Burning   ☐ Assistance Enroute ☐ Assistance On-Scene

☐ Flooding: ☐ Dewatering ☐ Lightering   ☐ Assistance Enroute ☐ Assistance On-Scene

☐ List: ☐ Port ☐ Starboard Degrees: ☐ Trim: ☐ Bow ☐ Stern Degrees:

ENVIRONMENTAL INFORMATION

Wind Speed: Knots	Wind Direction:	Air Temperature: F°	Water Temperature: F°	
Wave Height: Feet	Wave Direction:	Conditions:	Tide: ☐ Slack ☐ Flood ☐ Ebb	
Current: Knots	Current Direction:		High Tide at: Hours	
Swell Height: Feet	Swell Direction:		Low Tide at: Hours	

Prepared By: Date / Time Prepared	
	June 2000 **INITIAL INCIDENT INFORMATION**

Electronic version NOAA 1 0 June 1, 2000

Appendix C

HOSPITAL EMERGENCY INCIDENT COMMAND (HEICS) JOB ACTION SHEETS

In this appendix we have reproduced the Job Actions Sheets designed to compliment the Hospital Emergency Incident Command System (HEICS) designed and approved by the California Emergency Medical Services Authority (CA EMSA) for use in healthcare facilities in that state.

The intent of the sheets is to allow a person that has been trained in the HEICS system fundamentals and concepts to be able to fill any of the forty-nine specific functions used within the HEICS.

In practical use the content of the sheets can be modified with the exception of the job title and the reporting structure without loss of integrity of the HEICS system. This allows any hospital or other healthcare facility regardless of their nature to effectively use the system and these sheets in a HEICVS operational capacity.

As of the writing of this text the CA EMSA is currently working on HEICS IV, an update to make the HEICS system more NIMS complainant.

Note that each form in practice is designed to be either a one page or a two page form with the HEICS Organization chart printed on the back for reference.

Emergency Incident Commander
Public Information Officer
Liaison Officer
Safety and Security
Logistics Section Chief
Facility Unit Leader
Damage Assessment and ControlL Officer
Sanitation Systems Officer

Communications Unit Leader
Transportation Unit Leader
Materials Supply Unit Leader
Nutritional Supply Unit Leader
Planning Section Chief
Situation-Status (SITSTAT) Unit Leader
Labor Pool Unit Leader
Medical Staff Unit Leader
Nursing Unit Leader
Patientt Tracking Officer
Patient Information Officer
Finance Section Chief
Time Unit Leader
Procurement Unit Leader
Claims Unit Leader
Cost Unit Leader
Operations Section Chief
Medical Staff Director
Medical Care Director
In-Patient Areas Supervisor
Surgical Services Unit Leader
Maternal-Child Unit Leader
Critical Care Unit Leader
General Nursing Care Unit Leader
Out Patient Services Unit Leader
Treatment Areas Supervisor
Triage Unit Leader
Immediate Treatment Unit Leader
Delayed Treatment Unit Leader
Minor Treatment Unit Leader
Discharge Unit Leader
Morgue Unit Leader
Ancillary Services Director
Laboratory Unit Leader
Radiology Unit Leader
Pharmacy Unit Leader
Cardiopulmonary Unit Leader
Human Services Director
Staff Support Unit Leader
Psychological Support Unit Leader
Dependent Care Unit Leader

EMERGENCY INCIDENT COMMANDER

Mission: Organize and direct Emergency Operations Center (EOC). Give overall direction for hospital operations and if needed, authorize evacuation.

Immediate   ____ Initiate the Hospital Emergency Incident Command System by assuming role of Emergency Incident Commander.

____ Read this entire Job Action Sheet.

____ Put on position identification vest.

____ Appoint all Section Chiefs and the Medical Staff Director positions; distribute the four section packets which contain:
• Job Action Sheets for each position
• Identification vest for each position
• Forms pertinent to Section & positions

____ Appoint Public Information Officer, Liaison Officer, and Safety and Security Officer; distribute Job Action Sheets. (May be pre-established.)

____ Announce a status/action plan meeting of all Section Chiefs and Medical Staff Director to be held within 5 to 10 minutes.

____ Assign someone as Documentation Recorder/Aide.

____ Receive status report and discuss an initial action plan with Section Chiefs and Medical Staff Director. Determine appropriate level of service during immediate aftermath.

____ Receive initial facility damage survey report from Logistics Chief, if applicable, evaluate the need for evacuation.

____ Obtain patient census and status from Planning Section Chief. Emphasize proactive actions within the Planning Section. Call for a hospital-wide projection report for 4, 8, 24 & 48 hours from time of incident onset. Adjust projections as necessary.

____ Authorize a patient prioritization assessment for the purposes of designating appropriate early discharge, if additional beds needed.

EMERGENCY INCIDENT COMMANDER *(continued)*

     ____ Assure that contact and resource information has been established with outside agencies through the Liaison Officer.

Immediate   ____ Authorize resources as needed or requested by Section Chiefs.

     ____ Designate routine briefings with Section Chiefs to receive status reports and update the action plan regarding the continuance and termination of the action plan.

     ____ Communicate status to chairperson of the Hospital Board of Directors or the designee.

     ____ Consult with Section Chiefs on needs for staff, physician, and volunteer responder food and shelter. Consider needs for dependents. Authorize plan of action.

Extended   ____ Approve media releases submitted by P.I.O.

     ____ Observe all staff, volunteers and patients for signs of stress and inappropriate behavior. Report concerns to Psychological Support Unit Leader. Provide for staff rest periods and relief.

     ____ Other concerns:

PUBLIC INFORMATION OFFICER

Position Assigned To: _____

You Report To: _____ (Emergency Incident Commander)

Command Center: _____

Telephone: _____

Mission: Provide information to the news media.

Immediate   ____ Receive appointment from Emergency Incident Commander.

              ____ Read this entire Job Action sheet and review organizational
                    chart on back.

              ____ Put on position identification vest.

              ____ Identify restrictions in contents of news release information
                    from Emergency Incident Commander.

              ____ Establish a Public Information area away from E.O.C. and
                    patient care activity.

              ____ Ensure that all news releases have the approval of the Emergency
                    Incident Commander.

              ____ Issue an initial incident information report to the news media
                    with the cooperation of the Situation-Status Unit Leader. Relay
                    any pertinent data back to Situation-Status Unit Leader.

              ____ Inform on-site media of the physical areas which they have
                    access to, and those which are restricted. Coordinate with
                    Safety and Security Officer.

              ____ Contact other at-scene agencies to coordinate released infor-
                    mation, with respective P.I.O.s. Inform Liaison Officer of
                    action.

Extended   ____ Obtain progress reports from Section Chiefs as appropriate.

              ____ Notify media about casualty status.

PUBLIC INFORMATION OFFICER *(continued)*

Extended   ____ Direct calls from those who wish to volunteer to Labor Pool.
Contact Labor Pool to determine requests to be made to the
public via the media.

____ Observe all staff, volunteers and patients for signs of stress
and inappropriate behavior. Report concerns to Psychological
Support Unit Leader. Provide for staff rest periods and relief.

____ Other concerns:

LIAISON OFFICER

Position Assigned To: _____

You Report To: _____ (Emergency Incident Commander)

Command Center: _____

Telephone: _____

Mission: Function as incident contact person for representatives from other agencies.

Immediate
_____ Receive appointment from Emergency Incident Commander.

_____ Read this entire Job Action Sheet and review organizational chart on back.

_____ Put on position identification vest.

_____ Obtain briefing from Emergency Incident Commander.

_____ Establish contact with Communications Unit Leader in E.O.C. Obtain one or more aides as necessary from Labor Pool.

_____ Review county and municipal emergency organizational charts to determine appropriate contacts and message routing. Coordinate with Public Information Officer.

_____ Obtain information to provide the interhospital emergency communication network, municipal E.O.C. and/or county E.O.C. as appropriate, upon request. The following information should be gathered for relay:
  • The number of "Immediate" and "Delayed" patients that can be received and treated immediately (Patient Care Capacity).
  • Any current or anticipated shortage of personnel, supplies, etc.
  • Current condition of hospital structure and utilities (hospital's overall status).
  • Number of patients to be transferred by wheelchair or stretcher to another hospital.
  • Any resources which are requested by other facilities (i.e., staff, equipment, supplies).

LIAISON OFFICER *(continued)*

Immediate   ____ Establish communication with the assistance of the Communication Unit Leader with the interhospital emergency communication network, municipal E.O.C. or with county E.O.C./ County Health Officer. Relay current hospital status.

    ____ Establish contact with liaison counterparts of each assisting and cooperating agency (i.e., municipal E.O.C.). Keeping governmental Liaison Officers updated on changes and development of hospital's response to incident.

    ____ Request assistance and information as needed through the interhospital emergency communication network or municipal/ county E.O.C.

    ____ Respond to requests and complaints from incident personnel regarding inter-organization problems.

    ____ Prepare to assist Labor Pool Unit Leader with problems encountered in the volunteer credentialing process.

    ____ Relay any special information obtained to appropriate personnel in the receiving facility (i.e., information regarding toxic decontamination or any special emergency conditions).

Extended   ____ Assist the Medical Staff Director and Labor Pool Unit Leader in soliciting physicians and other hospital personnel willing to volunteer as Disaster Service Workers outside of the hospital, when appropriate.

    ____ Inventory any material resources which may be sent upon official request and method of transportation, if appropriate.

    ____ Supply casualty data to the appropriate authorities; prepare the following minimum data:
- Number of casualties received and types of injuries treated
- Number hospitalized and number discharged to home or other facilities
- Number dead
- Individual casualty data: name or physical description, sex, age, address, seriousness of injury or condition

LIAISON OFFICER *(continued)*

Extended   ____ Observe all staff, volunteers and patients for signs of stress and inappropriate behavior. Report concerns to Psychological Support Unit Leader. Provide for staff rest periods and relief.

                   ____ Other concerns:

SAFETY AND SECURITY OFFICER

Position Assigned To: _____

You Report To: _____ (Emergency Incident Commander)

Command Center: _____

Telephone: _____

Mission: Monitor and have authority over the safety of rescue operations and hazardous conditions. Organize and enforce scene/facility protection and traffic security.

Immediate   ____ Receive appointment from Emergency Incident Commander.

____ Read this entire Job Action sheet and review organizational chart on back.

____ Put on position identification vest.

____ Obtain a briefing from Emergency Incident Commander.

____ Implement the facility's disaster plan emergency lockdown policy and personnel identification policy.

____ Establish Security Command Post.

____ Remove unauthorized persons from restricted areas.

____ Establish ambulance entry and exit routes in cooperation with Transportation Unit Leader.

____ Secure the E.O.C., triage, patient care, morgue and other sensitive or strategic areas from unauthorized access.

____ Communicate with Damage Assessment and Control Officer to secure and post non-entry signs around unsafe areas. Keep Safety and Security staff alert to identify and report all hazards and unsafe conditions to the Damage Assessment and Control Officer.

____ Secure areas evacuated to and from, to limit unauthorized personnel access.

SAFETY AND SECURITY OFFICER *(continued)*

<u>Intermediate</u> ____ Initiate contact with fire, police agencies through the Liaison Officer, when necessary.

____ Advise the Emergency Incident Commander and Section Chiefs immediately of any unsafe, hazardous or security related conditions.

____ Assist Labor Pool and Medical Staff Unit Leaders with credentialing/screening process of volunteers. Prepare to manage large numbers of potential volunteers.

____ Confer with Public Information Officer to establish areas for media personnel.

____ Establish routine briefings with Emergency Incident Commander.

____ Provide vehicular and pedestrian traffic control.

____ Secure food, water, medical, and blood resources.

<u>Extended</u> ____ Inform Safety & Security staff to document all actions and observations.

____ Establish routine briefings with Safety & Security staff.

____ Observe all staff, volunteers and patients for signs of stress and inappropriate behavior. Report concerns to Psychological Support Unit Leader. Provide for staff rest periods and relief.

____ Other concerns:

LOGISTICS SECTION CHIEF

Position Assigned To: _____

You Report To: _____ (Emergency Incident Commander)

Logistics Command Center: _____

Telephone: _____

Mission: Organize and direct those operations associated with maintenance of the physical environment, and adequate levels of food, shelter and supplies to support the medical objectives.

Immediate   _____ Receive appointment from the Emergency Incident Commander. Obtain packet containing Section's Job Action Sheets, identification vests and forms.

        _____ Read this entire Job Action Sheet and review organizational chart on back.

        _____ Put on position identification vest.

        _____ Obtain briefing from Emergency Incident Commander.

        _____ Appoint Logistics Section Unit Leaders: Facilities Unit Leader, Communications Unit Leader, Transportation Unit Leader, Material's Supply Unit Leader, Nutritional Supply Unit Leader; distribute Job Action Sheets and vests. (May be pre-established.)

        _____ Brief unit leaders on current situation; outline action plan and designate time for next briefing.

        _____ Establish Logistics Section Center in proximity to E.O.C.

        _____ Attend damage assessment meeting with Emergency Incident Commander, Facility Unit Leader and Damage Assessment and Control Officer.

        _____ Obtain information and updates regularly from unit leaders and officers; maintain current status of all areas; pass status info to Situation-Status Unit Leader.

LOGISTICS SECTION CHIEF *(continued)*

_____ Communicate frequently with Emergency Incident Commander.

<u>Immediate</u>   _____ Obtain needed supplies with assistance of the Finance Section Chief, Communications Unit Leader and Liaison Unit Leader.

<u>Extended</u>   _____ Assure that all communications are copied to the Communications Unit Leader.

_____ Document actions and decisions on a continual basis.

_____ Observe all staff, volunteers and patients for signs of stress and inappropriate behavior. Report concerns to Psychological Support Unit Leader. Provide for staff rest periods and relief.

_____ Other concerns:

FACILITY UNIT LEADER

Position Assigned To: _____

You Report To: _____ (Logistics Section Chief)

Logistics Command Center: _____

Telephone: _____

Mission: Maintain the integrity of the physical facility to the best level. Provide adequate environmental controls to perform the medical mission.

Immediate ____ Receive appointment from Logistics Section Chief and Job Action Sheets for Damage Assessment and Control Officer, and Sanitation Systems Officer.

____ Read this entire Job Action Sheet and review organizational chart on back.

____ Put on position identification vest.

____ Meet with Logistics Section Chief to receive briefing and develop action plan; deliver preliminary report on the physical status of the facility if available.

____ Appoint Damage Assessment and Control Officer and Sanitation Systems Officer; supply the corresponding Job Action Sheets. Provide the Facility System Status Report Form to the Damage Assessment and Control Officer. (May be pre-established.)

____ Receive a comprehensive facility status report as soon as possible from Damage Assessment and Control Officer.

____ Facilitate and participate in damage assessment meeting between Emergency Incident Commander, Logistics Section Chief and Damage Assessment and Control Officer.

____ Prepare for the possibility of evacuation and/or the relocation of medical services outside of existing structure, if appropriate.

____ Receive continually updated reports from the Damage Assessment and Control Officer, and Sanitation Systems Officer.

FACILITY UNIT LEADER *(continued)*

Extended   ____ Forward requests of outside service providers/ resources to the
Materials Supply Unit Leader after clearing through the Logistics Section Chief.

____ Document actions and decisions on a continual basis. Obtain
the assistance of a documentation aide if necessary.

____ Observe all staff, volunteers and patients for signs of stress and
inappropriate behavior. Report concerns to Psychological
Support Unit Leader. Provide for staff rest periods and relief.

____ Other concerns:

DAMAGE ASSESSMENT AND CONTROL OFFICER

Position Assigned To: _____

You Report To: _____ (Facility Unit Leader)

Logistics Command Center: _____

Telephone: _____

Mission: Provide sufficient information regarding the operational status of the facility for the purpose of decision/policy making, including those regarding full or partial evacuation. Identify safe areas where patients and staff can be moved if needed. Manage fire suppression, search and rescue and damage mitigation activities.

Immediate

_____ Receive appointment, Job Action Sheet and Facility System Status Report form from Facility Unit Leader.

_____ Read this entire Job Action Sheet and review organizational chart on back.

_____ Put on position identification vest.

_____ Obtain briefing from Facility Unit Leader.

_____ Assign teams to check system components of entire facility, and report back within 5 minutes.

_____ Identify hazards, e.g., fire and assign staff to control and eliminate.

_____ Receive initial assessment/damage reports and immediately relay information in a briefing to Emergency Incident Commander, Logistics Section Chief and Facility Unit Leader; follow-up with written documentation.

_____ Notify Safety & Security Officer of unsafe areas and other security problems.

_____ Assemble light-duty search rescue team(s) to retrieve victims and deliver to Triage Area. Obtain Search and Rescue Team equipment pack from Materials Supply Unit Leader.

DAMAGE ASSESSMENT AND CONTROL OFFICER *(continued)*

_____ Notify Labor Pool of staffing needs.

Intermediate   _____ Identify areas where immediate repair efforts should be directed to restore critical services.

_____ Arrange to have structural engineer under contract report and obtain more definitive assessment if indicated.

_____ Inspect those areas of reported damage and photographically record.

_____ Identify areas where immediate salvage efforts should be directed in order to save critical services and equipment.

Extended   _____ Assign staff to salvage operations.

_____ Assign staff to repair operations.

_____ Brief Facility Unit Leader routinely to provide current damage/ recovery status.

_____ Observe and assist any staff who exhibit signs of stress and fatigue. Report concerns to Psychological Support Unit Leader. Provide for staff rest periods and relief.

_____ Other concerns:

SANITATION SYSTEMS OFFICER

Position Assigned To: _____

You Report To: _____ (Facility Unit Leader)

Logistics Command Center: _____

Telephone: _____

Mission: Evaluate and monitor the patency of existing sewage and sanitation systems. Enact pre-established alternate methods of waste disposal if necessary.

Immediate   ____ Receive appointment and Job Action Sheet from Facility Unit Leader.

    ____ Read this entire Job Action Sheet and review organizational chart on back.

    ____ Put on position identification vest.

    ____ Obtain briefing from Facility Unit Leader.

    ____ Coordinate the inspection of the hospital's sewage system with Damage Assessment and Control Officer.

    ____ Inspect the hazardous waste collection areas(s) to ensure patency of containment measures. Cordon off unsafe areas with assistance of the Safety & Security Officer.

    ____ Control observed hazards, leaks or contamination with the assistance of the Safety & Security Officer and the Damage Assessment and Control Officer.

    ____ Report all findings and actions to the Facility Unit Leader. Document all observations and actions.

    ____ Implement pre-established alternative waste disposal/collection plan, if necessary.

    ____ Assure that all sections and areas of the hospital are informed of the implementation of the alternative waste disposal/collection plan.

SANITATION SYSTEMS OFFICER *(continued)*

Intermediate   ____   Position portable toilets in accessible areas; away from patient care and food preparation.

____   Ensure an adequate number of handwashing areas are operational near patient care/food preparation areas, and adjacent to portable toilet facilities.

____   Inform Infection Control personnel of actions and enlist assistance where necessary.

Extended   ____   Monitor levels of all supplies, equipment and needs relevant to all sanitation operations.

____   Brief Facility Unit Leader routinely on current condition of all sanitation operations; communicate needs in advance.

____   Obtain support staff as necessary from Labor Pool.

____   Observe all staff, volunteers and patients for signs of stress and inappropriate behavior. Report concerns to Psychological Support Unit Leader. Provide for staff rest periods and relief.

____   Other concerns:

COMMUNICATIONS UNIT LEADER

Position Assigned To: _____

You Report To: _____ (Logistics Section Chief)

Logistics Command Center: _____

Telephone: _____

Mission: Organize and coordinate internal and external communications; act as custodian of all logged/documented communications.

Immediate   ____ Receive appointment from Logistics Section Chief.

              ____ Read this entire Job Action Sheet and review organizational chart back.

              ____ Put on position identification vest.

              ____ Obtain briefing from Emergency Incident Commander or Logistics Section Chief.

              ____ Establish a Communications Center in close proximity to E.O.C.

              ____ Request the response of assigned amateur radio personnel assigned to facility.

              ____ Assess current status of internal and external telephone system and report to Logistics Section Chiefs and Damage Assessment and Control Officer.

              ____ Establish a pool of runners and assure distribution of 2-way radios to pre-designated areas.

              ____ Use pre-established message forms to document all communication. Instruct all assistants to do the same.

              ____ Establish contact with Liaison Officer.

              ____ Receive and hold all documentation related to internal facility communications.

COMMUNICATIONS UNIT LEADER *(continued)*

Intermediate   ____ Monitor and document all communications sent and received
via the interhospital emergency communication network or
other external communication.

____ Establish mechanism to alert Code Team and Fire Suppression
Team to respond to internal patient and/or physical emergencies, i.e., cardiac arrest, fires, etc.

Extended   ____ Observe all staff, volunteers and patients for signs of stress
and inappropriate behavior. Report concerns to Psychological
Support Unit Leader. Provide for staff rest periods and relief.

____ Other concerns:

TRANSPORTATION UNIT LEADER

Position Assigned To: _____

You Report To: _____ (Logistics Section Chief)

Logistics Command Center: _____

Telephone: _____

Mission: Organize and coordinate the transportation of all casualties, ambulatory and non-ambulatory. Arrange for the transportation of human and material resources to and from the facility.

Immediate   ____ Receive appointment from Logistics Section Chief.

____ Read this entire Job Action Sheet and review the organizational chart on back.

____ Put on position identification vest.

____ Receive briefing from Logistics Section Chief.

____ Assess transportation requirements and needs for patients, personnel and materials; request patient transporters from Labor Pool to assist in the gathering of patient transport equipment.

____ Establish ambulance off-loading area in cooperation with the Triage Unit Leader.

____ Assemble gurneys, litters, wheelchairs and stretchers in proximity to ambulance off-loading area and Triage Area.

____ Establish ambulance loading area in cooperation with the Discharge Unit Leader.

____ Contact Safety & Security Officer on security needs of loading areas.

____ Provide for the transportation/shipment of resources into and out of the facility.

____ Secure ambulance or other transport for discharged patients.

TRANSPORTATION UNIT LEADER *(continued)*

Intermediate ____ Identify transportation needs for ambulatory casualties.

Extended ____ Maintain transportation assignment record in Triage Area, Discharge Area, and Material Supply Pool.

____ Keep Logistics Section Chief apprised of status.

____ Direct unassigned personnel to Labor Pool.

____ Observe and assist any staff who exhibits signs of stress or fatigue. Report concerns to Psychological Support Unit Leader. Provide for staff rest periods and relief.

____ Other concerns:

MATERIALS SUPPLY UNIT LEADER

Position Assigned To: _____

You Report To: _____ (Logistics Section Chief)

Logistics Command Center: _____

Telephone: _____

Mission: Organize and supply medical and non-medical care equipment and supplies.

Immediate   ____ Receive appointment from Logistics Section Chief.

          ____ Read this entire Job Action Sheet and review organizational chart on back.

          ____ Put on position identification vest.

          ____ Receive briefing from Logistics Section Chief.

          ____ Meet with and brief Materials Management and Central/Sterile Supply Personnel.

          ____ Establish and communicate the operational status of the Materials Supply Pool to the Logistics Section Chief, E.O.C. and Procurement Unit Leader.

          ____ Dispatch the pre-designated supply carts to Triage Area, Immediate Treatment Area, Delayed Treatment Area and the Minor Treatment Area, once these areas have been established. Enlist the assistance of the Transportation Unit Leader.

          ____ Release Search and Rescue Team equipment packs to those teams designated by the Damage Assessment and Control Officer.

          ____ Collect and coordinate essential medical equipment and supplies. (Prepare to assist with equipment salvage and recovery efforts.)

MATERIALS SUPPLY UNIT LEADER *(continued)*

Intermediate   ___   Develop medical equipment inventory to include, but not limited to the following:
- Bandages, dressings, compresses and suture material
- Sterile scrub brushes, normal saline, anti-microbial skin cleanser.
- Waterless hand cleaner and gloves
- Fracture immobilization, splinting and casting materials
- Backboard, rigid stretchers
- Non-rigid transporting devices (litters)
- Oxygen-ventilation-suction devices
- Advance life support equipment (chest tube, airway, major suture trays)

Extended   ___   Identify additional equipment and supply needs. Make requests/ needs known through Logistics Section Chief. Gain the assistance of the Procurement Unit Leader when indicated.

  ___   Determine the anticipated pharmaceuticals needed with the assistance of the Medical Care Director and Pharmacy Unit Leader to obtain/request items.

  ___   Coordinate with Safety & Security Officer to protect resources.

  ___   Observe and assist staff who exhibit signs of stress or fatigue. Report concerns to Psychological Support Unit Leader.

  ___   Other concerns:

NUTRITIONAL SUPPLY UNIT LEADER

Position Assigned To: _____

You Report To: _____ (Logistics Section Chief)

Logistics Command Center: _____

Telephone: _____

Mission: Organize food and water stores for preparation and rationing during periods of anticipated or actual shortage.

Immediate   ____ Receive appointment from Logistics Section Chief.

____ Read this entire Job Action Sheet and review organizational chart on back.

____ Put on position identification vest.

____ Receive briefing from Logistics Section Chief.

____ Meet with and brief Nutritional Services personnel.

____ Estimate the number of meals which can be served utilizing existing food stores; implement rationing if situation dictates.

____ Inventory the current emergency drinking water supply and estimate time when re-supply will be necessary. Implement rationing if situation dictates.

____ Report inventory levels of emergency drinking water and food stores to Logistics Section Chief.

____ Meet with Labor Pool Unit Leader and Staff Support Unit Leader to discuss location of personnel refreshment and nutritional break areas.

____ Secure nutritional and water inventories with the assistance of the Safety & Security Officer.

Intermediate   ____ Submit an anticipated need list of water and food to the Logistics Section Chief. Request should be based on current information concerning emergency events as well as projected needs for patients, staff and dependents.

NUTRITIONAL SUPPLY UNIT LEADER *(continued)*

<u>Extended</u>   ____ Meet with Logistics Section Chief regularly to keep informed of current status.

____ Observe and assist staff who exhibit signs of stress and fatigue. Report concerns to Psychological Support Unit Leader. Provide for staff rest period and relief.

____ Other Concerns:

PLANNING SECTION CHIEF

Position Assigned To: _____

You Report To: _____ (Emergency Incident Commander)

Planning Command Center: _____

Telephone: _____

Mission: Organize and direct all aspects of Planning Section operations. Ensure the distribution of critical information/data. Compile scenario/resource projections from all section chiefs and effect long range planning. Document and distribute facility Action Plan.

Immediate   _____ Receive appointment from Incident Commander. Obtain packet containing Section's Job Action Sheets.

    _____ Read this entire Job Action Sheet and review organizational chart on back.

    _____ Put on position identification vest.

    _____ Obtain briefing from Incident Commander.

    _____ Recruit a documentation aide from the Labor Pool

    _____ Appoint Planning unit leaders: Situation—Status Unit Leader, Labor Pool Unit Leader, Medical Staff Unit Leader, Nursing Unit Leader; distribute the corresponding Job Action Sheets and vests. (May be pre-established.)

    _____ Brief unit leaders after meeting with Emergency Incident Commander.

    _____ Provide for a Planning/Information Center.

    _____ Ensure the formulation and documentation of an incident-specific, facility Action Plan. Distribute copies to Incident Commander and all section chiefs.

    _____ Call for projection reports (Action Plan) from all Planning Section unit leaders and section chiefs for scenarios 4, 8, 24 & 48 hours from time of incident onset. Adjust time for receiving projection reports as necessary.

PLANNING SECTION CHIEF *(continued)*

Intermediate ____ Instruct Situation–Status Unit Leader and staff to document/ update status reports from all disaster section chiefs and unit leaders for use in decision making and for reference in post-disaster evaluation and recovery assistance applications.

____ Obtain briefings and updates as appropriate. Continue to update and distribute the facility Action Plan.

____ Schedule planning meetings to include Planning Section unit leaders, section chiefs and the Incident Commander for continued update of the facility Action Plan.

Extended ____ Continue to receive projected activity reports from section chiefs and Planning Section unit leaders at appropriate intervals.

____ Assure that all requests are routed/documented through the Communications Unit Leader.

____ Observe all staff, volunteers and patients for signs of stress and inappropriate behavior. Report concerns to Psychological Support Unit Leader. Provide for staff rest periods and relief.

____ Other concerns:

SITUATION-STATUS (SITSTAT) UNIT LEADER

Position Assigned To: _____

You Report To: _____ (Planning Section Chief)

Planning Command Center: _____

Telephone: _____

Mission: Maintain current information regarding the incident status for all hospital staff. Ensure a written record of the hospital's emergency planning and response. Develop the hospital's internal information network. Monitor the maintenance and preservation of the computer system.

Immediate   ____ Receive appointment from Planning Section Chief.

____ Read this entire Job Action Sheet and review organizational chart back.

____ Put on position identification vest.

____ Obtain briefing from Planning Section Chief.

____ Obtain status report on computer information system.

____ Assign recorder to document decisions, actions and attendance in E.O.C.

____ Establish a status/condition board in E.O.C. with a documentation aide. Ensure that this board is kept current.

____ Assign recorder to Communications Unit Leader to document telephone, radio and memo traffic.

____ Ensure that an adequate number of recorders are available to assist areas as needed. Coordinate personnel with Labor Pool.

____ Supervise backup and protection of existing data for main and support computer systems.

Intermediate   ____ Publish an internal incident informational sheet for employee information at least every 4-6 hours. Enlist the assistance of the Public Information Officer, Staff Support Unit Leader and Labor Pool Unit Leader.

SITUATION-STATUS (SITSTAT) UNIT LEADER *(continued)*

_____ Ensure the security and prevent the loss of medical record hard copies.

Extended   _____ Observe all staff, volunteers and patients for signs of stress and inappropriate behavior. Report concerns to Psychological Support Unit Leader. Provide for staff rest periods and relief.

_____ Other concerns:

LABOR POOL UNIT LEADER

Position Assigned To: _____

You Report To: _____ (Planning Section Chief)

Planning Command Center: _____

Telephone: _____

Mission: Collect and inventory available staff and volunteers at a central point. Receive requests and assign available staff as needed. Maintain adequate numbers of both medical and non-medical personnel. Assist in the maintenance of staff morale.

Immediate   ____ Receive appointment from Planning Section Chief.

    ____ Read this entire Job Action Sheet and review organizational chart on back.

    ____ Put on position identification vest.

    ____ Obtain briefing from the Planning Section Chief.

    ____ Establish Labor Pool area and communicate operational status to E.O.C. and all patient care and non-patient care areas.

    ____ Inventory the number and classify staff presently available. Use the following classifications and sub-classifications for personnel:

    I. MEDICAL PERSONNEL
      A. Physician (Obtain with assistance of Medical Staff Unit Leader.)
        1. Critical Care
        2. General Care
        3. Other
      B. Nurse
        1. Critical Care
        2. General Care
        3. Other
      C Medical Technicians
        1. Patient Care (aides, orderlies, EMTs, etc.)
        2. Diagnostic

LABOR POOL UNIT LEADER *(continued)*

Immediate   ____ Inventory the number and classify staff presently available. Use
                  the following classifications and sub-classifications for personnel:
                  II. NON-MEDICAL PERSONNEL
                  A. Engineering/Maintenance/Materials Management
                  B. Environmental/Nutritional Services
                  C. Business/Financial
                  D. Volunteer
                  E. Other

             ____ Establish a registration and credentialing desk for volunteers
                  not employed or associated with the hospital.

             ____ Obtain assistance from Safety & Security Officer in the
                  screening and identification of volunteer staff.

             ____ Meet with Nursing Unit Leader, Medical Staff Unit Leader and
                  Operations Section Chief to coordinate long term staffing needs.

             ____ Maintain log of all assignments.

             ____ Assist the Situation–Status Unit Leader in publishing an infor-
                  mational sheet to be distributed at frequent intervals to update
                  the hospital population.

             ____ Maintain a message center in Labor Pool Area with the coop-
                  eration of Staff Support Unit Leader and Situation–Status
                  Information Systems Unit Leader.

Extended   ____ Brief Planning Section Chief as frequently as necessary on the
                  status of labor pool numbers and composition.

             ____ Develop staff rest and nutritional area in coordination with
                  Staff Support Unit Leader and Nutritional Supply Unit Leader.

             ____ Document actions and decisions on a continual basis.

             ____ Observe all staff, volunteers and patients for signs of stress
                  and inappropriate behavior. Report concerns to Psychological
                  Support Unit Leader. Provide for staff rest periods and relief.

             ____ Other concerns:

MEDICAL STAFF UNIT LEADER

Position Assigned To: _____

You Report To: _____ (Planning Section Chief)

Planning Command Center: _____

Telephone: _____

Mission: Collect available physicians, and other medical staff, at a central point. Credential volunteer medical staff as necessary. Assist in the assignment of available medical staff as needed.

Immediate   _____ Receive assignment from Planning Section Chief.

    _____ Read this entire Job Action Sheet and refer to organizational chart on back.

    _____ Put on position identification vest.

    _____ Obtain briefing from Emergency Incident Commander or Planning Section Chief.

    _____ Establish Medical Staff Pool in predetermined location and communicate operational status to E.O.C. and Medical Staff Director. Obtain documentation personnel from Labor Pool.

    _____ Inventory the number and types of physicians, and other staff present. Relay information to Labor Pool Unit Leader.

    _____ Register and credential volunteer physician/medical staff. Request the assistance of the Labor Pool Unit Leader and Safety & Security Officer when necessary.

    _____ Meet with Labor Pool Unit Leader, Nursing Service Unit Leader and Operations Section Chief to coordinate projected staffing needs and issues.

    _____ Assist the Medical Staff Director in the assignment of medical staff to patient care and treatment areas.

Extended   _____ Establish a physician message center and emergency incident information board with the assistance of Staff Support Unit Leader and Labor Pool Unit Leader.

MEDICAL STAFF UNIT LEADER *(continued)*

Extended   Assist the Medical Staff Director in developing a medical staff rotation schedule.

        _____ Assist the Medical Staff Director in maintaining a log of medical staff assignments.

        _____ Brief Planning Section Chief as frequently as necessary on the status of medical staff pool numbers and composition.

        _____ Develop a medical staff rest and nutritional area in coordination with Staff Support Unit Leader and the Nutritional Supply Unit Leader.

        _____ Document actions and decisions on a continual basis.

        _____ Observe and assist medical staff who exhibit signs of stress and other fatigue. Report concerns to the Medical Staff Director and/or Psychological Support Unit Leader.

        _____ Other concerns:

NURSING UNIT LEADER

Position Assigned To: _____

You Report To: _____ (Planning Section Chief)

Planning Command Center: _____

Telephone: _____

Mission: Organize and coordinate nursing and direct patient care services.

Immediate   _____ Receive appointment from Planning Section Chief.

    _____ Read this entire Job Action Sheet and review organizational chart on back.

    _____ Put on position identification vest.

    _____ Obtain a briefing from Emergency Incident Commander or Planning Section Chief.

    _____ Appoint Patient Tracking Officer and Patient Information Officer and distribute the corresponding Job Action Sheets. Ensure the implementation of a patient tracking system.

    _____ Obtain current in-patient census and request a prioritization assessment (triage) of all in-house patients from the Medical Care Director.

    _____ Meet with Operations Chief, Medical Staff Director and Medical Care Director to assess and project nursing staff and patient care supply needs.

    _____ Recall staff as appropriate; assist the Labor Pool in meeting the nursing staff needs of the Medical Care Director.

    _____ Implement emergency patient discharge plan at the direction of the Emergency Incident Commander with support of the Medical Staff Director.

    _____ Meet regularly with the Patient Tracking Officer and Patient Information Officer.

NURSING UNIT LEADER *(continued)*

Immediate   ____ Meet with Labor Pool Unit Leader, Medical Care Director and
                            Operations Section Chief to coordinate long term staffing needs.

                  ____ Coordinate with the Labor Pool staff the number of nursing
                            personnel which may be released for future staffing or staffing
                            at another facility.

Extended   ____ Establish a staff rest and nutritional area in cooperation with
                            Labor Pool Unit Leader and Staff Support Unit Leader.

                  ____ Observe all staff, volunteers and patients for signs of stress
                            and inappropriate behavior. Report concerns to Psychological
                            Support Unit Leader. Provide for staff rest periods and relief.

                  ____ Other concerns:

PATIENT TRACKING OFFICER

Position Assigned To: _____

You Report To: _____ (Nursing Unit Leader)

Planning Command Center: _____

Telephone: _____

Mission: Maintain the location of patients at all times within the hospital's patient care system.

Immediate   ____ Receive appointment from Nursing Unit Leader.

   ____ Read this entire Job Action Sheet and review organizational chart on back.

   ____ Put on position identification vest.

   ____ Obtain a briefing from Nursing Unit Leader.

   ____ Obtain patient census from Nursing Unit Leader, Admitting personnel or other source.

   ____ Establish an area near the E.O.C. to track patient arrivals, location and disposition. Obtain sufficient assistance to document current and accurate patient information.

   ____ Ensure the proper use of the hospital disaster chart and tracking system for all newly admitted.

   ____ Meet with Patient Information Officer, Public Information Officer and Liaison Officer on a routine basis to update and exchange patient information and census data.

Extended   ____ Maintain log to document the location and time of all patients cared for.

   ____ Observe all staff, volunteers and patients for signs of stress and inappropriate behavior. Report concerns to Psychological Support Unit Leader. Provide for staff rest periods and relief.

   ____ Other concerns:

PATIENT TRACKING OFFICER

Position Assigned To: _____

You Report To: _____ (Nursing Unit Leader)

Planning Command Center: _____

Telephone: _____

Mission: Provide information to visitors and families regarding status and
location of patients. Collect information necessary to complete the
Disaster Welfare Inquiry process in cooperation with the American
Red Cross.

Immediate ____ Receive appointment from Nursing Unit Leader.

____ Read this entire Job Action Sheet and review organizational
chart back.

____ Put on position identification vest.

____ Obtain briefing on incident and any special instructions from
Nursing Unit Leader.

____ Establish Patient Information Area away from E.O.C.

____ Meet with Patient Tracking Officer to exchange patient related
information and establish regularly scheduled meetings.

____ Direct patient related news releases through Nursing Unit
Leader to the Public Information Officer.

____ Receive and screen requests about the status of individual
patients. Obtain appropriate information and relay to the
appropriate requesting party.

____ Request assistance of runners and amateur radio operators
from Labor Pool as needed.

Extended ____ Work with American Red Cross representative in development
of the Disaster Welfare Inquiry information.

PATIENT TRACKING OFFICER *(continued)*

Extended   ____ Observe all staff, volunteers and patients for signs of stress and inappropriate behavior. Report concerns to Psychological Support Unit Leader. Provide for staff rest periods and relief.

    ____ Other concerns:

FINANCE SECTION CHIEF

Position Assigned To: _____

You Report To: _____ (Emergency Incident Commander)

Finance Command Center: _____

Telephone: _____

Mission: Monitor the utilization of financial assets. Oversee the acquisition of supplies and services necessary to carry out the hospital's medical mission. Supervise the documentation of expenditures relevant to the emergency incident.

Immediate _____ Receive appointment from Emergency Incident Commander. Obtain packet containing Section's Job Action Sheets.

_____ Read this entire Job Action Sheet and review organizational chart on back.

_____ Put on position identification vest.

_____ Obtain briefing from Emergency Incident Commander.

_____ Appoint Time Unit Leader, Procurement Unit Leader, Claims Unit Leader and Cost Unit Leader; distribute the corresponding Job Action Sheets and vests. (May be pre-established.)

_____ Confer with Unit Leaders after meeting with Emergency Incident Commander; develop a section action plan.

_____ Establish a Financial Section Operations Center. Ensure adequate documentation/recording personnel.

_____ Approve a "cost-to-date" incident financial status report submitted by the Cost Unit Leader every eight hours summarizing financial data relative to personnel, supplies and miscellaneous expenses.

_____ Obtain briefings and updates from Emergency Incident Commander as appropriate. Relate pertinent financial status reports to appropriate chiefs and unit leaders.

FINANCE SECTION CHIEF *(continued)*

<u>Intermediate</u>   _____   Schedule planning meetings to include Finance Section unit leaders to discuss updating the section's incident action plan and termination procedures.

<u>Extended</u>   _____   Assure that all requests for personnel or supplies are copied to the Communications Unit Leader in a timely manner.

_____   Observe all staff, volunteers and patients for signs of stress and inappropriate behavior. Report concerns to Psychological Support Unit Leader. Provide for staff rest periods and relief.

_____   Other concerns:

TIME UNIT LEADER

Position Assigned To: _____

You Report To: _____ (Finance Section Chief)

Finance Command Center: _____

Telephone: _____

Mission: Responsible for the documentation of personnel time records. The monitoring and reporting of regular and overtime hours worked/ volunteered.

Immediate   ____ Receive appointment from Finance Section Chief.

              ____ Read this entire Job Action Sheet and review organizational chart on back.

              ____ Put on position identification vest.

              ____ Obtain briefing from Finance Section Chief; assist in the development of the section action plan.

              ____ Ensure the documentation of personnel hours worked and volunteer hours worked in all areas relevant to the hospital's emergency incident response. Confirm the utilization of the Emergency Incident Time Sheet by all section chiefs and/or unit leaders. Coordinate with Labor Pool Unit Leader.

              ____ Collect all Emergency Incident Time Sheets from each work area for recording and tabulation every eight hours, or as specified by the Finance Section Chief.

              ____ Forward tabulated Emergency Incident Time Sheets to Cost Unit Leader every eight hours.

Extended   ____ Prepare a total of personnel hours worked during the declared emergency incident.

              ____ Observe all staff, volunteers and patients for signs of stress and inappropriate behavior. Report concerns to Psychological Support Unit Leader. Provide for staff rest periods and relief.

TIME UNIT LEADER *(continued)*

<u>Extended</u> ____ Other concerns:

PROCUREMENT UNIT LEADER

Position Assigned To: _____

You Report To: _____ (Finance Section Chief)

Finance Command Center: _____

Telephone: _____

Mission: Responsible for administering accounts receivable and payable to contract and non-contract vendors.

Immediate   ____ Receive appointment from Finance Section Chief.

____ Read this entire Job Action Sheet and review organizational chart on back.

____ Put on position identification vest.

____ Obtain briefing from Finance Section Chief; assist in the development of the section action plan.

____ Ensure the separate accounting of all contracts specifically related to the emergency incident; and all purchases within the enactment of the emergency incident response plan.

____ Establish a line of communication with the Material Supply Unit Leader.

____ Obtain authorization to initiate purchases from the Finance Section Chief, or authorized representative.

____ Forward a summary accounting of purchases to the Cost Unit Leader every eight hours.

Extended   ____ Prepare a Procurement Summary Report identifying all contracts initiated during the declared emergency incident.

____ Observe all staff, volunteers and patients for signs of stress and inappropriate behavior. Report concerns to Psychological Support Unit Leader. Provide for staff rest periods and relief.

PROCUREMENT UNIT LEADER *(continued)*

<u>Extended</u>   ____ Other concerns:

CLAIMS UNIT LEADER

Position Assigned To: _____

You Report To: _____ (Finance Section Chief)

Finance Command Center: _____

Telephone: _____

Mission: Responsible for receiving, investigating and documenting all claims reported to the hospital during the emergency incident which are alleged to be the result of an accident or action on hospital property.

Immediate ____ Receive appointment from Finance Section Chief.

____ Read this entire Job Action Sheet and review the organizational chart on back.

____ Put on position identification vest.

____ Obtain briefing from Finance Section Chief; assist in the development of section action plan.

____ Receive and document alleged claims issued by employees and non-employees. Use photographs or video documentation when appropriate.

____ Obtain statements as quickly as possible from all claimants and witnesses.

____ Enlist the assistance of the Safety & Security Officer where necessary.

____ Inform Finance Section Chief of all alleged claims as they are reported.

____ Document claims on hospital risk/loss forms.

Extended ____ Report any cost incurred as a result of a claim to the Cost Unit Leader as soon as possible.

____ Prepare a summary of all claims reported during the declared emergency incident.

CLAIMS UNIT LEADER *(continued)*

<u>Extended</u>   ____ Observe all staff, volunteers and patients for signs of stress and inappropriate behavior. Report concerns to Psychological Support Unit Leader. Provide for staff rest periods and relief.

____ Other concerns:

COST UNIT LEADER

Position Assigned To: _____

You Report To: _____ (Finance Section Chief)

Finance Command Center: _____

Telephone: _____

Mission: Responsible for providing cost analysis data for declared emergency incident. Maintenance of accurate records of incident cost.

Immediate   _____ Receive appointment from Finance Section Chief.

_____ Read this entire Job Action Sheet and review the Organizational chart on back.

_____ Put on position identification vest.

_____ Obtain briefing from Finance Section Chief; assist in development of section action plan.

_____ Meet with Time Unit Leader, Procurement Unit Leader and Claims Unit Leader to establish schedule for routine reporting periods.

_____ Prepare a "cost-to-date" report form for submission to Finance Section Chief once every eight hours.

_____ Inform all section chief's of pertinent cost data at the direction of the Finance Section Chief or Emergency Incident Commander.

Extended   _____ Prepare a summary of all costs incurred during the declared emergency incident.

_____ Observe all staff, volunteers and patients for signs of stress and inappropriate behavior. Report concerns to Psychological Support Unit Leader. Provide for staff rest periods and relief.

_____ Other concerns:

OPERATIONS SECTION CHIEF

Position Assigned To: _____

You Report To: _____ (Emergency Incident Commander)

Operations Command Center: _____

Telephone: _____

Mission: Organize and direct aspects relating to the Operations Section. Carry out directives of the Emergency Incident Commander. Coordinate and supervise the Medical Services Subsection, Ancillary Services Subsection and Human Services Subsection of the Operations Section.

Immediate

_____ Receive appointment from Emergency Incident Commander. Obtain packet containing Section's Job Action Sheets.

_____ Read this entire Job Action Sheet and review organizational chart on back.

_____ Put on position identification vest.

_____ Obtain briefing from Emergency Incident Commander.

_____ Appoint Medical Staff Director, Medical Care Director, Ancillary Services Director and Human Services Director and transfer the corresponding Job Action Sheets. (May be pre-established.)

_____ Brief all Operations Section directors on current situation and develop the section's initial action plan. Designate time for next briefing.

_____ Establish Operations Section Center in proximity to E.O.C.

_____ Meet with the Medical Staff Director, Medical Care Director and Nursing Unit Leader to plan and project patient care needs.

_____ Designate times for briefings and updates with all Operations Section directors to develop/update section's action plan.

_____ Ensure that the Medical Services Subsection, Ancillary Services Subsection and Human Services Subsection are adequately staffed and supplied.

OPERATIONS SECTION CHIEF *(continued)*

Immediate   ____ Brief the Emergency Incident Commander routinely on the
                 status of the Operations Section.

Extended   ____ Assure that all communications are copied to the Communica-
                 tions Unit Leader; document all actions and decisions.

                 ____ Observe all staff, volunteers and patients for signs of stress
                 and inappropriate behavior. Report concerns to Psychological
                 Support Unit Leader. Provide for staff rest periods and relief.

                 ____ Other concerns

MEDICAL STAFF DIRECTOR

Position Assigned To: _____

You Report To: _____ (Operations Section Chief)

Operations Command Center: _____

Telephone: _____

Mission: Organize, prioritize and assign physicians to areas where medical care is being delivered. Advise the Incident Commander on issues related to the Medical Staff.

Immediate   ____ Receive appointment from the Operations Section Chief.

        ____ Read this entire Job Action Sheet and review organizational chart on back.

        ____ Put on position identification vest.

        ____ Meet with Operations Section Chief and other Operations Section directors for briefing and development of an initial action plan.

        ____ Meet with the Medical Staff Unit Leader to facilitate recruitment and staffing of Medical Staff. Assist in Medical Staff credentialing issues.

        ____ Document all physician assignments; facilitate rotation of physician staff with the assistance of the Medical Staff Unit Leader; where necessary, assist with physician orientation to in-patient and treatment areas.

        ____ Meet with Operations Chief, Medical Care Director and Nursing Unit Leader to plan and project patient care needs.

        ____ Provide medical staff support for patient priority assessment to designate patients for early discharge.

        ____ Meet with Incident Commander for appraisal of the situation regarding medical staff and projected needs. Establish meeting schedule with IC if necessary.

MEDICAL STAFF DIRECTOR *(continued)*

Immediate ____ Maintain communication with the Medical Care Director to co-monitor the delivery and quality of medical care in all patient care areas.

Extended ____ Ensure maintenance of Medical Staff time sheet; obtain clerical support from Labor Pool if necessary.

____ Meet as often as necessary with the Operations Section Chief to keep appraised of current conditions.

____ Observe all staff, volunteers and patients for signs of stress and inappropriate behavior. Report concerns to Psychological Support Unit Leader. Provide for staff rest periods and relief.

____ Other concerns:

MEDICAL CARE DIRECTOR

Position Assigned To: _____

You Report To: _____ (Operations Section Chief)

Operations Command Center: _____

Telephone: _____

Mission: Organize and direct the overall delivery of medical care in all areas of the hospital.

Immediate

_____ Receive appointment from the Operations Section Chief and receive the Job Action Sheets for the Medical Services Subsection.

_____ Read this entire Job Action Sheet and review organizational chart on back.

_____ Put on position identification vest.

_____ Meet with Operations Section Chief and other Operations Section directors for briefing and development of an initial action plan. Establish time for follow up meetings.

_____ Appoint the In-Patient Areas Supervisor and the Treatment Areas Supervisor and transfer the corresponding Job Action Sheets.

_____ Assist in establishing an Operations Section Center in proximity to the E.O.C.

_____ Meet with In-Patient Areas Supervisor and Treatment Areas Supervisor to discuss medical care needs and physician staffing in all patient care areas.

_____ Confer with the Operations Chief, Medical Staff Director and Nursing Unit Leader to make medical staff and nursing staffing/material needs known.

_____ Request Medical Staff Director to provide medical staff support to assist with patient priority assessment to designate those eligible for early discharge.

MEDICAL CARE DIRECTOR *(continued)*

Intermediate   ____   Establish 2-way communication (radio or runner) with In-Patient Areas Supervisor and Treatment Areas Supervisor.

____   Meet regularly with Medical Staff Director, In-Patient Areas Supervisor and Treatment Areas Supervisor to assess current and project future patient care conditions.

____   Brief Operations Section Chief routinely on the status/quality of medical care.

Extended   ____   Observe all staff, volunteers and patients for signs of stress and inappropriate behavior. Report concerns to Psychological Support Unit Leader. Provide for staff rest periods and relief.

____   Other concerns:

IN-PATIENT AREAS SUPERVISOR

Position Assigned To: _____

You Report To: _____ (Medical Care Director)

Operations Command Center: _____

Telephone: _____

Mission: Receive appointment from Medical Care Director and receive Job Action Sheets for the Surgical Services, Maternal - Child, Critical Care, General Nursing and Out Patient Services Unit Leaders.

        ____ Read this entire Job Action Sheet and review organizational chart on back.

        ____ Put on position identification vest.

        ____ Receive briefing from Medical Care Director; develop initial action plan with Medical Care Director, Treatment Areas Supervisor and Medical Staff Director.

        ____ Appoint Unit Leaders for:
- Surgical Services
- Maternal–Child
- Critical Care
- General Nursing Care
- Out Patient Services

        ____ Distribute corresponding Job Action Sheets, request a documentation aide/assistant for each unit leader from Labor Pool.

        ____ Brief unit leaders on current status. Designate time for follow-up meeting.

        ____ Assist establishment of in-patient care areas in new locations if necessary.

<u>Intermediate</u> ____ Instruct all unit leaders to begin patient priority assessment; designate those eligible for early discharge. Remind all unit leaders that all in-patient discharges are routed through the Discharge Unit.

IN-PATIENT AREAS SUPERVISOR *(continued)*

_____ Assess problems and treatment needs in each area; coordinate the staffing and supplies between each area to meet needs.

_____ Meet with Medical Care Director to discuss medical care plan of action and staffing in all in-patient care areas.

_____ Receive, coordinate and forward requests for personnel and supplies to the Labor Pool Unit Leader, Medical Care Director and Material Supply Unit Leader. Copy all communication to the Communications Unit Leader.

_____ Contact the Safety & Security Officer for any security needs. Advise the Medical Care Director of any actions/requests.

_____ Report equipment needs to Materials Supply Unit Leader.

_____ Establish 2-way communication (radio or runner) with Medical Care Director.

_____ Assess environmental services (housekeeping) needs in all in-patient care areas; contact Sanitation Systems Officer for assistance.

Extended   _____ Assist Patient Tracking Officer and Patient Information Officer in obtaining information.

_____ Observe and assist any staff who exhibit signs of stress and fatigue. Report any concerns to Psychological Support Unit Leader. Provide for staff rest periods and relief.

_____ Report frequently and routinely to Medical Care Director to keep apprised of situation.

_____ Document all action/decisions with a copy sent to the Medical Care Director.

_____ Other concerns:

SURGICAL SERVICES UNIT LEADER

Position Assigned To: _____

You Report To: _____ (In-Patient Areas Supervisor)

Operations Command Center: _____

Telephone: _____

Mission: Supervise and maintain the surgical capabilities to the best possible level in respect to current conditions in order to meet the needs of in-house and newly admitted patients.

Immediate ____ Receive appointment from In-Patient Areas Supervisor.

____ Read this entire Job Action Sheet and review organizational chart on back.

____ Put on position identification vest.

____ Receive briefing from In-Patient Areas Supervisor with other In-Patient Area unit leaders.

____ Assess current pre-op, operating suite and post-op capabilities. Project immediate and prolonged capacities to provide surgical services based on current data.

____ Begin patient priority assessment; designate those eligible for early discharge. Remind all staff that all in-patient discharges are routed through the Discharge Unit.

____ Develop action plan in cooperation with other In-Patient Area unit leaders and the In-Patient Areas Supervisor.

____ Request needed resources from the In-Patient Areas Supervisor.

____ Assign and schedule O.R. teams as necessary; obtain additional personnel from Labor Pool.

____ Identify location of Immediate and Delayed Treatment areas; inform patient transportation personnel.

SURGICAL SERVICES UNIT LEADER *(continued)*

Intermediate ____ Contact Safety & Security Officer of security and traffic flow
needs in the Surgical Services area. Inform In-Patient Areas
Supervisor of action.

____ Report equipment/material needs to Materials Supply Unit
Leader. Inform In-Patient Areas Supervisor of action.

Extended ____ Ensure that all area and individual documentation is current
and accurate. Request documentation/clerical personnel from
Labor Pool if necessary.

____ Keep In-Patient Areas Supervisor, Immediate Treatment and
Delayed Treatment Unit Leader apprised of status, capabilities
and projected services.

____ Observe and assist any staff who exhibit signs of stress and
fatigue. Report concerns to In-Patient Areas Supervisor. Provide
for staff rest periods and relief.

____ Review and approve the area documentation aide's recordings
of actions/decisions in the Surgical Services Area. Send copy
to the In-Patient Areas Supervisor.

____ Direct non-utilized personnel to Labor Pool.

____ Other concerns:

MATERNAL-CHILD UNIT LEADER

Position Assigned To: _____

You Report To: _____ (In-Patient Areas Supervisor)

Operations Command Center: _____

Telephone: _____

Mission: Supervise and maintain the obstetrical, labor & delivery, nursery, and pediatric services to the best possible level in respect to current conditions in order to meet the needs of in-house and newly admitted patients.

Immediate   ____ Receive appointment from In-Patient Areas Supervisor.

            ____ Read this entire Job Action Sheet and review organizational chart on back.

            ____ Put on position identification vest.

            ____ Receive briefing from In-Patient Areas Supervisor with other In-Patient Area unit leaders.

            ____ Assess current capabilities. Project immediate and prolonged capacities to provide all obstetrical and pediatric services based on current capabilities. (Give special consideration to the possibility of an increase in normal and premature deliveries due to environmental/emotional stress.)

            ____ Begin patient priority assessment; designate those eligible for early discharge. Remind all staff that all in-patient discharges are routed through the Discharge Unit.

            ____ Develop action plan in cooperation with other In-Patient Area unit leaders and the In-Patient Areas Supervisor.

            ____ Request needed resources from the In-Patient Areas Supervisor.

            ____ Assign delivery and patient teams as necessary; obtain additional personnel from Labor Pool.

MATERNAL-CHILD UNIT LEADER *(continued)*

Intermediate   ＿＿ Identify location of Immediate and Delayed Treatment areas; inform patient transportation personnel.

＿＿ Contact Safety & Security Officer of security and traffic flow needs. Inform In-Patient Areas Supervisor of action.

＿＿ Report equipment/material needs to Materials Supply Unit Leader. Inform In-Patient Areas Supervisor of action.

Extended   ＿＿ Ensure that all area and individual documentation is current and accurate. Request documentation/clerical personnel from Labor Pool if necessary.

＿＿ Keep In-Patient Areas Supervisor, Immediate Treatment and Delayed Treatment Unit Leader apprised of status, capabilities and projected services.

＿＿ Observe and assist any staff who exhibit signs of stress and fatigue. Report concerns to In-Patient Areas Supervisor. Provide for staff rest periods and relief.

＿＿ Review and approve the area documentation aide's recordings of actions/decisions in the Surgical Services Area. Send copy to the In-Patient Areas Supervisor.

＿＿ Direct non-utilized personnel to Labor Pool.

＿＿ Other concerns:

CRITICAL CARE UNIT LEADER

Position Assigned To: _____

You Report To: _____ (In-Patient Areas Supervisor)

Operations Command Center: _____

Telephone: _____

Mission: Supervise and maintain the critical care capabilities to the best possible level to meet the needs of in-house and newly admitted patients.

Immediate   ____ Receive appointment from In-Patient Areas Supervisor.

                         ____ Read this entire Job Action Sheet and review organizational chart on back.

                         ____ Put on position identification vest.

                         ____ Receive briefing from In-Patient Areas Supervisor with other In-Patient Area unit leaders.

                         ____ Assess current critical care patient capabilities. Project immediate and prolonged capabilities to provide services based on known resources. Obtain medical staff support to make patient triage decisions if warranted.

                         ____ Develop action plan in cooperation with other In-Patient Area unit leaders and the In-Patient Areas Supervisor

                         ____ Request the assistance of the In-Patient Areas Supervisor to obtain resources if necessary.

                         ____ Assign patient care teams as necessary; obtain additional personnel from Labor Pool.

                         ____ Identify location of Discharge Area; inform patient transportation personnel.

                         ____ Contact Safety & Security Officer of security and traffic flow needs in the critical care services area(s). Inform In-Patient Areas Supervisor of action.

CRITICAL CARE UNIT LEADER *(continued)*

<u>Intermediate</u> _____ Report equipment/material needs to Materials Supply Unit Leader. Inform In-Patient Areas Supervisor of action.

<u>Extended</u> _____ Ensure that all area and individual documentation is current and accurate. Request documentation/ clerical personnel from Labor Pool if necessary.

_____ Keep In-Patient Areas Supervisor, Immediate Treatment and Delayed Treatment Unit Leaders apprised of status, capabilities and projected services.

_____ Observe and assist any staff who exhibit signs of stress and fatigue. Report concerns to In-Patient Areas Supervisor. Provide for staff rest periods and relief.

_____ Review and approve the area document's recordings of actions/decisions in the Critical Care Area(s). Send copy to the In-Patient Areas Supervisor.

_____ Direct non-utilized personnel to Labor Pool.

_____ Other concerns:

GENERAL NURSING CARE UNIT LEADER

Position Assigned To: _____

You Report To: _____ (In-Patient Areas Supervisor)

Operations Command Center: _____

Telephone: _____

Mission: Supervise and maintain general nursing services to the best possible level to meet the needs of in-house and newly admitted patients.

Immediate   ____ Receive appointment from In-Patient Areas Supervisor.

____ Read this entire Job Action Sheet and review organizational chart on back.

____ Put on position identification vest.

____ Receive briefing from In-Patient Areas Supervisor with other In-Patient Area unit leaders.

____ Assess current capabilities. Project immediate and prolonged capacities to provide general medical/ surgical nursing services based on current data.

____ Begin patient priority assessment; designate those eligible for early discharge. Remind all staff that all in-patient discharges are routed through the Discharge Unit.

____ Develop action plan in cooperation with other In-Patient Area unit leaders and the In-Patient Areas Supervisor.

____ Request needed resources from the In-Patient Areas Supervisor.

____ Assign patient care teams as necessary; obtain additional personnel from Labor Pool.

____ Identify location of Immediate and Delayed Treatment areas; inform patient transportation personnel.

Intermediate   ____ Contact Safety & Security Officer of security and traffic flow needs. Inform In-Patient Areas Supervisor of action.

GENERAL NURSING CARE UNIT LEADER *(continued)*

<u>Intermediate</u> ____ Report equipment/material needs to Materials Supply Unit Leader. Inform In-Patient Areas Supervisor of action.

<u>Extended</u> ____ Ensure that all area and individual documentation is current and adhered. Request documentation/clerical personnel from Labor Pool if necessary.

____ Keep In-Patient Areas Supervisor, Immediate Treatment and Delayed Treatment Unit Leader apprised of status, capabilities and projected services.

____ Observe and assist any staff who exhibit signs of stress and fatigue. Report concerns to In-Patient Areas Supervisor. Provide for staff rest periods and relief.

____ Review and approve the area documenter's recordings of actions/decisions in the Surgical Services Area. Send copy to the In-Patient Areas Supervisor.

____ Direct non-utilized personnel to Labor Pool.

____ Other concerns:

OUT PATIENT SERVICES UNIT LEADER

Position Assigned To: _____

You Report To: _____ (In-Patient Areas Supervisor)

Operations Command Center: _____

Telephone: _____

Mission: Prepare any out patient service areas to meet the needs of in-house and newly admitted patients.

Immediate   ____ Receive appointment from In-Patient Areas Supervisor.

    ____ Read this entire Job Action Sheet and review organizational chart on back.

    ____ Put on position identification vest.

    ____ Receive briefing from In-Patient Areas Supervisor with other In-Patient Area unit leaders.

    ____ Assess current capabilities. Project immediate and prolonged capacities to provide nursing services based on current data.

    ____ Begin out patient priority assessment; designate those eligible for immediate discharge; admit those patients unable to be discharged. Remind all staff that all patient discharges are routed through the Discharge Unit.

    ____ Develop action plan in cooperation with other In-Patient Area unit leaders and the In-Patient Areas Supervisor.

    ____ Request needed resources from the In-Patient Areas Supervisor.

    ____ Assign patient care teams in configurations to meet the specific mission of the Out Patient areas; obtain additional personnel as necessary from Labor Pool.

    ____ Contact Safety & Security Officer of security and traffic flow needs. Inform In-Patient Areas Supervisor of action.

Intermediate   ____ Report equipment/material needs to Materials Supply Unit Leader. Inform In-Patient Areas Supervisor of action.

OUT PATIENT SERVICES UNIT LEADER *(continued)*

Extended   ____ Ensure that all area and individual documentation is current
and accurate. Request documentation/clerical personnel from
Labor Pool if necessary.

____ Keep In-Patient Areas Supervisor apprised of status, capabilities and projected services.

____ Observe and assist any staff who exhibit signs of stress and
fatigue. Report concerns to In-Patient Areas Supervisor. Provide for staff rest periods and relief.

____ Review and approve the area documenter's recordings of
actions/decisions in the Surgical Services Area. Send copy to
the In-Patient Areas Supervisor.

____ Direct non-utilized personnel to Labor Pool.

____ Other concerns:

TREATMENT AREAS SUPERVISOR

Position Assigned To: _____

You Report To: _____ (Medical Care Director)

Operations Command Center: _____

Telephone: _____

Mission: Initiate and supervise the patient triage process. Assure treatment of casualties according to triage categories and manage the treatment area(s). Provide for a controlled patient discharge. Supervise morgue service.

Immediate   ____ Receive appointment from Medical Care Director and Job Action Sheets for the Triage, Immediate-Delayed-Minor Treatment, Discharge and Morgue Unit Leaders.

   ____ Read this entire Job Action Sheet and review organizational chart on back.

   ____ Put on position identification vest.

   ____ Receive briefing from Medical Care Director and develop initial action plan with Medical Care Director, In-Patient Areas Supervisor and Medical Staff Director.

   ____ Appoint unit leaders for the following treatment areas:
   • Triage
   • Immediate Treatment
   • Delayed Treatment
   • Minor Treatment
   • Discharge
   • Morgue

   Distribute corresponding Job Action Sheets, request a documentation aide/assistant for each unit leader from Labor Pool.

   ____ Brief Treatment Area unit leaders. Designate time for follow-up meeting.

   ____ Assist establishment of Triage, Immediate, Delayed, Minor Treatment, Discharge and Morgue Areas in pre-established locations.

TREATMENT AREAS SUPERVISOR *(continued)*

Intermediate _____ Assess problem, treatment needs and customize the staffing and supplies in each area.

_____ Meet with Medical Care Director to discuss medical care plan of action and staffing in all triage/treatment/discharge/morgue areas. Maintain awareness of all in-patient capabilities, especially surgical services via the In-Patient Areas Supervisor.

_____ Receive, coordinate and forward requests for personnel and supplies to the Labor Pool Unit Leader, Medical Care Director and Material Supply Unit Leader. Copy all communication to the Communications Unit Leader.

_____ Contact the Safety and Security Officer for any security needs, especially those in the Triage, Discharge and Morgue areas. Advise the Medical Care Director of any actions/requests.

_____ Report equipment needs to Materials Supply Unit Leader.

_____ Establish 2-way communication (radio or runner) with Medical Care Director.

_____ Assess environmental services (housekeeping) needs for all Treatment Areas; contact Sanitation Systems Officer for assistance.

_____ Observe and assist any staff who exhibit signs of stress and fatigue. Report any concerns to Psychological Support Unit Leader. Provide for staff rest periods and relief.

_____ Assist Patient Tracking Officer and Patient Information Officer in obtaining information.

Extended _____ Report frequently and routinely to Medical Care Director to keep apprised of situation.

_____ Document all action/decisions with a copy sent to the Medical Care Director.

_____ Other concerns:

TRIAGE UNIT LEADER

Position Assigned To: _____

You Report To: _____ (Treatment Areas Supervisor)

Operations Command Center: _____

Telephone: _____

Mission: Sort casualties according to priority of injuries, and assure their disposition to the proper treatment area.

Immediate   ____ Receive appointment from Treatment Areas Supervisor.

        ____ Read this entire Job Action Sheet and review organizational chart on back.

        ____ Put on position identification vest.

        ____ Receive briefing from Treatment Areas Supervisor with other Treatment Area unit leaders.

        ____ Establish patient Triage Area; consult with Transportation Unit Leader to designate the ambulance off-loading area.

        ____ Ensure sufficient transport equipment and personnel for Triage Area.

        ____ Assess problem, triage-treatment needs relative to specific incident.

        ____ Assist the In-Patient Areas Supervisor with triage of internal hospital patients, if requested by Treatment Areas Supervisor.

        ____ Develop action plan, request needed resources from Treatment Areas Supervisor.

        ____ Assign triage teams.

        ____ Identify location of Immediate, Delayed, Minor Treatment, Discharge and Morgue areas; coordinate with Treatment Areas Supervisor.

TRIAGE UNIT LEADER *(continued)*

Intermediate ____ Contact Safety & Security Officer of security and traffic flow needs in the Triage Area. Inform Treatment Areas Supervisor of action.

Extended ____ Report emergency care equipment needs to Materials Supply Unit Leader. Inform Treatment Areas Supervisor of action.

____ Ensure that the disaster chart and admission forms are utilized. Request documentation/clerical personnel from Labor Pool if necessary.

____ Keep Treatment Areas Supervisor apprised of status, number of injured in the Triage Area or expected to arrive there.

____ Observe and assist any staff who exhibit signs of stress and fatigue. Report concerns to Treatment Areas Supervisor. Provide for staff rest periods and relief.

____ Review and approve the area documenter's recordings of actions/decisions in the Triage Area. Send copy to the Treatment Areas Supervisor.

____ Direct non-utilized personnel to Labor Pool.

____ Other concerns:

IMMEDIATE TREATMENT UNIT LEADER

Position Assigned To: _____

You Report To: _____ (Treatment Areas Supervisor)

Operations Command Center: _____

Telephone: _____

Mission: Coordinate the care given to patients received from the Triage Area; assure adequate staffing and supplies in the Immediate Treatment Area; facilitate the treatment and disposition of patients in the Immediate Treatment Area.

Immediate   ____ Receive appointment from Treatment Areas Supervisor.

     ____ Read this entire Job Action Sheet and review the organizational chart on back.

     ____ Put on position identification vest.

     ____ Receive briefing from Treatment Areas Supervisor with other Treatment Area unit leaders.

     ____ Assist Treatment Areas Unit Leader in the establishment of Immediate Treatment Area.

     ____ Assess situation/area for supply and staffing needs; request staff and supplies from the Labor Pool and Materials Supply Unit Leaders. Request medical staff support through Treatment Areas Supervisor.

     ____ Obtain an adequate number of patient transportation resources from the Transportation Unit Leader to ensure the movement of patients in and out of the area.

     ____ Ensure the rapid disposition and flow of treated patients from the Immediate Treatment Area.

     ____ Report frequently and routinely to the Treatment Areas Supervisor on situational status.

IMMEDIATE TREATMENT UNIT LEADER *(continued)*

Extended   ____ Observe and assist any staff who exhibits signs of stress and fatigue. Report any concerns to the Treatment Areas Unit Leader. Provide for staff rest periods and relief.

    ____ Review and approve the area documenter's recordings of actions/decisions in the Immediate Treatment Area. Send copy to the Treatment Areas Supervisor.

    ____ Direct non-utilized personnel to Labor Pool.

    ____ Other concerns:

DELAYED TREATMENT UNIT LEADER

Position Assigned To: _____

You Report To: _____ (Treatment Areas Supervisor)

Operations Command Center: _____

Telephone: _____

Mission: Coordinate the care given to patients received from the Triage Area. Assure adequate staffing and supplies in the Delayed Treatment Area. Facilitate the treatment and disposition of patients in the Delayed Treatment Area.

Immediate   _____ Receive appointment from Treatment Areas Supervisor.

    _____ Read this entire Job Action Sheet and review the organizational chart on back.

    _____ Put on position identification vest.

    _____ Receive briefing from Treatment Areas Supervisor with other Treatment Area unit leaders.

    _____ Assist Treatment Areas Supervisor in the establishment of Delayed Treatment Area.

    _____ Assess situation/area for supply and staffing need; request staff and supplies from the Labor Pool and Materials Supply Unit Leaders. Request medical staff support through Treatment Areas Supervisor.

    _____ Obtain an adequate number of patient transportation resources from the Transportation Unit Leader to ensure the movement of patients in and out of area.

    _____ Ensure the rapid disposition and flow of treated patients from the Delayed Treatment Area.

    _____ Report frequently and routinely to the Treatment Areas Supervisor on situational status.

DELAYED TREATMENT UNIT LEADER *(continued)*

Extended   ____ Observe and assist any staff who exhibits signs of stress and fatigue. Report any concerns to the Treatment Areas Supervisor. Provide for staff rest periods and relief.

____ Review and approve the area documenter's recordings of actions/decisions in the Delayed Treatment Area. Send copy to the Treatment Areas Supervisor.

____ Direct non-utilized personnel to Labor Pool.

____ Other concerns:

MINOR TREATMENT UNIT LEADER

Position Assigned To: _____

You Report To: _____ (Treatment Areas Supervisor)

Operations Command Center: _____

Telephone: _____

Mission: Coordinate the minor care of patients received from the Triage Area, and other areas of the hospital. Assure adequate staffing and supplies in the Minor Treatment. Facilitate the minor treatment of patients and disposition.

Immediate   _____ Receive appointment from the Treatment Areas Supervisor.

_____ Read this entire Job Action Sheet and review the organizational chart on back.

_____ Put on position identification vest.

_____ Receive briefing from Treatment Areas Supervisor with other Treatment Area unit leaders.

_____ Assist Treatment Areas Supervisor in the establishment of Minor Treatment Area.

_____ Assess situation/area for supply and staffing need; request staff and supplies from the Labor Pool and Materials Supply Unit Leaders. Request medical staff support through Treatment Areas Supervisor.

_____ Obtain an adequate number of patient transportation resources from the Transportation Unit Leader to ensure the movement of patients in and out of the area.

_____ Ensure a rapid, appropriate disposition of patients treated within Minor Treatment Area.

_____ Report frequently and routinely to the Treatment Areas Supervisor on situational status.

MINOR TREATMENT UNIT LEADER *(continued)*

Extended   ____ Observe and assist any staff who exhibit signs of stress or fatigue. Report any concerns to the Treatment Areas Supervisor. Provide for staff rest periods and relief.

    ____ Review and approve the area documenter's recordings of action/decisions in the Minor Treatment Area. Send copy to the Treatment Areas Supervisor.

    ____ Direct non-utilized personnel to Labor Pool.

    ____ Other concerns:

DISCHARGE UNIT LEADER

Position Assigned To: _____

You Report To: _____ (Treatment Areas Supervisor)

Operations Command Center: _____

Telephone: _____

Mission: Coordinate the controlled discharge, (possible observation and discharge) of patients received from all areas of the hospital. Facilitate the process of final patient disposition by assuring adequate staff and supplies in the Discharge Area.

Immediate ____ Receive appointment from the Treatment Areas Supervisor.

____ Read this entire Job Action Sheet and review the organizational chart on back.

____ Put on position identification vest.

____ Receive briefing from Treatment Areas Supervisor with other Treatment Areas unit leaders.

____ Assist Treatment Areas Supervisor in the establishment of Discharge Area. Coordinate with Human Services Director, Transportation Unit Leader and Safety & Security Officer.

____ Assess situation/area for supply and staffing need; request staff and supplies from the Labor Pool and Materials Supply Unit Leaders. Request medical staff support through Treatment Areas Supervisor. Prepare area for minor medical treatment and extended observation.

____ Request involvement of Human Services Director in appropriate patient disposition. Communicate regularly with Patient Tracking Officer.

Intermediate ____ Ensure that all patients discharged from area are tracked and documented in regards to disposition. Ensure a copy of the patient chart is sent with patient transfers. If copy service is not available, record chart number and destination for future

DISCHARGE UNIT LEADER *(continued)*

retrieval. (If other hospital areas are discharging patients, provide for accurate controls and documentation.) Provide for patient discharge services in Morgue Area.

_____ Report frequently and routinely to Treatment Areas Supervisor on situational status.

Extended _____ Observe and assist any staff or patient who exhibits sign of stress. Report concerns to the Treatment Areas Supervisor. Provide for staff rest periods and relief.

_____ Review and approve the area documenter's recordings of action/decisions in the Discharge Area. Send copy to the Treatment Areas Supervisor.

_____ Direct non-utilized personnel to Labor Pool.

_____ Other concerns:

MORGUE UNIT LEADER

Position Assigned To: _____

You Report To: _____ (Treatment Areas Supervisor)

Operations Command Center: _____

Telephone: _____

Mission: Collect, protect and identify deceased patients. Assist Discharge Area Unit Leader in appropriate patient discharge.

Immediate   ____ Receive appointment from the Treatment Areas Supervisor.

____ Read this entire Job Action Sheet and review the organizational chart on back.

____ Put on position identification vest.

____ Receive briefing from Treatment Areas Supervisor with other Treatment Area unit leaders.

____ Establish Morgue Area; coordinate with Treatment Areas Supervisor and Medical Care Director.

____ Request an on-call physician from the Treatment Areas Supervisor to confirm any resuscitatable casualties in Morgue Area.

____ Obtain assistance from the Transportation Unit Leader for transporting deceased patients.

____ Assure all transporting devices are removed from under deceased patients and returned to the Triage Area.

Extended   ____ Maintain master list of deceased patients with time of arrival for Patient Tracking Officer and Patient Information Officer.

____ Assure all personal belongings are kept with deceased patients and are secured.

____ Assure all deceased patients in Morgue Areas are covered, tagged and identified where possible.

MORGUE UNIT LEADER *(continued)*

Extended   ____ Keep Treatment Areas unit leaders apprised of number of deceased.

    ____ Contact the Safety & Security Officer for any morgue security needs.

    ____ Arrange for frequent rest and recovery periods, as well as relief for staff.

    ____ Schedule meetings with the Psychological Support Unit Leader to allow for staff debriefing.

    ____ Observe and assist any staff who exhibits signs of stress or fatigue. Report any concerns to the Treatment Areas Supervisor.

    ____ Review and approve the area documenter's recording of action/ decisions in the Morgue Area. Send copy to the Treatment Areas Supervisor.

    ____ Direct non-utilized personnel to Labor Pool.

    ____ Other concerns:

ANCILLARY SERVICES DIRECTOR

Position Assigned To: _____

You Report To: _____ (Operations Section Chief)

Operations Command Center: _____

Telephone: _____

Mission: Organize and manage ancillary medical services. To assist in providing for the optimal functioning of these services. Monitor the use and conservation of these resources.

Immediate   ____ Receive appointment from Operation Section Chief and subsection's Job Action Sheets.

    ____ Read this entire Job Action Sheet and review organizational chart on back.

    ____ Put on position identification vest.

    ____ Meet with Operations Section Chief and other Operations Section directors for a briefing and development of initial action plan. Designate time for next meeting.

    ____ Appoint unit leaders for:
- Laboratory Services
- Radiology Services
- Pharmacy Services
- Cardiopulmonary Services

    Distribute corresponding Job Action Sheets; request a documentation aide/assistant for each unit leader from the Labor Pool.

    ____ Brief all unit leaders. Request an immediate assessment of each service's capabilities, human resources and needs. Designate time for follow-up meeting.

    ____ Receive, coordinate and forward requests for personnel and materials to the appropriate individual.

    ____ Report routinely to the Operations Section Chief the actions, decisions and needs of the Ancillary Services Section.

ANCILLARY SERVICES DIRECTOR *(continued)*

Intermediate   ____ Track the ordering and receiving of needed supplies.

            ____ Supervise salvage operations within Ancillary Services when indicated.

            ____ Meet routinely with Ancillary Services unit leaders for status reports, and relay important information to Operation Section Chief.

Extended   ____ Observe and assist any staff who exhibits signs of stress or fatigue. Report any concerns to Psychological Support Unit Leader. Provide for staff rest periods and relief.

            ____ Review and approve the documenter's recordings of actions/ decisions in the Ancillary Services Section. Send copy to the Operations Section Chief.

            ____ Direct non-utilized personnel to Labor Pool.

            ____ Other concerns:

LABORATORY UNIT LEADER

Position Assigned To: _____

You Report To: _____ (Ancillary Services Director)

Operations Command Center: _____

Telephone: _____

Mission: Maintain Laboratory services, blood and blood products at appropriate levels. Prioritize and manage the activity of the Laboratory Staff.

Immediate   _____ Receive appointment from Ancillary Services Director.

    _____ Read this entire Job Action Sheet and review organizational chart on back.

    _____ Put on position identification vest.

    _____ Receive briefing from Ancillary Services Director with other subsection unit leaders; develop a subsection action plan.

    _____ Inventory available blood supply and designate those units of blood, if any, which may be released for use outside the facility. Report information to Ancillary Services Director and Communications Unit Leader.

    _____ Evaluate Laboratory Service's capacity to perform:
    • Hematology studies
    • Chemistry studies
    • Blood Bank services

    _____ Ascertain the approximate "turn around" time for study results. Report capabilities and operational readiness to Ancillary Services Director.

    _____ Assign a phlebotomies and runner with adequate blood collection supplies to the Immediate Treatment and Delayed Treatment Areas.

Intermediate   _____ Contact Materials Supply Unit Leader in anticipation of needed supplies.

LABORATORY UNIT LEADER *(continued)*

Intermediate   _____   Prepare for the possibility of initiating blood donor services.

        _____   Send any unassigned personnel to Labor Pool.

        _____   Inform patient care areas of currently available service.

        _____   Communicate with Patient Tracking Officer to ensure accurate routing of test results.

Extended   _____   Provide for routine meetings with Ancillary Services Director.

        _____   Review and approve the documenter's recordings of actions/decisions in the Laboratory Services area. Send copy of to the Ancillary Services Director.

        _____   Observe and assist any staff who exhibit signs of stress and fatigue. Report concerns to Ancillary Services Director. Provide for staff rest periods and relief.

        _____   Other concerns:

RADIOLOGY UNIT LEADER

Position Assigned To: _____

You Report To: _____ (Ancillary Services Director)

Operations Command Center: _____

Telephone: _____

Mission: Maintain radiology and other diagnostic imaging services at appropriate levels. Ensure the highest quality of service under current conditions.

Immediate ____ Receive appointment from Ancillary Services Director.

____ Read this entire Job Action Sheet and review the organizational chart on back.

____ Put on position identification vest.

____ Receive briefing from Ancillary Services Director with other subsection unit leaders; develop a subsection action plan.

____ Evaluate Radiology Service's capacity to perform x-ray and other appropriate procedures:
- Number of Operational X-ray suites
- Number of operational portable X-ray units
- Number of hours of film processing available
- Availability of CT scan or MRI
- Availability of fluoroscopy
- Report status to Ancillary Services Director.

____ Provide radiology technician and portable X-ray unit to Immediate and Delayed Treatment Areas, if available.

____ Contact Materials Supply Unit Leader in anticipation of needed supplies.

____ Send any unassigned personnel to Labor Pool.

____ Inform patient care areas of currently available radiology services.

RADIOLOGY UNIT LEADER *(continued)*

Intermediate   ____ Communicate with Patient Tracking Officer to ensure accu-
                      rate routing of test results.

Extended   ____ Provide for routine meetings with Ancillary Services Director.

                 ____ Review and approve the documenter's recordings of action/
                      decisions in the Radiology Services Area. Send copy to Ancil-
                      lary Services Director.

                 ____ Observe and assist any staff who exhibit signs of stress and
                      fatigue. Report concerns to Ancillary Services Director. Pro-
                      vide for staff rest periods and relief.

                 ____ Other concerns:

PHARMACY UNIT LEADER

Position Assigned To: _____

You Report To: _____ (Ancillary Services Director)

Operations Command Center: _____

Telephone: _____

Mission: Ensure the availability of emergency, incident specific, pharmaceutical and pharmacy services.

Immediate   ____   Receive appointment from Ancillary Services Director.

____ Read this entire Job Action Sheet and review the organizational chart on back.

____ Put on position identification vest.

____ Receive briefing from Ancillary Services Director with other subsection unit leaders; develop a subsection action plan.

____ Assign pharmacist to Immediate and Delayed Treatment Areas, when appropriate.

____ Inventory most commonly utilized pharmaceutical items and provide for the continual update of this inventory.

____ Identify any inventories which might be transferred upon request to another facility and communicate list to the Ancillary Services Director.

____ Communicate with the Materials Supply Unit Leader to assure a smooth method of requisitioning and delivery of pharmaceutical inventories within the hospital.

Extended   ____   Provide for routine meetings with Ancillary Services Director.

____ Review and approve the documenter's recordings of actions/ decisions in the Pharmacy Service Area. Send copy to Ancillary Services Director.

PHARMACY UNIT LEADER *(continued)*

<u>Extended</u>　　____　Observe and assist any staff who exhibit signs of stress and fatigue. Report any concerns to Ancillary Services Director. Provide for staff rest periods and relief.

　　　　　　　____　Other concerns:

CARDIOPULMONARY UNIT LEADER

Position Assigned To: _____

You Report To: _____ (Ancillary Services Director)

Operations Command Center: _____

Telephone: _____

Mission: Provide the highest level of Cardiopulmonary services at levels sufficient to meet the emergency incident needs.

Immediate   ____ Receive appointment from Ancillary Services Director.

              ____ Read this entire Job Action Sheet and review the organizational chart on back.

              ____ Put on position identification vest.

              ____ Receive briefing from Ancillary Services Director with other subsection unit leaders; develop a subsection action plan.

              ____ Evaluate Cardiopulmonary service's capacity to supply/perform:
              • Operational ventilatory equipment
              • Arterial blood gas analysis (ABG's)
              • Electrocardiograph study (EKG)
              • In-wall oxygen, nitrous oxide and other medical gases (confer with Damage Assessment and Control Officer)
              • Size and availability of gas cylinders

              Report status/information of Ancillary Services Director.

              ____ Assign respiratory therapist technician and EKG technician to the Immediate Treatment Area, when appropriate.

              ____ Consider the possibility of requesting additional Cardiopulmonary resources vs. developing a list of resources which may be loaned out of the facility.

Extended   ____ Monitor levels of all medical gases.

             ___ Provide for routine meetings with Ancillary Services Director.

CARDIOPULMONARY UNIT LEADER *(continued)*

Extended   _____ Review and approve the documenter's recordings of actions/
                    decisions in the Cardiopulmonary Services area. Send copy to
                    the Ancillary Services Director.

                    _____ Observe and assist any staff who exhibit signs of stress and
                    fatigue. Report any concerns to Ancillary Services Director.
                    Provide for staff rest periods and relief.

                    _____ Other concerns:

HUMAN SERVICES DIRECTOR

Position Assigned To: _____

You Report To: _____ (Operations Section Chief)

Operations Command Center: _____

Telephone: _____

Mission: Organize, direct and supervise those services associated with the social and psychological needs of the patients, staff and their respective families. Assist with discharge planning.

Immediate

_____ Receive appointment from Operations Section Chief. Obtain packet containing subsection Job Action Sheets.

_____ Read this entire Job Action Sheet and review organizational chart on back.

_____ Put on position identification vest.

_____ Obtain briefing from Operations Section Chief with other section directors and assist with development of the Operations Section's action plan. Designate time for follow up meeting.

_____ Appoint Staff Support Unit Leader, Psychological Support Unit Leader and Dependent Care Unit Leader. Distribute corresponding Job Action Sheets and identification vests.

_____ Brief unit leaders on current situation; outline action plan for subsection and designate time for next briefing.

_____ Establish Human Services Center near Discharge Area or near staff rest/rehabilitation area.

_____ Assist with establishment of Discharge Area. Lend support personnel to assist with patient discharge process.

_____ Assist in the implementation of patient early discharge protocol on the direction of Operations Section Chief. Secure the aid of Nursing Unit Leader.

HUMAN SERVICES DIRECTOR *(continued)*

Intermediate   _____ Assist Psychological Support Unit Leader in securing a debrief-
                ing area.

                _____ Meet regularly with unit leaders to receive updates and requests.

                _____ Communicate frequently with Operations Section Chief.

Extended   _____ Document action and decisions on a continual basis.

                _____ Observe and assist anyone who exhibits signs of stress and
                fatigue. Provide for staff rest and relief.

                _____ Other concerns:

STAFF SUPPORT UNIT LEADER

Position Assigned To: _____

You Report To: _____ (Human Services Director)

Operations Command Center: _____

Telephone: _____

Mission: Assure the provision of logistical and psychological support of the hospital staff.

Immediate ____ Receive assignment from Human Services Director.

____ Read this entire Job Action Sheet and review the organizational chart on back.

____ Put on position identification vest.

____ Obtain briefing from Human Services Director with other subsection unit leaders; assist in development of subsection action plan. Designate time for follow up meeting.

____ Anticipate staff needs as they might relate to the specific disaster.

Intermediate ____ Establish a staff rest and nutritional area in a low traffic area. Provide for a calm relaxing environment provide overall disaster information updates (bulletins) for rumor control.

____ Provide for nutritional support and sleeping arrangements; contact Nutritional Supply Unit Leader and Labor Pool Unit Leader for assistance.

____ Establish a staff Information Center with the help of Communications Unit Leader, Nursing Unit Leader and Labor Pool Unit Leader. Provide overall disaster info updates (bulletins) for rumor control.

____ Arrange for routine visits/evaluations by the Psychological Support Unit Leader. Assist in establishment of separate debriefing area.

STAFF SUPPORT UNIT LEADER *(continued)*

Extended   _____   Observe all staff closely for signs of stress and fatigue; intervene appropriately. Provide for personal staff rest periods and relief.

_____   Assist staff with logistical and personal concerns; act as facilitator when appropriate.

_____   Report routinely to the Human Services Director.

_____   Document all actions, decisions and interventions.

_____   Other concerns:

PSYCHOLOGICAL SUPPORT UNIT LEADER

Position Assigned To: _____

You Report To: _____ (Human Services Director)

Operations Command Center: _____

Telephone: _____

Mission: Assure the provision of psychological, spiritual and emotional support to the hospital staff, patients, dependents and guests. Initiate and organize the Critical Stress Debriefing process.

Immediate   ____ Receive appointment from Human Services Director.

                  ____ Read this entire Job Action Sheet and review the organizational chart on back.

                  ____ Put on position identification vest.

                  ____ Receive briefing from Human Services Director; assist in development of subsection action plan. Designate time for follow up meeting.

                  ____ Establish teams composed of staff, clergy and other mental health professionals to support the psycho-social needs of the staff, patients and guests.

                  ____ Designate a secluded debriefing area where individual and group intervention may take place. Coordinate with Staff Support Unit Leader.

                  ____ Appoint psychological support staff to visit patient care and non-patient care areas on a routine schedule.

                  ____ Meet regularly with all members of the Human Services Subsection.

                  ____ Assist the Staff Support Unit Leader in establishment of staff information/status board (situation, disaster update, hospital activities).

PSYCHOLOGICAL SUPPORT UNIT LEADER

<u>Extended</u> ＿＿＿ Advise psychological support staff to document all contacts.

＿＿＿ Observe psychological support staff for signs of stress and fatigue. Arrange for frequent, mandatory rest periods and debriefing sessions.

＿＿＿ Schedule and post the dates and times for critical stress debriefing sessions during and after the immediate disaster period.

＿＿＿ Document all actions, decisions and interventions.

＿＿＿ Other concerns:

DEPENDENT CARE UNIT LEADER

Position Assigned To: _____

You Report To: _____ (Human Services Director)

Operations Command Center: _____

Telephone: _____

Mission: Initiate and direct the sheltering and feeding of staff and volunteer dependents.

Immediate
    ____ Receive appointment from Human Services Director.

    ____ Read this entire Job Action Sheet and review the organizational chart on back.

    ____ Put on position identification vest.

    ____ Obtain briefing from Human Services Director; participate in development of subsection action plan. Designate time for follow up meeting.

    ____ Establish a controlled, comfortable area where patients and visitors may wait for disposition home.

    ____ Establish a Dependent Care Area removed from any patient care areas.

    ____ Obtain volunteers from the Labor Pool to assist with child and/or adult care. Make tentative plans for extended care.

    ____ Monitor the area continuously for safety and dependant needs with a minimum of two hospital employees.

    ____ Implement a positive I.D. system for all children cared for under age of 10 years of age. Provide matching I.D. for retrieving guardian to show upon release of child.

    ____ Document care and all personnel in the area.

    ____ Contact the Safety & Security Officer for assistance.

DEPENDENT CARE UNIT LEADER *(continued)*

Immediate   _____ Contact Materials Supply Unit Leader and Nutritional Supply
              Unit Leader for supplies and food; advise Situation–Status Unit
              Leader and Labor Pool Unit Leader of any extended plans.

Extended   _____ Assure that those dependents taking medications have sufficient
              supply for estimated length of stay.

              _____ Arrange for the Psychological Support Unit Leader to make rou-
              tine contact with dependents in the shelter, as well as respond-
              ing when necessary.

              _____ Observe staff and dependents for signs of stress and fatigue.
              Provide for staff rest periods and relief.

              _____ Report routinely to Human Services Director. Document all
              actions/decisions.

              _____ Other concerns:

Appendix D

EXAMPLES OF TACTICAL WORKSHEETS FOR INCIDENT COMMAND

THE CONCEPT BEHIND THE USE TACTICAL WORKSHEETS FOR INCIDENT COMMAND

The basic concept for the use of a tactical command sheet is that some jobs or task within a specialty realm such as Aircraft Rescue and Firefighting (ARFF or Hazardous Materials (HAZMAT) response are of a nature that lends themselves to the development of checklists to prompt an Incident Commander to make certain that some "routine" tasks or assignments are taken care of in a timely fashion at such incidents.

The generic worksheets in this Appendix are meant to serve as guidelines only for the development of agency specific worksheets for this purpose. These worksheets are not meant to be all inclusive of such worksheets but again serve as guidelines for agencies wishing to develop such response tools for their own needs. Also the topics represented are again a sampling of possible titles for such worksheets, you will note some redundancy between the examples included here. This is intentional as the concept of generic templates such as this is common place in emergency services.

The topics of the worksheets in this appendix are:

WORKSHEET TITLE
Aircraft Incident
Confined Space Incident
HAZMAT Incident
Lowrise Incident
Multi-Victim (MVI) or Mass Casualty (MCI) Incident
Railroad Incident
Tanker Truck Incident
Terrorism Incident

AIRCRAFT INCIDENT		OFFENSE
INCIDENT COMMAND SHEET		DEFENSE

DISTRICT	PLOT PLAN #	ALARM #	DATE / /

ADDRESS	TIME OF ALARM	TIME CONTAIN

AIRCRAFT I.D./ TYPE	AIRLINE NAME	TIME CONTROL

UNITS

RESPONDING	ASSIGNED

ASSIGNMENTS

I.C.
OPERATIONS
MEDICAL GROUP
RIC
SAFETY OFFICER
LIAISON
N.T.S.B.
HAZ-MAT
ORDINANCE DISPOSAL
CRASH RESCUE
MILITARY

REMINDERS

	2nd ALARM		RESPONSIBLE PARTY
	3rd ALARM		RED CROSS
	RESCUE/SEARCH		P.I.O. PRESS
	EXPOSURES		RELIEF CREWS
	AIRPORT		BARRICADES
	ELECTRICITY		PLYWOOD
	GAS		TOILET TRAILER
	DOMESTIC WATER		INVESTIGATORS
	SALVAGE		C.S.I.
	FIREBRAND PATROL		FOOD/DRINK
	N.T.S.B./MILITARY		COMMAND OFFICERS
	ORDINANCE		FOAM
	LIGHTING		UTILITIES REP
	AIR-LIGHT UNIT		CHAPLAIN
	S.A.P.D. CROWD		CORONER
	PARAMEDICS		
	AMBULANCE		
	HAZ-MAT		
	SCENE PRESERVED		

AIRCRAFT STABILIZATION
SHUT OFF FUEL PUMP, BATTERIES, MAGNETO
AFFF FOAM BLANKET / REPEAT
NON-SPARKING FORCIBLE ENTRY TOOLS

CONFINED SPACE INCIDENT INCIDENT COMMAND SHEET				
DATE	TIME	INCIDENT #		PERMIT AVAILABLE YES NO
ADDRESS		TIME OF ALARM		TIME FINISHED
ISOLATE	ESTABLISH HOT ZONE		ESTABLISH WARM ZONE	

FAILURE SOURCE

	SECURE	LOCKOUT	TAGOUT	BLOCKOUT	SAFE

FAILURE SOURCE

	SECURE	LOCKOUT	TAGOUT	BLOCKOUT	SAFE

FAILURE SOURCE

	SECURE	LOCKOUT	TAGOUT	BLOCKOUT	SAFE

IDENTIFY	PRODUCT	PROCESS		MSDS
	PHYSICAL HAZARD	TEMP	DOT GUIDE #	

NOTIFY	SITE MANAGER	US&R	HAZ-MAT	OSHA

	SITUATION
	VICTIM PROFILE
PLAN OF ACTION	RESCUE OPERATIONS
	RESOURCES NEEDED

ENTRY TERMINATION TIME		JUSTIFICATION
COMMENTS		

ASSIGNMENTS		REMINDERS	
I.C.		PHASE 1...SIZE-UP	VENTILATE
		WITNESSES	LOCKOUT/TAGOUT
SAFETY		PRIME HAZARDS	PHASE 3...RESCUE
		LOCATE VICTIMS	ACTION PLAN
RESCUE TEAM MANAGER		ENTRY PERMIT	ENTRY TEAM
		TYPE OF SPACE	BACK-UP TEAM
ATTENDANT		PRODUCT	PROT. EQUIPMENT
		SEC. HAZARDS	LIGHTING
ENTRANT #1		DIAGRAM	COMMUNICATIONS
		STABILITY	RESPIRATORY
ENTRANT #2		PROPER EQUIPMENT	AIR MONITOR
		RESCUE	FALL PROTECT
BACK-UP TEAM MEMBER #1		PHASE 2...PRE-ENTRY	RESCUE HARNESS
		PERIMETER	PT. PACKAGING
BACK-UP TEAM MEMBER #2		EVACUATE	PHASE 4...TERMINATE
		TRAFFIC	ACCOUNTABILITY
AIR SUPPLY		LOBBY CONTROL	REMOVE EQUIP
		ACCOUNTABILITY	SECURE SCENE
MEDIC #1		ATMOSPHERE (IDLH)	DEBRIEFING
MEDIC #2			

				MONITOR						
TECHNICAL EXPERT	TIME									
	O2	19.59%								
	LEL	10% LEL								
RIC	CO	35 PPM								
	H2S	10 PPM								

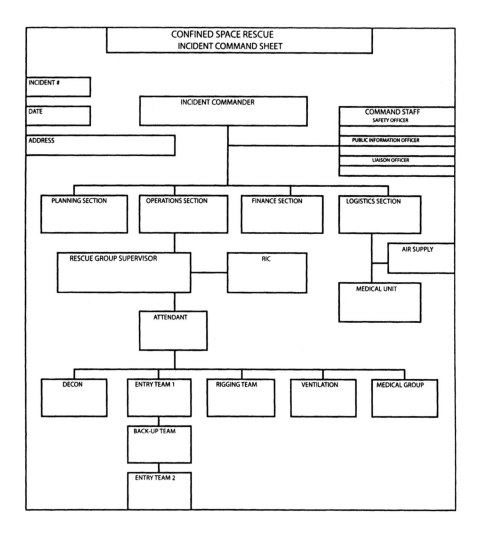

POSITION PROGRESS NEEDS	PLAN OF ACTION		RESCUE / RECOVERY	Other

SITUATION: _____

VICTIM PROFILE: _____

RESCUE OPERATIONS: _____

RESOURCES ORDERED: _____

MONITOR

TIME												
OXYGEN	O₂	19.50%										
FLAMMABLE (CORRECTED)	(LEL)	10%										
CARBON MONOXIDE	(CO)	35 PPM										
HYDROGEN SULFIDE	(H₂S)	10 PPM										
OTHER												

Entry Termination Time: _____ Justification: _____

Comments: _____

GRADE "D" AIR	AMOUNT ON SITE	EXPECTED DURATION	REQUIRED RESUPPLY	EQUIPMENT ORDERED	OTHER
SCBA					
CYLINDERS ( )					
LIGHT / AIR					
AIR PUMP					
OTHER					

MANIFOLD / AIR CART

NUMBER					
TIME					
PRESSURE					
PRESSURE					
PRESSURE					
PRESSURE					

ENTRANT / S AIR SUPPLY

NAME	SCBA	SAR	HOSE I.D.	SOURCE	ESCAPE	TIME-IN	WORK	ETA OUT	EXIT

HAZ-MAT INCIDENT			OFFENSIVE
INCIDENT COMMAND SHEET			DEFENSIVE

DISTRICT	PLOT PLAN #	ALARM #	DATE / /

ADDRESS		TIME OF ALARM	TIME CONTAINED

OCCUPANCY	OWNER/OCCUPANT	TIME CONTROL

UNITS

RESPONDING	ASSIGNED

MATERIAL INFORMATION

ID #	GUIDE #	MATERIAL NAME

FLASH POINT	SPEC. GRAVITY	EVACUATION DISTANCE
		FT. X FT.

WEATHER INFORMATION

TIME	TIME	TIME	TIME
WIND DIR	WIND DIR	WIND DIR	WIND DIR
SPEED	SPEED	SPEED	SPEED

NFPA 704

ASSIGNMENTS

HAZ-MAT I.C.	
OPERATIONS	
DIVISION	
VENTILATION GROUP	
SALVAGE GROUP	
INCIDENT BASE	
SAFETY OFFICER	
MEDICAL TEAM	
LIAISON	
RIC	

REMINDERS

UPHILL/UPWIND	OC WATER QUALITY
STATIC OR MOVING	A.Q.M.D.
RESCUE	SCHOOL DISTRICT
FULL PROTECTIVE	PESTICIDE/OC AGR
ISOLATE CONT. AREA	FRWY CHP/ CALTRANS
DIKING	O.C. HEALTH
IDENTIFY MATERIAL	EVACUATION SITE
HAZ-MAT TEAMS	RED CROSS
PARAMEDICS	CITY SEWER SYSTEM
RESPONSIBLE PARTY	O.S.H.A.
MSDS	CHAPLAIN
SHIPPING PAPERS	CORONER
CROWD CONTROL	P.I.O. PRESS
AIR-LIGHT UNIT	FOOD/DRINK
FOAM	TOILET TRAILER
AMBULANCES	
RELIEF CREWS	
COMMAND OFFICERS	
UTILITIES	

MISC

LOWRISE INCIDENTS INCIDENT COMMAND SHEET	OFFENSIVE
	DEFENSIVE

DISTRICT	PLOT PLAN #	ALARM #	DATE / /

ADDRESS	TIME OF ALARM	TIME CONTAINED

OCCUPANCY	OWNER/OCCUPANT	TIME CONTROL

UNITS		SKETCH
RESPONDING	ASSIGNED	

ASSIGNMENTS

I.C.
OPERATIONS
DIVISION
VENTILATION GROUP
SALVAGE GROUP
INCIDENT BASE
RIC
SAFETY OFFICER
MEDICAL TEAM
LIAISON

REMINDERS

2nd ALARM		RESPONSIBLE PARTY
3rd ALARM		RED CROSS
RESCUE/SEARCH		P.I.O. PRESS
EXPOSURES		RELIEF CREWS
VENTILATION		BARRICADES
ELECTRICITY		PLYWOOD
GAS		TOILET TRAILER
DOMESTIC WATER		INVESTIGATORS
SALVAGE		C.S.I.
FIREBRAND PATROL		FOOD/DRINK
SPRINKLER CHARGE		COMMAND OFFICER
STANDPIPE CHARGE		FOAM
LIGHTING		UTILITIES REP.
AIR-LIGHT UNIT		CHAPLAIN
SAPD CROWD		CORONER
PARAMEDICS		
AMBULANCES		
HAZ-MAT		
SCENE PRESERVED		

REMINDERS

LIGHTING	
CROWD CONTROL	
P.I.O.	
RED CROSS	
INVESTIGATORS	
COMMAND OFFICERS	
REFERENCE GUIDE	
UTILITIES	
EVACUATION	
HAZ-MAT	
CHAPLAIN	
CORONER	
PRESS	

MISC

ASSIGNMENTS

I.C.	
OPERATIONS	
MEDICAL COMM MANAGER	
MEDICAL GROUP	
TRIAGE TEAM	
TREATMENT TEAM	
MORGUE TEAM	
INCIDENT BASE	
LIAISON	

| | | RAILROAD INCIDENTS | | | OFFENSIVE |
| | | INCIDENT COMMAND SHEET | | | DEFENSIVE |

| DISTRICT | PLOT PLAN # | ALARM # | DATE   /   / |

| ADDRESS/RAIL MARKER | TIME OF ALARM | TIME CONTAINED |

| RAILROAD I.D. | SHUT DOWN RAIL TRAFFIC | TIME CONTROL |

UNITS		SKETCH
RESPONDING	ASSIGNED	

ASSIGNMENTS

I.C.

OPERATIONS

DIVISION

MEDICAL GROUP

HAZ-MAT GROUP

TECHNICAL RESCUE GROUP

SAFETY OFFICER

LIAISON

REMINDERS

2nd ALARM	SCENE PRESERVED
3rd ALARM	RESPONSIBLE PARTY
RESCUE/SEARCH	RED CROSS
EXPOSURES	PRESS
FUEL SHUT-OFF	RELIEF CREWS
RAIL NOTIFY	BARRICADES
BATTERY TURN-OFF	CONSIST PAPERS
RIC	TOILET TRAILER
UTILITIES	INVESTIGATORS
OVERHEAD WIRES	FOOD/DRINK
MVI - MCI	COMMAND OFFICER
FOAM	UTILITIES REP.
LIGHTING	CHAPLAIN
AIR-LIGHT UNIT	P.I.O.
CROWD	
PARAMEDICS	
AMBULANCES	
HAZ-MAT	
CORONER	

MISC

Glossary

Advanced Life Support (ALS) - A service capable of delivering advanced skills performed by Emergency Medical Services (EMS) practitioners (e.g., intravenous [IV] fluids and drug administration).

Advanced Life Support (ALS) Ambulance - An ambulance capable of delivering ALS service performed by Emergency Medical Services (EMS) practitioners (e.g., intravenous [IV] fluids and drug administration).

After Action Report (AAR) - A report covering response actions, application of IMS, modifications to plans and procedures, training needs, and recovery activities.

Agency Executive or Administrator - Chief Executive Officer (or designee) of the agency or jurisdiction that has responsibility for the incident.

Agency Representative - An individual assigned to an incident from an assisting or cooperating agency who has been delegated authority to make decisions on matters affecting that agency's participation at the incident. Agency Representatives report to the Incident Liaison Officer.

Allocated Resources - Resources dispatched to an incident that have not yet checked-in with the Incident Communications Center.

Ambulance Strike Team - An Ambulance Strike Team is a group of five ambulances of the same type with common communications and a leader. It provides an operational grouping of ambulances complete with supervisory elements for organization command and control. The strike teams may be all ALS or all BLS.

Ambulance Task Force - An Ambulance Task Force is a group of any combination of ambulances, within the span of control, with common communications and a leader.

Animal Health Incident Management Team - Team provides overall management of animal-related volunteers and donations.

Animal Rescue Team - A team proficient in animal handling and capture and management (minimum teams of two). Environments include water (swift water and flood), wildfire, and hazardous materials (HazMat) conditions. Operations include communications and/or evacuations to effect animal rescue.

Animal Health Technician - Technician performs variety of animal healthcare duties to assist veterinarians in settings such as veterinarians' clinics, zoos, research laboratories, kennels, and commercial facilities. Prepares treatment room for examination of animals and holds or restrains animals during examination, treatment, or inoculation.

Animal Sheltering Team - A team proficient in animal handling, animal care, and animal shelter management and manages the setup, management, and staffing of temporary animal shelters.

Area Command - Area Command is an expansion of the incident command function primarily designed to manage a very large incident that has multiple incident management teams assigned. However, an Area Command can be established at any time that incidents are close enough that oversight direction is required among incident management teams to ensure conflicts do not arise.

Assigned Resources - Resources checked-in and assigned work tasks on an incident.

Assistant - Title for subordinates of the Command Staff positions. The title indicates a level of technical capability, qualifications, and responsibility subordinate to the primary positions. Assistants may also be used to supervise unit activities at camps.

Assisting Agency - An agency directly contributing suppression, rescue, support, or service resources to another agency.

Available Resources - Resources assigned to an incident and available for an assignment.

Base - That location at which the primary logistics functions are coordinated and administered. (Incident name or other designator will be added to the term "Base.") The Incident Command Post may be collocated with the Base. There is only one Base per incident.

Basic Life Support (BLS) - A service capable of delivering basic emergency interventions performed by Emergency Medical Services (EMS) practitioners trained and credentialed to do so (e.g., splinting, bandaging, oxygen administration).

Basic Life Support (BLS) Ambulance - An ambulance service capable of delivering basic emergency interventions performed by Emergency Medical Services (EMS) practitioners trained and credentialed to do so (e.g., splinting, bandaging, oxygen administration).

Biological Agent - Living organisms or the materials derived from them (such as bacteria, viruses, fungi, and toxins) that cause disease in or harm to humans, animals, or plants, or cause deterioration of material.

Bomb Squad/Explosives Ordinance Disposal (EOD) Teams - A police unit specializing in the investigation and disarming of suspected explosive devices and otheer explosives.

Branch - That organizational level having functional or geographic responsibility for major parts of incident operations. The Branch level is organizationally between Section and Division/Group in the Operations Section, and between Section and Units in the Logistics Section. Branches are identified by the use of Roman Numerals or by functional name (e.g., medical, security, etc.).

Breathing Apparatus Support Vehicle (SCBA Support; Breathing Air, Firefighting) - A mobile unit designed and constructed for the purpose of providing specified level of breathing air support capacity and personnel capable of refilling Self-Contained Breathing Apparatus (SCBA) at remote incident locations (Compressor systems or cascade type).

Camp - A geographical site, within the general incident area, separate from the Base, equipped and staffed to provide food, water, and sanitary services to incident personnel.

Chemical Warfare Agent - A chemical substance (such as a nerve agent, blister agent, blood agent, choking agent, or irritating agent) used to kill, seriously injure, or incapacitate people through its physiological effects.

Chain of Command - A series of management positions in order of authority.

Clear Text - The use of plain English in radio communications transmissions. No "Ten Codes", or agency-specific codes are used when using Clear Text.

Command - The act of directing, ordering and/or controlling resources by virtue of explicit legal, agency, or delegated authority.

Command Staff - The Command Staff consists of the Information Officer, Safety Officer, and Liaison Officer, who report directly to the Incident Commander.

Chain of Command - A series of management positions in order of authority.

Communications Unit - An organizational unit in the Logistics Section responsible for providing communication services at an incident. A Communications Unit may also be a facility (e.g., a trailer or mobile van) used to provide the major part of an Incident Communications Center.

Compensation Unit/Claims Unit - Functional unit within the Finance/Administration Section responsible for financial concerns resulting from property damage, injuries or fatalities at the incident or within an incident.

Cooperating Agency - An agency supplying assistance other than direct suppression, rescue, support, or service functions to the incident control effort (e.g. American Red Cross, law enforcement agency, telephone company, etc.).

Coordination - The process of systematically analyzing a situation, developing relevant information, and informing appropriate command authority (for its decision) of viable alternatives for selection of the most effective combination of available resources to meet specific objectives. The coordination process (which can be either intra- or interagency) does not in and of itself involve command dispatch actions. However, personnel responsible for coordination may perform command or dispatch functions within limits as established by specific agency delegations, procedures, legal authority, etc.

Coordination Center - Term used to describe any facility that is used for the coordination of agency or jurisdictional resources in support of one or more incidents.

Cost Unit - Functional unit within the Finance/Administration Section responsible for tracking costs, analyzing cost data, making cost estimates, and recommending cost-saving measures.

Crew Transport - Any vehicle capable of transporting a specified number of crew personnel in a specified manner.

Critical Care Transport (CCT) Ambulance - An ambulance transport of a patient from a scene or a clinical setting whose condition warrants care commensurate with the scope of practice of a physician or registered nurse (e.g., capable of providing advanced hemodynamic support and monitoring, use of ventilators, infusion pumps, advanced skills, therapies, and techniques).

Debris Management Team - The team that facilitates and coordinates the removal, collection, and disposal of debris following a disaster, to mitigate against any potential threat to the health, safety, and welfare of the impacted citizens, and expedite recovery efforts in the impacted area, and address any threat of significant damage to improved public or private property. Team mobilization will vary depending on the team selection, need, and or emergency. Debris removal process will vary depending on the team selection and need.

Decontamination - The physical or chemical process of reducing and preventing the spread of contaminants from persons and equipment used at a Hazardous Materials (HazMat) incident.

Demobilization Unit - Functional Unit within the Planning Section responsible for assuring orderly, safe, and efficient demobilization of an incident and the assigned resources.

Deployment - Departure of team or personnel from home unit or Base.

Deputy - A fully qualified individual who, in the absence of a superior, could be delegated the authority to manage a functional operation or perform a specific task. In some cases, a Deputy could act as relief for a superior and therefore must be fully qualified in the position. Deputies can be assigned to the Incident Commander, General Staff, and Branch Directors.

Deputy Incident Commander (Section Chief or Branch Director) - A fully qualified individual who, in the absence of a superior, could be delegated the authority to manage a functional operation or perform a specific task. In some cases, a Deputy could act as relief for a superior and therefore must be fully qualified in the position.

Disaster - A sudden calamitous emergency event bringing great damage loss or destruction.

Disaster Assessment Team - Governed by type and magnitude of the disaster, the structure of the team consists of people most knowledgeable about the collection or material inventory of the disaster site, and assessing the magnitude and extent of impact on both the population and infrastructure of society. Trained specifically for disaster assessment techniques, team members are multidisciplinary and can include health

personnel, engineering specialists, logisticians, environmental experts, and communications specialists. Responsibilities include recording observations and decisions
made by the team, photographing and recording disaster site damage, and investigating where damage exists. Teams also analyze the significance of affected infrastructures, estimate the extent of damages, and establish initial priorities for recovery.
Disaster assessment teams can perform an initial assessment that comprises situational and needs assessments in the early, critical stages of a disaster to determine
the type of relief needed for an emergency response, or they may carry out a much
more expedited process termed a rapid assessment.

**Disaster Medical Assistance Team (DMAT)—Basic, National Disaster Medical
System (NDMS)** - A DMAT is a volunteer group of medical and nonmedical individuals, usually from the same state or region of a state, which has formed a response
team under the guidance of the NDMS (or under similar state or local auspices).
Usually includes a mix of physicians, nurses, nurse practitioners, physician's assistants, pharmacists, emergency medical technicians, other allied health professionals, and support staff. A Standard DMAT has 35 deployable personnel.

**Disaster Medical Assistance Team (DMAT)—Burn Specialty, National Disaster
Medical System (NDMS)** - A Burn Specialty DMAT is a volunteer group of medical and nonmedical individuals, usually from the same state or region of a state,
that has formed a response team under the guidance of the NDMS (or state or local
auspices), and whose personnel have specific training/skills in the acute management of burn trauma patients. Members of the burn team are trained surgeons, nurses,
and support personnel that include physical and occupational therapists, social
workers, child life specialists, psychologists, nutrition and pharmacy consultants,
respiratory therapists, chaplains, and volunteers. Team composition is usually determined ad hoc, based on the projected mission at hand.

**Disaster Medical Assistance Team (DMAT)—Crush Injury Specialty, National
Disaster Medical System (NDMS)** - A Crush Injury Specialty DMAT is a volunteer group of medical and nonmedical individuals, usually from the same state or
region of a state, that has formed a response team under the guidance of the NDMS
(or state or local auspices), and whose personnel have specific training/skills in the
management of crush injury patients. Crush teams deal with crush and penetrating
injuries. Usually includes a mix of physicians, nurses, nurse practitioners, physician's
assistants, pharmacists, emergency medical technicians, other allied health professionals, and support staff. Team composition is usually determined ad hoc, based
on the mission at hand.

**Disaster Medical Assistance Team (DMAT)—Mental Health Specialty, National
Disaster Medical System (NDMS)** - A Mental Health Specialty DMAT is a volunteer group of medical and nonmedical individuals, usually from the same state or
region of a state, that has formed a response team under the guidance of the NDMS
(or state or local auspices), and whose personnel have specific training/skills in the
management of psychiatric patients. A multidisciplinary staff of specially trained and
licensed mental health professionals provides emergency mental health assessment

and crisis intervention services. Usually includes a mix of physicians, nurses, nurse practitioners, physician's assistants, pharmacists, emergency medical technicians, other allied health professionals, and support staff. Team composition is usually determined ad hoc, based on the mission at hand.

Disaster Medical Assistance Team (DMAT)—Pediatric Specialty, National Disaster Medical System (NDMS) - A Pediatric Specialty DMAT is a volunteer group of medical and nonmedical individuals, usually from the same state or region of a state, that has formed a response team under the guidance of the NDMS (or state or local auspices), and whose personnel have specific training/skills in the management of pediatric patients. Usually includes a mix of physicians, nurses, nurse practitioners, physician's assistants, pharmacists, emergency medical technicians, other allied health professionals, and support staff. Team composition is usually determined ad hoc, based on the projected mission at hand.

Disaster Mortuary Operational Response Team (DMORT), National Disaster Medical System (NDMS) - A DMORT is a volunteer group of medical and forensic personnel, usually from the same geographic region, that has formed a response team under the guidance of the NDMS (or state or local auspices), and whose personnel have specific training/skills in victim identification, mortuary services, and forensic pathology and anthropology methods. Usually includes a mix of medical examiners, coroners, pathologists, forensic anthropologists, medical records technicians, fingerprint technicians, forensic odentologists, dental assistants, radiologists, funeral directors, mental health professionals, and support personnel. DMORTs are mission-tailored on an ad-hoc basis, and usually deploy only with personnel and equipment specifically required for the current projected mission.

Disaster Mortuary Operational Response Team (DMORT)—Weapons of Mass Destruction (WMD), National Disaster Medical System (NDMS) - Same as DMORT except adds additional capability to deal with deceased persons residually contaminated by chemical, biological, or radiological agents.

Dispatch - The implementation of a Command decision to move a resource or resources from one place to another.

Dispatch Center - A facility from which resources are directly assigned to an incident.

Division - Divisions are used to divide an incident into geographical areas of operation. Divisions are identified by alphabetic characters for horizontal applications and, often, by numbers when used in buildings.

Division or Group Supervisor - The position title for individuals responsible for command of a Division or Group at an Incident.

Documentation Unit - Functional unit within the Planning Section responsible for collecting, recording, and safeguarding all documents relevant to an incident.

Donations Coordinator - The Donations Coordinator is a subsection of a Donations Management Team and has working knowledge of the Individual Assistance and Public Assistance functions under FEMA/State agreement. A Donations Coordinator also has working knowledge of establishing long-term recovery committees on local levels following events. A Donations Coordinator possesses an operational knowledge of all aspects of donations coordination, including management of solicited and unsolicited funds, goods and services from concerned citizens and private organizations following a catastrophic disaster situation.

Donations Management Team - A donations management team consists of one or two persons trained and experienced in all aspects of donations management. The team will be deployed to a disaster-affected jurisdiction after impact to assist in the organization and operations of state or local donations management in support of the affected jurisdiction.

Emergency - A condition of disaster or extreme peril to the safety of persons and property caused by such conditions as air pollution, fire, flood, hazardous material incident, storm, epidemic, riot, drought, sudden and severe energy shortage, plant or animal infestations or disease.

Emergency Management Coordinator, Emergency Management Director, or Emergency Services Director - The individual within each political subdivision or jurisdiction that has overall responsibility for jurisdiction emergency management and or day-to-day responsibility for the development and maintenance of all emergency management coordination efforts. For cities and counties, this responsibility is commonly assigned by local ordinance.

Emergency Medical Task Force - An Emergency Medical Task Force is any combination (within span of control) of resources (Ambulances, Rescues, Engines, Squads, etc.) assembled for a medical mission, with common communications, and a leader (supervisor). Self-sufficient for 12-hour Operational Periods, although it may be deployed longer, depending on need.

Emergency Medical Technician (EMT) - A health-care specialist with particular skills and knowledge in pre-hospital emergency medicine. A practitioner credentialed by a State to function as an EMT by a State Emergency Medical Services (EMS) system.

Emergency Operations Center (EOC) - A location from which centralized emergency management can be performed. EOC facilities are established by an agency or jurisdiction to coordinate the overall agency or jurisdictional response and support to an emergency.

Emergency Operations Plan (EOP) - The plan that each jurisdiction has and maintains for responding to appropriate hazards.

Emergency Response Agency - Any organization responding to an emergency, or providing mutual aid support to such an organization, whether in the field, at the scene of an incident, or to an operations center.

EMS Strike Team - A team comprised of five resources or less of the same type with a supervisor and common communications capability. Whether it is five resources or less, a specific number must be identified for the team. For instance, a basic life support (BLS) strike team would be five BLS units and a supervisor or, for example, an advanced life support (ALS) strike team would be comprised of five ALS units and a supervisor.

EMS Task Force - A team comprised of five resources or less of different types with a supervisor and common communications capability. Whether it is five resources or less, a specific number must be identified for the team. For instance, an EMS Task Force might be comprised of two ALS teams and three BLS teams and a supervisor.

Engine, Fire (Engine Company) - Any ground vehicle providing specified levels of pumping, water, hose capacity, and staffed with a minimum number of personnel.

Emergency Operations Center (EOC) Management Support Team - Team provides support to an Incident Commander (IC). An IC is an optional member of the team, because it is assumed that an Incident Command/lead has already been established under which these support functions will operate. Typically comprised of an Information Officer, Liaison Officer, Safety Officer, Logistics Officer, and Administrative Aide.

Emergency Traffic - A term used to clear designated channels used at an incident to make way for important radio traffic for an emergency situation or an immediate change in tactical operations.

EOC Finance/Administration Section Coordinator - An EOC Finance/Administration Section Coordinator is an individual at the EOC responsible for tracking incident costs and reimbursement accounting, and coordinating/administering support for EOC personnel during disaster operations. This function is part of the standardized ICS structure per the National Incident Management System. If situation warrants, chief/coordinator oversees subunits of this function, including Compensation/Claims, Procurement, Cost, and Time.

EOC Operations Section Chief - An EOC Operations Section Chief is an individual at the EOC responsible for managing tactical operations at the incident site directed toward reducing the immediate hazard, saving lives and property, establishing situation control, and restoring normal conditions; responsible for the delivery and coordination of disaster assistance programs and services, including emergency assistance, human services assistance, and infrastructure assistance; and oversight of subunits of Operations Section, including Branches (up to five), Division/Groups (up to 25) and Resources as warranted.

EOC Planning Section Chief - The EOC Planning Section Chief is an individual at the EOC who oversees all incident-related data gathering and analysis regarding incident operations and assigned resources, develops alternatives for tactical operations, conducts planning meetings, and prepares the IAP for each operational period.

Event - A planned, non-emergency activity. IMS can be used as the management system for a wide range of events, e.g., parades, concerts or sporting events.

External Resources - Resources that fall outside a team's particular agency, including other agency resources or commercially contracted resources.

Facilities Unit - Functional Unit within the Support Branch of the Logistics Section that provides fixed facilities for the incident. These facilities may include the Incident Base, feeding areas, sleeping areas, sanitary facilities, etc.

Field Operations Guide (FOG) - A pocket-size manual of instructions on the application of the Incident Command System.

Finance/Administration Section - One of the five primary functions found in all IMS systems which is responsible for all costs and financial considerations. At the incident the Section can include the Time Unit, Procurement Unit, Compensation/Claims Unit and Cost Unit.

Food Dispenser Unit - Any vehicle capable of dispensing food to incident personnel.

Food Unit - Functional Unit within the Service Branch of the Logistics Section responsible for providing meals for incident personnel.

General Staff - The group of incident management personnel comprised of: The Incident Commander, The Operations Section Chief, The Planning/Intelligence Section Chief, The Logistics Section Chief, The Finance/Administration Section Chief.

Ground Ambulance (Medical Transport) - A ground transport vehicle configured, equipped, and staffed to respond to, care for, and transport patients.

Ground Support Unit - Functional unit within the Support Branch of the Logistics Section in all IMS systems that is responsible for the fueling, maintaining and repairing of vehicles, and the transportation of personnel and supplies.

Group - Groups are established to divide the incident into functional areas of operation. Groups are composed of resources assembled to perform a special function not necessarily within a single geographic division. (*See* Division.) Groups are located between Branches (when activated) and Resources in the Operations Section.

Hazardous Materials (HazMat) - Any material that is explosive, flammable, poisonous, corrosive, reactive, or radioactive, or any combination thereof, and requires special care in handling because of the hazards it poses to public health, safety, and/or the environment. Any hazardous substance under the Clean Water Act, or any element, compound, mixture, solution, or substance designated under the Comprehensive Environmental Response, Compensation, and Liability Act (CERCLA); any hazardous waste under the Resource Conservation and Recovery Act (RCRA); any toxic pollutant listed under pretreatment provisions of the Clean Water Act; any hazardous pollutant under Section 112 of the Clean Air Act; or any imminent hazardous chemical substance for which the administrator has taken action under the Toxic Substances Control Act (TSCA) Section 7. (Section 101[14] CERCLA)

Hazardous Material Response Team (HMRT) - An organized group of individuals that is trained and equipped to perform work to control actual or potential leaks, spills, discharges, or releases of HazMat, requiring possible close approach to the material. The team/equipment may include external or contracted resources.

Hazardous Materials Incident - Uncontrolled, unlicensed release of HazMat during storage or use from a fixed facility or during transport outside a fixed facility that may impact public health, safety, and/or the environment.

HazMat Task Force (HMTF) - A group of resources with common communications and a leader. A HazMat Task Force may be preestablished and sent to an incident, or formed at the incident.

HazMat Trained and Equipped - To the level of training and equipment defined by the Occupational Safety and Health Administration (OSHA) and the National Fire Protection Association (NFPA).

Helibase - A location within the general incident area for parking, fueling, maintenance, and loading of helicopters.

Helicopters, Firefighting (Helicopter or Copter) - An aircraft that depends principally on the lift generated by one or more rotors for its support in flight. Capable of the delivery of firefighters, water, or chemical retardants (either a fixed tank or bucket system), and internal or external cargo.

Helispot - A location where a helicopter can take off and land. Some helispots may be used for temporary retardant loading.

Helitanker - A helicopter equipped with a fixed tank, Air Tanker Board Certified, capable of delivering a minimum of 1,100 gallons of water, retardant, or foam.

Hierarchy of Command - *See* Chain of Command.

Incident - An occurrence or event, either human-caused or by natural phenomena, that requires action by emergency response personnel to prevent or minimize loss of life or damage to property and/or natural resources.

Incident Action Plan (IAP) - The plan developed at the field response level which contains objectives reflecting the overall incident strategy and specific tactical actions and supporting information for the next operational period. The plan may be oral or written. When written the Incident Action Plan contains objectives reflecting the overall incident strategy and specific control actions for the next operational period. When complete, the Incident Action Plan will have a number of attachments. Contains: ICS-202, ICS-203, ICS-204, ICS-205, ICS-206, Incident Traffic Plan, and Incident map.

Incident Base - Location at the incident where the primary logistics functions are coordinated and administered. (Incident name or other designator will be added to the term "Base.") The Incident Command Post may be co-located with the Base. There is only one Base per incident.

Incident Commander (IC) - The individual responsible for the command of all functions at the field response level.

Incident Communications Center (ICC) - The location of the Communications Unit and the Incident Message Center.

Incident Command Post (ICP) - That location at which the primary command functions are executed and usually co-located with the incident base.

Incident Command System (ICS) - The combination of facilities, equipment, personnel, procedures, and communications operating within a common organizational structure with responsibility for the management of assigned resources to effectively accomplish stated objectives pertaining to an incident.

Incident Objectives - Statements of guidance and direction necessary for the selection of appropriate strategy(s), and the tactical direction of resources. Incident Objectives are based on realistic expectations of what can be accomplished when all allocated resources have been effectively deployed. Incident Objectives must be achievable and measurable, yet flexible enough to allow for strategic and tactical alternatives.

Incident Management Team (IMT) - A command team comprised of the Incident Commander (IC), appropriate command, and general staff personnel assigned to an incident. (*Source:* FIRESCOPE)

Incident Management Team, Animal Protection - An Animal Protection Incident Management Team, when deployed, will assess the emergency situation and determine the number of operational strike teams that will be required for the rescuing, transporting, and sheltering of animals.

Incident Management Team, Firefighting - An Incident Management Team is an interagency organization under the auspices of NWCG composed of the Incident Commander (IC) and appropriate general and command staff personnel assigned to an incident, trained and certified to the Type I level. Type I level personnel possess the highest level of training available and are experienced in the management of complex incidents.

Information Officer (IO) - A member of the Command Staff responsible for interfacing with the public and media or with other agencies requiring information directly from the incident. There is only one Information Officer per incident. The Information Officer may have Assistants. This position is at times referred to as Public Affairs Officer or Public Information Officer (PIO) in some disciplines.

Initial Action - The actions taken by resources which are the first to arrive at an incident.

Initial Response - Resources initially committed to an incident.

International Medical Surgical Response Team (IMSuRT), National Disaster Medical System (NDMS) - An IMSuRT is a volunteer group of medical and non-medical individuals, usually from the same State or region of a State, which has formed a response team under the guidance of the NDMS and the State Department, and whose personnel and equipment give it deployable medical and surgical treatment capability worldwide. It is the only NDMS medical team with surgical operating room capability. Full team consists of roughly 26 personnel, which is a mix of physicians, nurses, medical technicians, and allied personnel.

Jurisdictional Agency - The agency having jurisdiction and responsibility for a specific geographical area.

Jurisdiction - The range or sphere of authority. Public agencies have jurisdiction at an incident related to their legal responsibilities and authority for incident mitigation.

Jurisdictional authority at an incident can be political/geographical (e.g., special district city, county, state or federal boundary lines), or functional (e.g., police department, health department, etc.). (*See* Multijurisdiction.)

LCES Checklist - In the wildland fire environment, Lookouts, Communications, Escape Routes, Safety Zones (LCES) is key to safe procedures for firefighters. The elements of LCES form a safety system used by wildland firefighters to protect themselves. This system is put in place before fighting the fire: select a lookout or lookouts, set up a communication system, choose escape routes, and select a safety zone or zones.

Leader - The ICS title for an individual responsible for a functional Unit, Task Forces, or Strike Teams.

Liaison Officer - A member of the Command Staff in all IMS systems that is responsible for coordinating with representatives from cooperating and assisting agencies.

Life-Safety - Refers to the joint consideration of both the life and physical well-being of individuals.

Logistics Section - One of the five primary functions found at all IMS systems. The Section responsible for providing facilities, services, and materials for the incident or at an EOC.

Management by Objectives - In IMS, the field and in the EOC level of emergency management. This is a top-down management activity which involves a three-step process to achieve the desired goal. The steps are: establishing the objectives, selection of appropriate strategy(s) to achieve the objective, and the direction or assignments associated with the selected strategy.

Management Support Team (MST), National Disaster Medical System (NDMS) - An MST is a command and control team that provides support and liaison functions for other NDMS teams in the field. MSTs are usually staffed by a mix of Federal employees and are constituted on an ad-hoc, mission-specific basis. An MST (perhaps as small as one or two individuals) always accompanies an NDMS unit on a deployment.

Medical Unit - Functional Unit within the Service Branch of the Logistics Section at the IMS Field levels responsible for the development of the Medical Emergency Plan, and for providing emergency medical treatment of incident personnel.

Message Center - The Message Center is part of the Communications Center and is collocated or placed adjacent to it. It receives, records, and routes information about resources reporting to the incident, resource status, and administration and tactical traffic.

Mobile Communications Center (MCC) (Mobile Emergency Operations Center (MEOC); Mobile Command Center (MCC); Continuity of Operations Vehicle) A vehicle that serves as a self-sustaining mobile operations center capable of operating in an environment with little to no basic services, facilitating communications between multiple entities using an array of fixed and/or wireless communications equipment, providing appropriate work space for routine support functions, and providing basic services for personnel in short-term or long-term deployments.

Mobile Feeding Kitchen (Mobile Field Kitchen; Rapid Deployment Kitchen) - A containerized kitchen that can be positioned forward in fulfillment of Emergency Support Function (ESF) #11—Food and Water. The units are used to support feeding operations at emergency incidents.

Mobile Kitchen Unit - A unit designed and constructed to dispense food for incident personnel providing a specified level of capacity.

Mobilization - The process and procedures used by all organizations, federal, state and local, for activating, assembling, and transporting all resources that have been requested to respond to or support an incident.

Mobilization Center - An off-incident location at which emergency service personnel and equipment are temporarily located pending assignment, release, or reassignment.

Multi-Agency or Inter-Agency Coordination - The participation of agencies and disciplines involved at any level of the IMS organization working together in a coordinated effort to facilitate decisions for overall emergency response activities, including the sharing of critical resources and the prioritization of incidents.

Multi-Agency Coordination System (MACS) - The combination of personnel, facilities, equipment, procedures, and communications integrated into a common system. When activated, MACS has the responsibility for coordination of assisting agency resources and support in a multi-agency or multijurisdictional environment. A MAC Group functions within the MACS. The combination of facilities, equipment, personnel, procedures, and communications integrated into a common system with responsibility for coordination of assisting agency resources and support to agency emergency operations.

Multi-Agency Incident - An incident where one or more agencies assist a jurisdictional agency or agencies. The incident may be managed under single or unified command.

Multi-Jurisdiction Incident - An incident requiring action from multiple agencies that have a statutory responsibility for incident mitigation. In IMS these incidents will be managed under Unified Command.

Mutual Aid Agreement - Written agreement between agencies and/or jurisdictions in which they agree to assist one another upon request by furnishing personnel and equipment.

Mutual Aid Coordinator - An individual at local government, operational area, region or state level that is responsible to coordinate the process of requesting, obtaining, processing, and using mutual aid resources. Mutual Aid Coordinator duties will vary depending upon the mutual aid system.

National Urban Search and Rescue (US&R) Incident Support Team (IST) - ISTs are components of ERT-As that provide Federal, State, and local officials with technical assistance in the acquisition and use of search and rescue resources through advice, Incident Command assistance, management, and coordination of US&R task forces and obtaining logistic support.

National Strike Force, U.S. Coast Guard - The U.S. Coast Guard National Strike Force was created in 1973 as a Coast Guard special force under the National Contingency Plan (NCP/see 40 CFR 300.145) to respond to oil and hazardous chemical incidents. The NSF consists of three interoperable regionally based Strike Teams: Atlantic, Gulf and Pacific, and the Public Information Assist Team (PIAT). The NSF supports USCG and EPA Federal On-Scene Coordinators (FOSCs) to protect public health, welfare, and the environment. In recent years, the capabilities have been expanded to include response to weapons of mass destruction (WMD) incidents, as well as incident management assistance.

Occupational Health & Safety Specialists (Occupational Physicians; Occupational Health Nurses; Industrial Hygienists; Occupational Safety Specialists; Occupational Safety & Health Technicians; Health and Safety Inspectors; Industrial Hygienists) - Personnel with specific training in occupational safety and health and topics such as workplace assessment or occupational medicine. Occupational health and safety specialists and technicians help keep workplaces safe and workers in good health unscathed. They promote occupational health and safety within organizations by developing safer, healthier, and more efficient ways of working. They analyze work environments and design programs to control, eliminate, and prevent disease or injury caused by chemical, physical, and biological agents or ergonomic factors. They may conduct inspections and enforce adherence to laws, regulations, or employer policies governing worker health and safety.

Operations Section - One of the five primary functions found in all IMS systems. The Section responsible for all tactical operations at the incident, or for the coordination of operational activities at an EOC. The Operations Section in an IMS can include Branches, Divisions and/or Groups, Task Forces, Teams, Single Resources and Staging Areas as necessary because of span of control considerations.

Operations Coordination Center (OCC) - The primary facility of the Multi-Agency Coordination System. It houses the staff and equipment necessary to perform the MACS functions.

Operational Period - The period of time scheduled for execution of a given set of operation actions as specified in the Incident Action Plan.

Out-of-Service Resources - Resources assigned to an incident but unable to respond for mechanical, rest, or personnel reasons.

Overhead Personnel - Personnel who are assigned to supervisory positions which includes Incident Commander, Command Staff, General Staff, Directors, Supervisors and Unit Leaders.

Paramedic - A practitioner credentialed by a State to function at the advanced life support (ALS) level in the State Emergency Medical Services (EMS) system.

Personnel Accountability - The ability to account for the whereabouts and welfare of personnel. It is accomplished when supervisors ensure that ICS principles and processes are functional and personnel are working within these guidelines.

Personal Protective Equipment (PPE) - Equipment and clothing required to shield or isolate personnel from the chemical, physical, thermal, and biological hazards that may be encountered at a Hazardous Materials (HazMat) incident.

Planning Meeting - A meeting, held as needed throughout the duration of an incident, to select specific strategies and tactics for incident control operations and for service and support planning.

Planning Section (Also referred to as Planning/Intelligence) - One of the five primary functions found in all IMS systems that is responsible for the collection, evaluation, and dissemination of information related to the incident or an emergency, and for the preparation and documentation of Incident Action Plans. The section also maintains information on the current and forecasted situation, and on the status of resources assigned to the incident. The Section will include the Situation, Resource, Documentation, and Demobilization Units, as well as Technical Specialists.

Procurement Unit - Functional Unit within the Finance/Administration Section responsible for financial matters involving vendor contracts.

Public Information Officer (PIO) - The individual in the IMS Command Staff (usually the IO or his Designee) that has been delegated the authority to prepare public information releases and to interact with the media.

Radiological Material - Any material that spontaneously emits ionizing radiation.

Recorders - Individuals within IMS organizational units who are responsible for recording information. Recorders may be found in Planning/Intelligence, Logistics and Finance/Administration Units.

Reinforced Response - Those resources requested in addition to the initial response.

Release - Any spilling, leaking, pumping, pouring, emitting, emptying, discharging, injecting, escaping, leaching, dumping, or disposing into the environment (including the abandonment or discharging of barrels, containers, and other closed receptacles containing any hazardous substance or pollutant or contaminant).

Rescue - To access, stabilize, and evacuate distressed or injured individuals by whatever means necessary to ensure their timely transfer to appropriate care or to a place of safety.

Resources - Personnel and equipment available, or potentially available, for assignment to incidents. Resources are described by kind and type, and may be used in tactical support or supervisory capacities at an incident.

Resources Unit - Functional Unit within the Planning/Intelligence Section in all IMS systems responsible for recording the status of resources committed to the incident. The Unit also evaluates resources currently committed to the incident, the impact that additional responding resources will have on the incident, and anticipated resource needs.

Responder Rehabilitation - Also known as "rehab"; resting and treatment of incident personnel who are suffering from the effects of strenuous work and/or extreme conditions.

Rest and Recuperation (R&R) - Time away from work assignment to give personnel proper rest so they remain productive, physically capable, and mentally alert to perform their jobs safely.

Safety Officer (SO) - A member of the Command Staff at the incident responsible for monitoring and assessing safety hazards or unsafe situations, and for developing measures for ensuring personnel safety. The Safety Officer may have Assistants as needed.

Search - To locate an overdue or missing individual, individuals, or objects.

Section - That organization level with responsibility for a major functional area of the incident, e.g., Operations, Planning/Intelligence, Logistics, Administration/Finance.

Section Chief - The ICS title for individuals responsible for command of functional sections: Operations, Planning/Intelligence, Logistics and Finance Administration.

Service Branch - A Branch within the Logistics Section responsible for service activities at the incident. Includes the Communications, Medical, and Food Units.

Shelter Management Team - Team provides managerial and operational support for a shelter during an emergency. Responsibilities of the team may include all or some of the following: operating the shelter; establishing security; ensuring the availability of adequate care, food, sanitation, and first aid; selecting and training personnel to perform operational tasks; monitoring contamination; performing decontamination; establishing exposure control and monitoring; monitoring overpressure and filtration systems; performing post-event reconnaissance; and directing egress.

Sheltering Team, Large Animal, Animal Protection - An Animal Protection Large Animal Sheltering Team will deploy for a minimum of 7 days and will be responsible for advising and supporting local efforts in setting up a large animal shelter.

Sheltering Team, Small Animal, Animal Protection - An Animal Protection Small Animal Sheltering Team will deploy for a minimum of 7 days and will be responsible for advising and supporting local efforts in setting up a small animal shelter.

Single Resource - An individual, a piece of equipment and its personnel complement, or a crew or team of individuals with an identified work supervisor that can be used on an incident.

Situation Unit - Functional Unit within the Planning/Intelligence Section responsible for the collection, organization and analysis of incident status information, and for analysis of the situation as it progresses. Reports to the Planning/Intelligence Section Chief.

Span of control - The supervisory ratio maintained within an IMS organization. A span of control of five-positions reporting to one supervisor is considered optimum. The IC (or his designee) determines appropriate span of control for a given task.

Staging Area - That location where incident personnel and equipment are assigned on a three (3) minutes to available status.

Staging Area Managers - Individuals within ICS organizational units that are assigned specific managerial responsibilities at Staging Areas. (Also Camp Manager.)

Standardized Emergency Management System (SEMS) - A system utilizing ICS principles including the five elements of Command, Operations, Planning/Intelligence, Logistics, and Finance/Administration. SEMS is used in California at five levels of government: Field Response, Local Government, Operational Areas, Regions, and State.

Strategy - The general plan or direction selected to accomplish incident objectives.

Supply Unit - Functional Unit within the Support Branch of the Logistics Section responsible for ordering equipment and supplies required for incident operations.

Support Branch - Branch within the Logistics Section responsible for providing personnel, equipment and supplies to support incident operations. Includes the Supply, Facilities and Ground Support Units.

Support Resources - Non-tactical resources under the supervision of the Logistics, Planning/Intelligence, Finance/Administration Sections or the Command Staff.

Supporting Materials - Refers to the several attachments that may be included with an Incident Action Plan, e.g., Communications Plan, map, Safety Plan, Traffic Plan, and Medical Plan.

Strike Team - Specified combinations of the same kind and type of resources, with common communications and a Leader.

Sustainability - Ability to continue response operations for the prescribed duration necessary.

Special Weapons and Tactics (SWAT)/Tactical Teams - SWAT teams are specially trained to handle high-risk situations and specialized tactical needs. Team members have advanced skills beyond that of typical patrol officers.

Strike Team, Large Animal Rescue, Animal Protection - An Animal Protection Large Animal Rescue Strike Team is a six-member team capable of completing an average of one rescue every 30 minutes in a suburban setting and one rescue every hour in rural settings.

Strike Team, Small Animal Rescue, Animal Protection - An Animal Protection Small Animal Rescue Strike Team is a six-member team capable of completing an average of one rescue every 30 minutes in a suburban setting and one rescue every hour in rural settings.

Tactics - Deploying and directing resources on an incident to accomplish the objectives designated by strategy.

Task Force - A combination of single resources assembled for a particular tactical need, with common communications and a Leader.

Technical Specialists - Personnel with special skills who are activated only when needed. Technical Specialists may be needed in the areas of fire behavior, water resources, environmental concerns, resource use, training areas, geographic information systems, and damage inspection.

Tender, Foam (Firefighting Foam Tender) - The apparatus used to mix concentrate with water to make solution, pump, and mix air and solution to make foam, and transport and apply foam.

Tender, Fuel (Fuel Tender) - Any vehicle capable of supplying fuel to ground or airborne equipment.

Tender, Helicopter (Helicopter Tender) - A ground service vehicle capable of supplying fuel and support equipment to helicopters.

Time Unit - Functional Unit within the Finance/Administration Section responsible for recording time for incident personnel and hired equipment.

Transport Team, Large Animal, Animal Protection - An Animal Protection Large Animal Transport Team will deploy for a minimum of 7 days and will be responsible for transporting large animals from a disaster site. All required vehicles will accompany team.

Transport Team, Small Animal, Animal Protection - An Animal Protection Small Animal Transport Team will deploy for a minimum of 7 days and will be responsible for transporting large animals from a disaster site. All required vehicles will accompany team.

Type - Refers to resource capability. A Type 1 resource provides a greater overall capability due to power, size, capacity, etc., than would be found in a Type 2 resource. Resource typing provides managers with additional information in selecting the best resource for the task.

Unified Command (UC) - In an IMS, Unified Command is a unified team effort which allows all agencies with responsibility for the incident, either geographical or functional, to manage an incident by establishing a common set of incident objectives and strategies. This is accomplished without losing or abdicating agency authority, responsibility or accountability.

Unified Area Command (UAC) - A Unified Area Command is established when incidents under an Area Command are multijurisdictional. (*See* Area Command and Unified Command.)

Unit - An organizational element having functional responsibility. Units are commonly used in Incident Planning/Intelligence, Logistics, or Finance/Administration sections and can be used in operations for some applications.

Unity of Command - The concept by which each person within an organization reports to one and only one designated person.

Urban Search and Rescue (US&R) - US&R involves the location, rescue (extrication), and initial medical stabilization of victims trapped in confined spaces.

Urban Search and Rescue (US&R) Task Force (US&R Team) - Federal asset that conducts physical search and rescue in collapsed buildings; provides emergency medical care to trapped victims; assesses and controls gas, electrical services, and Hazardous Materials (HazMat); and evaluates and stabilizes damaged structures.

Veterinary Epidemiologist - A practitioner who studies factors influencing existence and spread of diseases among humans and animals, particularly those diseases transmissible from animals to humans. Required to hold degree of Doctor of Veterinary Medicine.

Veterinary Medical Assistance Team (VMAT), National Disaster Medical System (NDMS) - VMATs are volunteer teams of veterinarians, technicians, and support personnel, usually from the same region, that have organized a response team under the guidance of the American Veterinary Medical Association (AVMA) and the NDMS, and whose personnel have specific training in responding to animal casualties and/or animal disease outbreaks during a disaster. They help assess medical needs of animals and conduct animal disease surveillance, hazard mitigation, biological and chemical terrorism surveillance, and animal decontamination. Usually includes a mix of veterinarians, veterinary technicians, support personnel, microbiologists, epidemiologists, and veterinary pathologists.

Volunteer Agency Liaison (VAL) - The Volunteer Agency Liaison serves as the central point between government entities and volunteer organizations in the coordination of information and activities of VOADs (Volunteer Organizations Active in Disasters) responding in times of disaster.

Weapons of Mass Destruction (WMD) - (1) Any destructive device as defined in section 921 of this title ("destructive device" defined as any explosive, incendiary, or poison gas, bomb, grenade, rocket having a propellant charge of more than 4 ounces, missile having an explosive or incendiary charge of more than $1/4$ ounce, mine or device similar to the above); (2) any weapon that is designed or intended to cause serious bodily injury through the release, dissemination, or impact of toxic or poisonous chemicals, or their precursors; (3) any weapon involving a disease organism; or (4) any weapon that is designed to release radiation or radioactivity at a level dangerous to human life. (United States Code, Title 18-Crimes and Criminal Procedure, Part I-Crimes, Chapter 113B-Terrorism, Sec. 2332a)

WMD Chem/Bio - A short-hand phrase for "weapons of mass destruction, chemical/biological," in reference to those substances that were developed by military institutions to create widespread injury, illness, or death.

INDEX

CPSIA information can be obtained at www.ICGtesting.com
Printed in the USA
BVOW021557291211

279074BV00001B/1/P